The Biology
of Aging

A Publication of the American Institute of Biological Sciences

The Biology of Aging

Edited by
John A. Behnke
Editor, *BioScience*

Caleb E. Finch
Professor of Biology and Gerontology
University of Southern California
Los Angeles, California

and

Gairdner B. Moment
National Institute of Child Health and Human Development
and Professor Emeritus of Biological Sciences
Goucher College
Baltimore, Maryland

Plenum Press · New York and London

Library of Congress Cataloging in Publication Data

Main entry under title:

The Biology of aging.

Includes bibliographies and index.
1. Aging. I. Behnke, John A. II. Finch, Caleb Ellicott. III. Moment, Gairdner
Bostwick, 1905-   [DNLM: 1. Aging. 2. Time. WT104.3 B615]
QP86.B518   574.3'72   78-19012
ISBN 0-306-31139-9

First Printing – November 1978
Second Printing – August 1979

© 1978 Plenum Press, New York
A Division of Plenum Publishing Corporation
227 West 17th Street, New York, N.Y. 10011

Printed in the United States of America

Contributors

William H. Adler • Gerontology Research Center, Clinical Physiology Branch, Baltimore City Hospitals, Baltimore, Maryland 21224 of the National Institute on Aging, National Institutes of Health, Bethesda, Maryland 20014

Edwin D. Bransome, Jr. • Section of Metabolic and Endocrine Disease, Medical College of Georgia, Augusta, Georgia 30901

Douglas E. Brash • Departments of Biophysics and Radiology, The Ohio State University, Columbus, Ohio 43210

Mary Anne Brock • Gerontology Research Center, Clinical Physiology Branch, Baltimore City Hospitals, Baltimore, Maryland 21224 of the National Institute on Aging, National Institutes of Health, Bethesda, Maryland 20014

Vincent J. Cristofalo • The Wistar Institute of Anatomy and Biology, Philadelphia, Pennsylvania 19104

Richard G. Cutler • Gerontology Research Center, Baltimore City Hospitals, Baltimore, Maryland 21224 of the National Institute on Aging, National Institutes of Health, Bethesda, Maryland 20014

Paul J. Davis • Endocrinology Division, Department of Medicine, Medical School of the State University of New York at Buffalo, Buffalo, New York 14215

Michael J. Eckardt • Clinical and Biobehavioral Research Branch, Division of Intramural Research, National Institute on Alcohol Abuse and Alcoholism, Rockville, Maryland 20857

Caleb E. Finch • The Andrus Gerontology Center, University of Southern California, Los Angeles, California 90007

David E. Harrison • The Jackson Laboratory, Bar Harbor, Maine 04609

Ronald W. Hart • Departments of Biophysics and Radiology, The Ohio State University, Columbus, Ohio 43210

Kenneth H. Jones • Gerontology Research Center, Clinical Physiology Branch, Baltimore City Hospitals, Baltimore, Maryland 21224 of the National Institute on Aging, National Institutes of Health, Bethesda, Maryland 20014

A. Carl Leopold • Boyce Thompson Institute for Plant Research, Cornell University, Ithaca, New York 14853

Gairdner B. Moment • Goucher College, Baltimore, Maryland 21204, and Gerontology Research Center, Baltimore City Hospitals, Baltimore, Maryland 21224 of the National Institute on Aging, National Institutes of Health, Bethesda, Maryland 20014

Morris H. Ross • The Institute for Cancer Research, Philadelphia, Pennsylvania 19111

George S. Roth • Endocrinology Section, Clinical Physiology Branch, Gerontology Research Center, Baltimore City Hospitals, Baltimore, Maryland 21224 of the National Institute on Aging, National Institutes of Health, Bethesda, Maryland 20014

Elizabeth S. Russell • The Jackson Laboratory, Bar Harbor, Maine 04609

Edward L. Schneider • Laboratory of Cellular and Comparative Physiology, Gerontology Research Center, Baltimore City Hospitals, Baltimore, Maryland 21224 of the National Institute on Aging, National Institutes of Health, Bethesda, Maryland 20014

Roy J. Shephard • Department of Preventive Medicine and Biostatistics, University of Toronto, Toronto, Canada

Tracy M. Sonneborn • Department of Biology, Indiana University, Bloomington, Indiana 47401

Betzabé M. Stanulis • The Wistar Institute of Anatomy and Biology, Philadelphia, Pennsylvania 19104

Robert D. Terry • Department of Pathology, Albert Einstein College of Medicine, New York, N.Y. 10461

William F. Van Heukelem • Pacific Biomedical Research Center, Kewalo Marine Laboratory, University of Hawaii, Honolulu, Hawaii 96813

Harold W. Woolhouse • Department of Plant Sciences, University of Leeds, Leeds, LS2 9JT, England

Vernon R. Young • Department of Nutrition and Food Science, Massachusetts Institute of Technology, Cambridge, Massachusetts 02139

Preface

Egocentricity is characteristically human. It is natural for our prime interest to be ourselves and for one of our major concerns to be what affects us personally. Aging and death — universal and inevitable — have always been of compelling concern. Mystical explanations were invented when scientific answers were lacking.

As scientific knowledge developed, anatomy and gross physiological processes were explained, and the roles of the endocrine glands were revealed. Since the sex hormones obviously lose some of their potency with age, it was logical to assume that they played the major role in declining general well-being. The puzzle of aging would now be solved. The Ponce de Leon quest would soon be fulfilled.

Pseudoscientists and quacks rushed in where most scientists feared to tread. By the time the glowing promises of perpetual youth through gland transplants and injections had proved illusory, serious study of the aging process had been set back for years. The field had lost "respectability," and most capable scientists shunned it. Those who did continue to seek answers to its tough questions deserve special recognition.

Actually, real progress in understanding the aging processes required more sophisticated approaches than were available in the monkey-gland era. Endocrinology had to become a much more sophisticated discipline and needed the modern companion sciences of genetics, molecular biology, biochemistry, and neurology. Fascination with the causes of aging had never waned, and a new generation of scientists with highly specialized skills and no hang-ups with the past have pitched in and are seeking answers on the cellular and biochemical level. Inevitably, they are an interdisciplinary group, closely aligned with the exciting field of development.

The goal of extending the healthy life-span seems likely to take its place among the host of other biological freedoms achieved through scientific insight and technical solutions. This possibility has already begun to tantalize many, although most scientists agree that major shifts in human life-span cannot be predicted at present. Possibly, clues to regulating aging will come from ongoing studies of the major diseases of aging such as cardiovascular disease, maturity-onset diabetes, and senile dementia. Many think that it is not time alone that sets the stage of such diseases, but some still mysterious physiological and cellular changes. Clearly, there is more to aging than the diseases associated with aging. Nature, through genetics, has already shown control over some of these changes, since neural and immunological events of aging occur 30 times more slowly in humans than they do in mice.

The study of other forms of life — plant and animal — is providing important clues. But even these researches raise fascinating questions, more challenging than the topography of the other side of the moon! Why is the life-span of a flying squirrel about twice as long as the three years of a rat when both are studied in captivity? Why should a chicken be aged at 12 years while parrots live for 20 or more?

The American Institute of Biological Sciences and the Editorial Board of its magazine *BioScience* decided to summarize some of the major research developments in the study of the aging process. Since, as is so often true in scientific studies, answers to basic questions are more likely to be found from research on organisms other than man, the articles in a special issue of the magazine reported work on a number of animals and even plants. The ultimate extrapolations to *Homo sapiens*, however, were not overlooked.

From this auspicious beginning, we decided to expand the coverage to a far wider series of topics, including more of the findings on the human level. This book is the result. However, the material is pitched at a less technical level than the articles with the aim of making the content intelligible to a more general audience, including use in various courses. We had the good fortune of bringing together a very distinguished group of specialists to do exactly that.

The chapters that follow represent many (but not all) of the current concepts, conclusions, and curiosities of research on aging. The observant reader will detect signs of a major current debate on the nature of aging: namely, whether each cell of the body has its built-in clock, or, on the other hand, whether a limited group of cells function as pacemakers of aging for the rest of the body. Come back in a decade, if you can, and see how the journey along the Ponce de Leon trail has progressed.

A special word of thanks is due Martha Kresheskey, whose dedication to this effort while secretary to J.A.B. helped immeasurably in bringing this project to completion.

John A. Behnke
Caleb E. Finch
Gairdner B. Moment

Contents

IV • AGING IN HUMANS AND OTHER MAMMALS

V • HORMONES AND AGING

VI • AGING IN PERSPECTIVE

1

The Ponce de Leon Trail Today

GAIRDNER B. MOMENT

Aging is both a complex and challenging scientific problem and a fact of universal concern. Raymond Pearl began his classic *The Biology of Death* by asserting that "probably no subject so deeply interests human beings as that of the duration of human life." This concern is extremely ancient. Despite the belief of some that the "cult of youth" is a modern invention, even a product of American advertising, no one need look very far to be reminded otherwise. It was in the sixteenth century that the Spanish conquistador Ponce de Leon searched for the Fountain of Youth. In fact, preliterate peoples attributed to their gods, whether in Valhalla or on Mt. Olympus, magical ways of postponing old age forever.

The goals of modern aging research are considerably more modest. But public interest in aging is growing steadily as more and more people swell the ranks of those over 65, and younger people realize more keenly than ever before two stark facts. First, the burden of supporting this increasing number of senior citizens, whether through taxation or private efforts, is already stupendous and will increase. The number of Americans over 65 is already 22,000,000 and is destined to reach an incredible 50,000,000 early in the next century. Already, the fastest-growing segment of our population is the group over 85. The second fact is that today's young and middle-aged adults realize that they themselves will join those over 65 with a probability far greater than at any time in the past.

Old age can be a healthy, productive, even a creative period. Verdi composed what is generally regarded as his greatest opera, *Otello*, when he was 75. Churchill was 66 when he became prime minister at the start of World War II. The history-making Pope John XXIII was 77 when he

GAIRDNER B. MOMENT, Ph.D. ● Goucher College, Baltimore, Maryland 21204, and Gerontology Research Center, Baltimore City Hospitals, Baltimore, Maryland 21224 of the National Institute on Aging, National Institutes of Health, Bethesda, Maryland 20014

became pope. Among scientists, botanists are notable for lives that are productive far into old age. Outstanding American examples include Asa Gray, Darwin's champion at Harvard, and Liberty Hyde Bailey at Cornell, the only biologist known to have mentioned Mendel's work on heredity before it was rediscovered three decades later. Darwin's friend and contemporary, the botanist Sir Joseph Hooker, continued to publish for almost 30 years after Darwin was under the floor of Westminster Abbey.

All such individuals are happy exceptions. Their chief importance here is as evidence revealing the human potential inherent in our later years. For far too many, old age is a wretched time characterized by a sense of uselessness, by physical and social neglect, and by increasingly severe physical and mental deterioration. The urgency of this problem is evident in countless personal tragedies, in the loss to society of the potential for self-reliance and productiveness of millions of our citizens, and indeed in the problems of our federal Social Security and Medicare Programs.

THE NEW NATIONAL INSTITUTE ON AGING

To help correct this intolerable situation, the Congress of the United States established in 1974 a National Institute on Aging as one of the National Institutes of Health. Its stated mission is to gain the knowledge necessary to alleviate "the physical infirmities resulting from advanced age [and] the economic, social and psychological factors associated with aging which operate to exclude millions of older Americans from a full life and place in our society." The thrust of the new institute, on the biological side, will be to seek to extend the healthy middle years into the later decades.

Fortunately, biomedical investigators have not been idly waiting for answers. For many years, under the leadership of Nathan Shock, the NIH Gerontology Research Center, now the nucleus of the National Institute on Aging, has pursued a vigorous program, documenting the aging changes that occur in individuals and looking into the physiological, psychological, cytological, and biochemical facts of the aging process. Approximately 650 men — and now women will be added — present themselves every 2 years for 2.5 days of a complete physical examination and laboratory testing. This longitudinal program is making it possible to know how age changes actually occur in individuals rather than in averages taken from groups of individuals, each group in a different period of life-span but without records of individual past histories. Thus, it is providing a necessary base line of knowledge on which future advances can be built.

The director of the new institute is Robert N. Butler, M.D., best known to the general public as author of the Pulitzer Prize-winning book, *Why Survive? Growing Old in America.* The new institute will push forward on a broad front. Emphasis in the biomedical areas will be placed on major problems of the later years, including cardiovascular physiology, the influence of diet on aging and possible changing dietary needs with age, arthritis and related conditions, immune functions and changing susceptibility to infections, altered responses with age to alcohol and other drugs (including those commonly used as medications), and, second to none in importance, neurological changes with age. The general aim may be said to resemble that of Oliver Wendell Holmes's famous deacon, who, it will be recalled, set out to construct a wonderful one-horse shay so well made that no part would wear out before any other part, and the entire vehicle would run perfectly right up to the final total catastrophe. Perhaps a shocking termination but far better than seeing a perfectly healthy mind and body destroyed by a malfunctioning heart or, worse yet, a functionless mind leading a vegetable existence in a healthy body.

DIE-HARD IDEAS: PRECURSORS TO DISCOVERY OR BLIND ALLEYS?

Before discussing some of the theories about aging and possible lines of investigation or the implications and even possible surprises from fundamental research in the field of aging, it is surely worthwhile to look at and perhaps clear away a number of ideas which are still very much alive in the popular mind. Some contain elements of truth, others are dead ends. The most talked about at the present time may be termed the Shangri-La phenomenon. Tucked away in remote mountain fastnesses are small groups of people, many of whom are said to live very active lives for 140 years or longer. The most studied region lies up in the Caucasus mountains between the Black Sea and the Caspian Sea in southern Russia. Another is the remote village of Vilcabamba in the high Equadorian Andes, and the third area lies in a high valley in the Karakorum mountains of northernmost Kashmir at the extreme western end of the Himalayas.

All three regions have been visited by learned observers, including Alexander Leaf of the Harvard medical faculty, who has described them in a recent book. All three regions have much in common. Life there involves much hard physical labor. The diets are well below European and American standards as to the number of calories eaten per day and probably also for the amount of fats and proteins. The aged continue to be active and respected members of their communities.

What is the hard evidence that people actually live to such extraordinary ages in these places? Leaf relies on ancient church records in Vilcabamba. But it is difficult to place much faith in records made well over a century ago in a remote village of illiterate peasants. Furthermore, in many Latin towns there are recurring favored names as common as John Smith in this country. Nor would one expect to find the best-educated priests in remote and inaccessible villages. Consequently, it seems only prudent to take these claims with a very large grain of salt. At an international conference in Bethesda in the spring of 1978 attended by Alexander Leaf and the Minister of Health of Ecuador, it was agreed, on the basis of far more careful scrutiny of old records and studies of bone characteristics, that the oldest individual in Vilcabamba was less than 100 years of age.

The evidence is no better in the Caucasus. Sula Benet, Emeritus Professor of Anthropology, Hunter College, lived among these people and came away convinced that some of the inhabitants are at least 125 years of age. When challenged as to how she could be certain of those claims, she said that she had found the people friendly and completely reliable, which is no doubt true, and added that there are many families with five generations living together. However, if it is assumed that the parents in these families had averaged 19 years of age at the birth of their first child and that the fifth generation were children 10 years old or younger, surely not unreasonable assumptions, then the oldest member of such a family would be 86, a ripe old age but quite a long way from 106, not to mention 125. Zhores Medvedev, a leading Russian investigator of aging, has also studied these people. He reports that falsification of birth dates and ages in that region has been going on for a very long time, under both Tsars and Soviets, to avoid military service. Medvedev concludes that "the best way to get a reputation for being extremely old is to lie a little." I know of instances in remote villages near Erzurum in far eastern Turkey where local villagers delight to fill the city slicker from Istanbul with all manner of fantastic tales about village life, and I have seen the same game played by my own relatives in northwestern Connecticut with the tourist from New York City.

A similar lack of firm evidence obtains for the very interesting Hunza people in the remotest part of the Kashmir. It would be instructive to compare the life expectancies of the whole populations of these tiny villages with those of towns and villages in other parts of the world. Such vital statistics do not seem to be available.

One of the oldest of the well-publicized theories about the cause of aging was proposed in 1889 by Brown-Séquard of the University of Paris. He believed that the aging of the gonads is the key factor and that he had demonstrated that testicular extracts have a general rejuvenating effect. In the 1920s, Serge Voronoff, Collège de France, and others in Europe and the United States reported success by grafting sex glands from young nonhuman primates into aging humans. These claims must also be taken *cum grano salis*. To begin with, psychological and other factors were not properly controlled. Furthermore, as Robert Butler has noted, the idea that elderly men and women are without sex is a myth, provided they enjoy good health, are free of psychological impediments, and are not influenced by various drugs, including tranquilizers. The likelihood that any such foreign grafts would have been rather rapidly rejected is great. A vast amount of modern experience has shown the enormous difficulty of getting transplanted hearts or kidneys to remain healthy and functional even when grafted between two individuals carefully matched for blood types and with the immune system of the host knocked out. In any case, if aging were under gonadal control in either male or female animals, one would expect the life-span of eunuchs and of castrated or spayed animals to be notably different from that of intact individuals. Several studies confirm the common observation that such is not the case.

At the turn of the century, Élie Metchnikoff, a younger associate of Pasteur in Paris and the discoverer of phagocytosis, proposed the theory that aging is due to the chronic effects of intestinal toxins. As a corrective, he advocated a diet rich in *Lactobacillus bulgaricus*, which is present in, for example, yogurt. There are a number of such foods, and in all probability they are very healthful, but any extensions of life expectancy among those consuming them would appear to be due to added freedom from disease because maximum life-span is not increased.

Noting the flabby muscles and sagging skin of the aged, an otherwise obscure Russian, A. A. Bogomolets, proposed that aging is due to the deterioration of the connective tissue and claimed to have revitalized this tissue with some sort of undefined serum. Nothing much has ever come of this idea, in part because the materials and methods he used were very vaguely defined. However, the possibility remains that there may have been, just possibly, antigens or antibodies, in the serum he prepared which stimulated the host's own immune system in some beneficial way. Such precursor or pseudodiscoveries are commonplace in the history of science. Today major efforts are under way in many parts of the world to explore the role of the immune system in aging.

For the past two decades, Anna Aslan in Romania has been claiming significant rejuvenating results in human patients from injections of a special procaine ("novocaine") solution which she calls Gerovital H3. Its alleviating action on the symptoms of senescence was an accidental discovery noticed in the course of treating patients for peripheral vascular disease, a not uncommon ailment of the aging. The procaine used is modified by the action of benzoic acid and potassium metabisulfate. Investigators in this country have found that the final solution is a weak inhibitor of monoamine oxidase. Mono- and other amines play an important role in the central nervous system, including within the brain, specifically within the hypothalamus, which some suspect may be the site of an "aging clock," if such exists. Consequently, it is conceivable that this procaine might affect a pacemaker for aging. Aslan herself states that her modified procaine does not extend the life-span. This indicates that in all probability whatever beneficial effects it has achieved are attained by some means other than action on the underlying aging mechanism, which is not to say that it is worthless.

If it seems strange, even a bit scandalous, that no thorough tests of this procaine treatment have been made in any of our research institutions as a possible method to alleviate "the physical infirmities resulting from advanced age . . . which operate to exclude millions of older Americans from a fuil life and place in our society" (1974 National Institute on Aging guidelines), it should be remembered that a long time lag between the first proposal of an idea and its final testing and acceptance or rejection is commonplace in the history of science and nowhere more so than in medicine. The probable primary role of the nervous system in aging, indeed any role except a passive one, was not even suspected until recently. Finally, the time and funds available for research on aging are strictly limited. Every investigator has to choose what looks to him or her to be the most promising line of exploration. No one wants to waste time and money on a fool's errand.

New nostrums continue to appear. Often they are treated as commercial secrets, a practice which prevents scientific tests for effectiveness or even for safety, and which opens the door to fraud. Any preparation which has not been fully described and made available for testing on subhuman mammals and for subsequent double-blind tests on human patients must be regarded with considerable suspicion.

THE SEARCH FOR FUNDAMENTAL KNOWLEDGE

Unfortunately, there is no generally accepted theory about the cause or causes of aging. In fact, human understanding of the biology of aging is in much the same position that knowledge of the causes of contagious disease was before the germ theory was established. This means that all that anyone can do for the present is to treat the ailments of the later years piecemeal without knowing what the controlling events are. This does not mean that such tinkering is worthless. Far from it. Malaria, for example, was successfully treated with quinine for over 200 years before there was any inkling that malaria is due to a protozoan blood parasite transmitted by the bite of a particular kind of mosquito. What it does indicate is the need for the kind of basic knowledge that will finally enable us to know what is really going on and, therefore, be able to develop rational regimens for living out our lives with a maximum of satisfaction.

The contemporary student of aging is faced with more theories than a centipede has legs. Some are conflicting and some mutually supporting and appear to deal with different aspects of the same general events. Most can claim, at least, some degree of credibility and some are backed by considerable masses of evidence, but in no case is there anything like definite proof. However, all these theories fall into two general groups recognized over half a century ago by C. M. Child of the University of Chicago in his *Senescence and Rejuvenescence*. According to one view, termed "epiphenomenalist" by Alex Comfort, a leading English student of aging, but also called the "extrinsic" or "random" theory, aging results from the contingencies of living rather than from a programmed development.

Put simply, aging results from some form of wear and tear. Perhaps one organ or organ system becomes worn or damaged, and this throws added strains on some other system and that, now under duress, strains a third, so an interacting downward spiral is produced. Possibly there is an accumulation of waste products, or various chemical changes occur which irreparably damage cells or the DNA- and protein-synthesizing machinery.

The other general view holds that aging, at least in birds and mammals and perhaps in most other animal groups as well, is genetically programmed by some kind of a pacemaker or biological death clock. Comfort terms this the "fundamentalist" view. Others sometimes speak of "intrinsic" or "controlled" theories. Some investigators place an intrinsic aging chronometer in every cell, others placing a controlling clock in a single center usually somewhere in the brain. In such a case, the pacemaker would be extrinsic to most of the cells of the body although still intrinsic to the organism.

As will become evident later, the distinction between these two types of theories may turn out not to be nearly as sharp as has been generally supposed.

AGING BY WEAR AND TEAR

One of the most outspoken advocates of an epiphenomenalist view is Hans Selye at the University of Montreal, a Canadian investigator well known for his pioneer work on adaptation to stress. He holds that "It has always been assumed as self-evident that aging has a specific cause and that, consequently, we might discover what this is and perhaps find a way to block it. There is no justification for such an assumption."

The most widely discussed of the newer theories of the general wear-and-tear type was proposed by Leo Szilard from the University of Chicago, a man who helped develop the first nuclear reactor and who received an "Atoms for Peace" award. According to this theory, aging is due to the background radiation to which we are all subjected from the sun and from the rocks of the earth itself. Random hits — "aging hits" — on the chromosomes damage large portions of chromosomes or single genes, inactivating them or producing harmful mutations. It has been known for many decades that radiation causes mutations (i.e., errors in the DNA which interfere with its proper functioning). Leslie Orgel, an English biochemist and theorist now at The Salk Institute, Medvedev, and others have subsequently refined and extended this idea to show how an error, once made anywhere by any agent in the DNA–protein-synthesizing apparatus could be magnified, producing faulty templates which, in turn, would serve as faulty models for the production of more faulty enzymes, resulting in more faulty templates — a process leading to a gradually accelerating cumulative increase of errors until the final "error catastrophe" — death — occurs. Since his original proposal, Orgel has published, in 1970, a correction to his calculations, pointing out that his famous "error catastrophe" is not inevitable with time but depends on a number of contingencies which may or may not occur.

If aging is due to an accumulation of errors in the DNA- and protein-synthesizing machinery, it should be possible to identify an accumulation of abnormal enzymes and other abnormal proteins in aged animals. Some proteins of slightly altered properties have been found, but whether or not they are part of a buildup to an error catastrophe remains unknown. Collagen in the tendons of old animals is somewhat different from that in young, but collagen freshly produced by old animals closely resembles that of very young ones. Chromosomal aberrations in rat liver cells increase with the age of the rat but not sufficiently to prevent normal liver regeneration. The RNA of fetal mouse liver is different from that of adult mouse liver. Yet in the first stages of regeneration in the liver of an adult mouse, the RNA is typical fetal-type RNA and only later becomes replaced by the adult type. Thus, it is clear that genes which have been inactive for long periods during an animal's life have not become seriously damaged.

It has been suggested that redundancy in the genetic information possessed by any species of animal would be a safety factor in some degree against random radiation damage and wear on the code from other agents or continual use. However, Richard Cutler of Texas and the NIA Gerontology Research Center, working with presently available techniques, found "no clear correlation between the percentage of re-iterated nucleotide sequences" (i.e., redundant genetic information) and the rate of aging in different species of mammals. There was some increase with increasing life-span, and this relatively small increase may turn out to be significant. It is too soon to tell.

For many years, it has been recognized that the macromolecules characteristic of biological material such as hides, rubber, and proteins can be stabilized ("tanned") and profoundly changed in their properties by chemical cross-linkages. The accumulation of such damaging chemical links between molecules with time seems slowed or even largely prevented by various antioxidants. A number of investigators, including Cutler and Johan Bjorksten, an independent researcher in Madison, Wisconsin, have looked at aging from the point of view of the effects of cross-linkages, especially at the possible but unproven alleviating effects of vitamin E, which is a well-known antioxidant.

DIET AND AGING

The relation of diet to aging is now undergoing a vigorous reinvestigation. It has been known since the work of C. M. McCay of Cornell University 40 years ago that the life of rats can be prolonged up to 25% by restricting the diet, especially early in life. Morris Ross of the Institute for Cancer Research in Philadelphia has recently demonstrated that the onset of age-related diseases in rats can be "readily" controlled by dietary manipulation and that in old age, dietary restriction may shorten life. The great importance of all this work in a hungry world is obvious. There are, however, problems. Most of the work has been done on rats, and rats commonly die of either lung or kidney diseases. If a given diet prevents or postpones these diseases, it will prolong the life of the rats. But the question remains — "Have we been tampering with the basic aging process itself?" — to quote Roy Walford, a veteran investigator of aging at the University of California, Los Angeles. It may be no different in principle from increasing life expectancy by eliminating some childhood disease. From the practical human point of view, this question is largely irrelevant. Not so irrelevant is the fact that many of the rats on the restricted diets appear sickly while others look healthy but are sterile. Very early dietary restriction in children seems to result in mental retardation. Clearly, it will require years of carefully controlled testing on various subhuman species before such methods are applicable to human beings.

The explanation for any increase in the duration of life by dietary restriction is a mystery. It may be that the use of the genetic code is slowed as fewer enzymes and other proteins are required and hence there is less wear and tear on the DNA. It might also be that some aging clock is slowed.

RATE OF LIVING

As long ago as 1908, a German physiologist, Max Rubner, proposed that the length of life in mammals is a reflection of the rate of physiological living and pointed out that the total number of calories of energy burned per gram of body weight during the lifetime of mammals of greatly different sizes and life-spans is approximately the same. It is indeed an intriguing fact that at the end of the life-span, the very rapidly beating heart of a 3.5-year-old mouse and the slowing beating heart of a 70-year-old elephant patriarch will have beaten about the same number of times, roughly 1.1×10^9 and 1.0×10^9, respectively. The significance of these facts has been argued about ever since, usually with negative conclusions as far as relevance to aging is concerned. The correlations are only approximate. Furthermore, if any animal is to maintain a constant body temperature above that of its surroundings, the smaller it is, the more calories it will metabolize per gram of body weight to maintain that temperature, and the reverse will be true for larger and larger animals. This is a consequence of the fact that the smaller a body, the greater its surface area is in proportion to its weight and the obvious fact that heat can only be lost to the surroundings from its surface.

However, this might be at least part of the reason that, in general, the larger the animal, the longer its life-span. The genetic code might be worn out less rapidly or some hypothetical time recording process slowed. It has been known since the work of Raymond Pearl at Johns Hopkins University and of others subsequently that the life-spans of fruit flies, *Drosophila*, can be more than doubled or cut in half by either lowering or raising the temperature at which they live. Subsequent work has shown that the time in the life-span in which a fly is exposed to a given temperature influences the result. Evidently, the problem is not as simple as it first appeared. It is of interest here that women, on the average, live about 8 years longer than men, who have a basal metabolic rate about 6% higher than women and more red blood cells per milliliter of blood. New interest in this aspect of aging is sure to be aroused by the report that Bernard Strehler, an enterprising and imaginative investigator at the University of Southern California, has found that mice with lower body temperatures than usual live up to twice as long.

AGING BY PROGRAM

Among the more articulate of the advocates of the theory that aging results from some definite program are the Australian A. V. Everitt, who speaks of an "aging clock"; Caleb Finch of the University of Southern California, who has written of "pacemakers" of aging; and W. D. Denckla at the Harvard School of Medicine, who argues for a special "aging hormone" secreted by the brain or pituitary. The arguments for some sort of a timekeeper are highly persuasive. As Charles Sedgwick Minot, a nineteenth-century American student of development and aging wrote: "Aging begins at birth." All the years must be counted equally from the first. Were this not true, a man of 40 would be the same as one of 20, and no one could ever attain old age. The continuity of aging is shown by the onset of sexual maturity and by menopause. Aging is a continuous developmental process shown with especial clarity in chickens. A leghorn hen lays the maximum number of eggs she will ever lay in a single year during her first year. In every subsequent year, she will lay approximately 80% of the number of eggs she laid the previous year.

The site presently favored for a localized aging chronometer is in the base of the brain, specifically in the hypothalamus. Here are the centers which control the production of growth hormone by the pituitary and the development and activities of the gonads, the thyroid, and other glands. Further indications of involvement of the nervous system in aging include evidence that this system is largely responsible for the atrophy of muscles in the very old and the demonstration that ovarian activity can be revived in very old rats, permanently without reproductive cycles, by very precise electrical stimulation of a specific location within the hypothalamus.

What is the mechanism of the "clock"? One suggestion is that aging is due to, or at least accompanied by (the old horse vs. cart problem), a gradual elevation of the threshold of sensitivity of the hypothalamus to the normal feedback suppression from the endocrine glands by which the metabolic stability of the body is maintained. It is as though a furnace became less and less responsive to a thermostat, with the result that it poured out more and more heat before it turned off. It has been supposed for many years that the metamorphosis of a tadpole into a frog, which is most certainly due to an increase in thyroid hormone, occurs because of a progessive loss of sensitivity of the thyroid-stimulating cells in the brain to the regulatory inhibitory feedback from the thyroid. It has sometimes been thought that perhaps the thymus, a gland in the chest that atrophies with the onset of adolescence, might be involved in the regulation of aging. There is now an explosion of knowledge about the thymus and the immune system in general, but its causal relationship to aging is highly problematical.

Ultimately, any timekeeper must reside in the biochemical properties of the cells. One guess is that the timekeeper lies in cell membranes, as some suppose to be true for the 24-hour or circadian rhythms of many plants and animals. If there is a similarity in mechanisms, knowledge of these rhythms may throw some light on the aging clock. In the fruit fly *Drosophila*, Seymour Benzer of the California Institute of Technology has shown that the circadian rhythms of activity are controlled by not more than three or four genes apparently located close together on the same chromosome. In genetic mosaics, where, by accident during early development, different portions of the body have different genetic compositions, the rhythm of the entire fly follows the genetic constitution of the head. This not only indicates that the circadian clock is under genetic control, but suggests that the site of the control mechanism lies in the brain, which is known to be true in moths from brain-transplantation experiments. It is still entirely possible that the chronometer is the amount of protective redundancy of genetic information already discussed or in the effectiveness of DNA repair enzymes, both of which would certainly be under genetic control.

A specific genetic program which determines the life-span of animals was first proposed by August Weismann of the University of Freiburg-im-Breisgau, founder of the germ plasm theory — the generally accepted theory that individuals are the expression of the genetic information contained in the continuous line of reproductive cells generation after generation. At the meeting of the Association of German Naturalists in Mozart's little city of Salzburg in 1881, Weismann argued that the characteristic life-span of different species has been determined by natural selection, like the other traits of any organism, such as length of ears or the season of birth of the young. Specifically, he suggested "that the restriction [of life-span] may conceivably follow from the limitation in number of cell generations for the somatic [body] cells of each organ and tissue." Weismann himself disclaimed any "ability to indicate the molecular properties of cells" upon which such a restriction in the possible number of cell generations might depend.

Weismann's ideas about aging have been under investigation ever since. Shortly after he announced them, several investigators — notably Emile Maupas in Algiers and Richard Hertwig in Germany — neatly combined the function of sex and the limitation in clonal life-span. It will be recalled that a clone is a group of cells or of plants or animals all descended from a single individual by asexual methods. These workers found that a clone of cells, specifically free-living protozoans such as paramecium, had a limited life-span at the end of which general vigor declines, and the rate of cell division gradually falls to zero, much as Weismann had predicted. However, if sexual interchange occurs before this happens, the cells are reinvigorated and another series of asexual cell divisions begins and continues until the limit is again reached. These discoveries were thrown into almost total eclipse among tissue and cell culturists by the famous work of Alexis Carrel with his supposedly immortal culture of chicken heart cells. In contrast, a number of protozoologists, notably Tracy Sonneborn and his associates at Indiana University, continued work on this problem and have repeatedly demonstrated the existence of a limited clonal life-span in several species of ciliated protozoans and the way it can be overcome by sexual interchange. For over 40 years, biologists have been asking how sexual reproduction achieves this result.

Within the past decade, Leonard Hayflick of Stanford University and many others have established to the satisfaction of probably a majority of investigators that normal human diploid cells cultured *in vitro* possess a limited clonal life-span not unlike that of certain protozoans. If true, such a limitation would impose some kind of upper limit on life-span, although there is no evidence that animals reach that limit and then die. Within the past year, R. R. Kohn, Case-Western Reserve University, writing in *Science,* has summarized the results of several workers which do not conform to the Hayflick model. Furthermore, disquieting thoughts arise about the many horticultural varieties of plants which have been propagated for long periods of time by grafting and other asexual methods. The Concord grape, for example, has been so propagated since Ephraim Bull, a Massachusetts farmer, introduced it in 1853. Cells from carrot roots are said to have been continuously cultured since 1937. Perhaps in this respect plants differ from animals, but in the light of all that is known about the biochemical genetics of both plants and animals, this seems highly improbable, although not inconceivable. It has been well said that in any scientific problem the last word is often long in coming.

DOES HEREDITY SET THE HUMAN LIFE-SPAN?

In the sense that a man lives longer than a mouse or a dog, the answer is clearly yes. What of a difference between different human individuals? The old dictum of Pearl, based on a massive study, has become proverbial — the best way to achieve a long life is to choose long-lived grandparents. A different kind of study by F. J. Kallman of College of Physicians and Surgeons, Columbia University, and L. W. Sander of Boston University gives the same answer. They compared over 1000 pairs of twins, some one-egg or monozygotic, which share exactly the same genes, and some two-egg or dizygotic, which are no more closely related than any two siblings. They found that the average difference in age at death between the two members of the two-egg or fraternal pairs was about twice that between the monozygotic pairs. It is very hard to deny an important hereditary factor in the light of this result. In 1974, M. H. Abbott and her collaborators at Johns Hopkins published the results of a follow-up of the descendants of Pearl's nonagenarians. They found only a very slight correlation between the age at death of the descendants and of their grandparents.

These results raise one of the most bothersome problems in aging studies. Is any observed increase in length of life due to improved health, enabling an animal to live out more of its hereditary potential life-span, or is it due to an actual lengthening of that inherent life-span? Advances in medicine have greatly increased life expectancy by the virtual elimination of many children's diseases and the successful treatment of many of the diseases of the middle years. Thus, it may be that the reason the descendants of both the shorter- and longer-lived subjects in the Pearl study have lived about the same time is simple. The genes for lack of resistance to various diseases which were inherited from the shorter-lived ancestors have been compensated for by improved modern medicine.

WHILE WE WAIT

While waiting for answers to the fundamental question of why we all age, whether we are men, elephants, or parrots, and how the later years of life can be made more rewarding for each individual and for society at large, it is worth remembering that far-reaching and practical results have often come from research on plants and animals far removed from the human species. The laws of heredity were discovered by studying the garden pea and how the human heart beat is controlled from experiments on the hearts of frogs and horseshoe crabs.

There are already some commonsense answers to the insistent question, "What can I do to live a long and healthy life?" Clearly in order is a sensible diet and more-or-less regular exercise somewhat short of singles tennis in the sun. Unhappily for many people, cigarette smoking induces lung cancer and heart damage, both important life-shorteners. Alcohol is a major factor in the causation of cirrhosis of the liver, which is responsible for over half of all deaths between the ages of 45 and 65. In addition, alcohol has been shown to injure the brain and heart. To return to the positive side, there is evidence that keeping active interests and a purpose promotes long life.

POSSIBLE OUTCOMES AND SURPRISES

There is no reason to doubt a continuing flow of beneficient discoveries from modern aging research. The goal of making it possible to age well will be achieved so that a satisfying old age, both physically and mentally, can be the good fortune of men and women everywhere.

What of the possibility of a radical discovery, extending not just life expectancy so that more people will live the full proverbial three score years and ten, but extending the life span to 200, 500, or even 10,000 years? Such a discovery seems highly unlikely, but the history of science has been full of surprises, as the physicists found out with the totally unexpected discovery of x-rays and then atomic fission. What is important to remember about such discoveries is the Columbus Principle. Just as such surprise discoveries cannot be predicted; so they can scarcely be prevented. You do not have to be seeking a new continent to find one. You only have to sail in a certain direction, and if the continent exists, you will run into it. A discovery that would enable us to slow or stop the aging process might, for example, come out of studies on how best to treat the circadian rhythms that produce jet lag for east–west travelers, or it might be an unintentional spin-off from work on the biochemistry of mental illness.

Some thinking has already been begun by Strehler and others on the problems that would arise in the event of such a discovery. More is needed, for we would be faced by a watershed in human history of unparalleled dimensions. The population problem would immediately assume incredible urgency. Enormous economic, social, and emotional problems would challenge our best thinking.

To begin with, at what age would you choose to stop the clock? Would the age of 9 or 10 seem the truly best years, or the late teens, or perhaps young adulthood, or maybe you would prefer a comfortable middle age? It is to be hoped that few would opt for 800 years of shuffleboard and bingo. The available options should include joining those old botanists and other fortunate men and women mentioned earlier who can enjoy an age of serious purposes and lively delight in the inexhaustible wonders of this world. How could anyone make an intelligent choice before experiencing each age firsthand? In addition, there may be limits to how much the human brain can assimilate and remember and for how long.

Furthermore, if you chose to remain a teenager, what sort of a 16-year-old would you really be after you had lived 500 years? Sir William Osler, one of the four founders of the Johns Hopkins Medical School, was not the first to note that, with rare exceptions, radical new ideas and great achievements have sprung from young men. Would stopping the aging process also stop the flow of truly innovative ideas and new outlooks? Would a world without children be like a world without the fresh winds of dawn or even a world without mornings?

Yet, in some very fundamental ways, life might not seem very different. As the ancient psalmist explained, a thousand years in the sight of God are as an evening past. In comparison to the immense stretches of astronomical time open to our investigation, 10,000 times 10,000 is but an instant. Thus, we can still take courage from the words of Cicero recorded in his *De Senectute* over 2000 years ago: "The short period of life is yet long enough for living well and honorably."

BIBLIOGRAPHY

General

Child, C. M. 1915. *Senescence and rejuvenescence.* The University of Chicago Press, Chicago.

Comfort, A. 1956. *The biology of senescence.* Routledge & Kegan Paul Ltd., London.

Comfort, A. 1976. *A good age.* Crown Publishers, Inc., New York.

Galston, A. W. 1975. In search of the antiaging cocktail. *Natural History* 84: 14–19.

Rosenfeld, A. 1967. *Prolongevity.* Alfred A. Knopf, Inc., New York.

Segerberg, O., Jr. 1974. *The immortality factor.* E. P. Dutton & Co., Inc., New York.

Strehler, B. L. 1975. Implications of aging research for society. *Federation Proceedings* 34: 5–8.

Technical

Burnet, M. 1974. *Intrinsic mutagenesis: a genetic approach to ageing.* John Wiley & Sons, Inc., New York.

Everitt, A. V. 1973. The hypothalamic–pituitary control of ageing and age-related pathology. *Experimental Gerontology* 8: 265–277.

Finch, C. E., and L. Hayflick, eds. 1977. *Handbook of the biology of aging.* Van Nostrand Reinhold Company, New York.

Hamilton, J. B., and G. E. Mestler. 1969. Mortality and survival: comparison of eunuchs with intact men and women in a mentally retarded population. *Journal of Gerontology* 24: 395–411.

Kallman, F. J., and G. Sander, 1948. Twin studies on ageing and longevity. *Journal of Heredity* 39: 349–357.

Macfarlane, M. D. 1975. Procaine HCl (Gerovital H3): a weak reversible, fully competitive inhibitor of monoamine oxidase. *Federation Proceedings* 34: 108–110.

Orgel, L. E. 1970. The maintenance of the accuracy of protein synthesis and its relevance to ageing: a correction. *Proceedings of the National Academy of Sciences of the USA* 67:1477.

Rockstein M., ed. 1974. *Theoretical aspects of aging*. Academic Press, Inc., New York.

Roth, G. S., and R. C. Adelman. 1974. Age related changes in hormonal binding by target cells and tissues: possible role in altered adaptive reponsiveness. *Experimental Gerontology* 10: 1–11.

Sonneborn, T. M. 1974. Genetics of the 14 species of *Paramecium aurelia*. In R. C. King, ed. *Handbook of genetics*. Plenum Press, New York.

Van Heukelem, W. F. 1973. Growth and lifespan of *Octopus cyanea*. *Journal of Zoology (London)* 169: 299–315.

2

Cell Aging: A Model System Approach

VINCENT J. CRISTOFALO and BETZABÉ M. STANULIS

Although there is a large literature describing age-related physiological changes at the cell, tissue, organ, and organism level, fundamental questions about the aging process remain unanswered. In fact, it can be said that the experimental design required to answer them is in many ways obscure. For example, we still need answers to such questions as: (1) is aging intrinsic or extrinsic to the organism; (2) is aging intrinsic to an individual cell or an integrated function; (3) are there mechanisms which control the rate of aging; (4) is there a kind of pacemaker cell that signals the rest of the cells to degenerate and die, or does each cell have its own clock-like machinery by which aging proceeds; and (5) in either case, would such a clock mechanism be contained in the nucleus or in the cytoplasm? Definitive answers to all these questions must precede elucidation of the biology of aging.

With reference to the first question, it seems clear that the maximal life-span is genetically regulated since, within limits, it is characteristic for each individual species; so, at least in part, aging must be intrinsic to the organism.

The second question then becomes the key one. We know that at least some of the cells in an aged organism must be senescent. It is also clear that functional failures must occur in cells at different rates since different functional capacities in the old organism decline at different rates. We do not know, however, whether the changes that we see in individual cells are intrinsic to the cells or are the effects of decline in integrative function.

This question of whether aging is supracellular has been asked by many scholars over many decades. It was proposed at the turn of the

VINCENT J. CRISTOFALO, Ph.D. and BETZABÉ M. STANULIS, Ph.D. • The Wistar Institute of Anatomy and Biology, Philadelphia, Pennsylvania 19104

century that aging was the price cells paid for differentiation and that isolated cells growing outside the constraints of a highly complex organism would be able to live forever. Alexis Carrel, a biologist active in the early part of this century, and his co-workers believed that they had demonstrated this essential immortality of isolated cells by keeping cultures of chick heart cells growing for 34 years before terminating their experiments. Support for the observation of Carrel and his colleagues came from the discoveries in the 1940s and 1950s that cell lines such as "L" cells, a cell culture derived from mouse tissue, and HeLa cells, a cell culture isolated from a human cervical tumor, could be kept growing in culture without a decline in proliferative vigor.

In contrast, in the late 1950s and early 1960s, Leonard Hayflick and Paul Moorhead, two scientists then at the Wistar Institute in Philadelphia, observed that cells from a variety of normal human tissues would proliferate in culture for various periods of time but would eventually degenerate and die. Under the culture conditions used, normal cells living outside the body were clearly not immortal, and they proposed that this was the expression of aging at the cellular level.

THE CELL SYSTEM

To evaluate the evidence for these opposing views in any depth requires an understanding of the procedures involved in cell culture. Briefly, to prepare cultures of human cells, pieces of fresh tissue are disrupted with trypsin, a proteolytic enzyme, to isolate individual cells; the isolated cells are placed in a container with a nutritive medium designed to resemble human blood serum. The cells attach to the surface of the container and multiply, forming a mat which eventually covers the surface. The cells then stop multiplying and are subcultivated by removal from the surface with trypsin and placed into two new containers with fresh medium. When the growing surfaces of these two fresh containers are covered with cells, the population is said to have undergone one doubling. One measurement of a culture's age is that of the cumulative number of doublings the population has undergone. Subcultivation of normal cells can be done serially for various periods but cannot be done indefinitely; after a specific finite number of subcultivations, the population can no longer be passaged. At any point in the subcultivation, however, cells can be frozen in liquid nitrogen and, when thawed, will grow again. The total number of doublings they can attain, however, remains the same. Such procedures have been carried out for cell populations derived from both fetal and adult tissues.

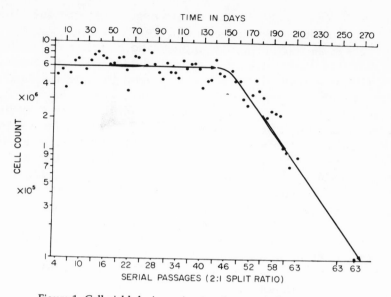

Figure 1. Cell yield during aging in a human diploid cell culture.

Interestingly, cells from adult tissues had a shorter life-span than those from fetuses.

Figure 1 is a typical growth curve for fetal lung cells in culture obtained by Hayflick and Moorhead. The points represent the cell yield at each subcultivation. As can be seen, after establishment in culture, there is a period during which proliferation is rapid and subcultivations can be carried out frequently. This is followed by a period of declining proliferative capacity during which many of the cells become granular and accumulate debris until, eventually, the culture is lost. The average number of population doublings before senescence was found to be 50 (range 35 to 63) for cells derived from fetal lung and 20 (range 14 to 29) for cells derived from adult lung. In both types, the cells retained their normal diploid character until late in the life-span, at which time nuclear abnormalities appeared.

Hayflick and Moorhead also carried out some simple but ingenious experiments to examine whether this limited life-span was environmentally induced. Exploiting the fact that cells from a male lack the Barr bodies (condensed sex chromatin) visible in the interphase nuclei of cells from a female, these workers mixed "old" male cells with "young" female cells and then measured how long it took each component in the mixture to die in comparison with unmixed controls. The results showed that the "old" cells in the mixture stopped growing after the same number of population doublings as did "old" controls, while the "young" cells in the mixture grew vigorously and stopped growing after the same number of population doublings as did the "young" controls. Thus, in the same medium, the old cells stopped growing at a time at which the young cells were proliferating vigorously. This experiment would seem to rule out any simple limitation on the life-span of the culture from deficiencies in the medium, from microbial contamination, or from the presence of toxic products produced by the cells. Additionally, when populations of cells were frozen in liquid nitrogen at a particular doubling level and later thawed and recultured, the cells "remembered" the level at which they had stopped growing and continued to divide from that point on until they had achieved the average number of doublings characteristic for their strain.

Therefore, it would seem clear that the regulation of senescence in human diploid cells in culture is intrinsic to the cells and precisely timed. However, we must still explain the findings of Carrel and others that purport to show the immortality of at least some cell lines. First, no one has been able to duplicate the Carrel experiments. Workers, in a systematic series of studies exactly duplicating the conditions used by Carrel, have been able to keep chick cells continuously proliferating for no more than 2 years. One possible explanation for Carrel's apparent long-term maintenance of these cells may be that live chick cells were inadvertently introduced into the experimental populations every time cultures were routinely fed with fresh chick embryo extract.

The immortality of the HeLa and L lines, however, is representative of a large number of cell types which do indeed appear to be essentially immortal. These cells occasionally appear to arise "spontaneously" in tissue culture. They may also be produced by infecting normal, mortal cell cultures with viruses or by exposing them to certain chemicals or to radiation. Such cells will continue to grow and multiply indefinitely if subcultured. Thus far, all "immortal" cell lines have been shown to be abnormal in one or more ways. They differ in their chromosome number and chromosomal morphology from the cells of the donor tissue. Many of these immortal cell lines not only were derived from tumors but give rise to tumors when introduced into appropriate laboratory host animals. Cell lines like the human fetal lung fibroblast-like cell lines are distinguished from indefinitely proliferating cell lines by the retention of features characteristic of the donor tissue. They are normal diploid cells and are not capable of initiating malignancy in laboratory animals.

The life history of cells in culture, therefore, can be summarized as follows: Cell cultures derived from normal tissues typically undergo a period of vigorous proliferation and then enter a senescent phase characterized by a reduction in proliferative ability that ultimately leads to death. At any point in their life history, cells can change and become abnormal. This change can be induced by viruses or other agents, or it can apparently occur spontaneously. These abnormal cells acquire an indefinite life-span.

The proclivity of cells in culture to transform and give rise to cells with unlimited proliferative potential seems to depend on the species from which they were derived. Fibroblast-like cells derived from chick embryos have been found to have a stable, limited life-span and to not give rise to permanent cell lines either spontaneously or after treatment with viruses. Mouse fibroblasts, on the other hand, after a period of rapid growth followed by proliferative decline, almost always give rise to permanent cell lines. Human fibroblasts lie between those of chick and mouse but are much closer to chick cells in respect to their general inability to transform "spontaneously."

RELATIONSHIP TO AGING *IN VIVO*

Although intrinsic timing mechanisms appear to exist in at least some types of cells in culture, it is important to consider whether or not such mechanisms are evident *in vivo* and whether aging in culture bears any relevance to aging *in situ*.

An old organism contains old cells. However, populations of cells must age at different rates because some tissues retain better functional capacities in old organisms than others. *In situ* cells look and behave differently. These cell populations *in vivo* can be divided into those such as skin, gut lining, and the blood-forming elements that are continuously proliferating, those such as cartilage and liver that are reverting postmitotic cells (i.e., cells which are not usually dividing but retain the capacity to divide throughout life), and those such as nervous tissue and muscle that have lost the capacity to divide and are called fixed postmitotic cells. It is important to recognize that all these types of cell populations show age changes; aging is not limited to the nondividing, fixed postmitotic cells. Several studies have indicated that proliferating tissues, like gut lining and regenerative liver, lose proliferative capacity as the animal ages.

In addition, studies in which normal somatic tissues have been serially transplanted to young inbred hosts appear to support the tissue culture studies and the concept that cells have an intrinsically regulated and limited proliferative capacity. For example, it was found that in mice of one genotype the survival of skin transplanted from mice of one age to mice of another age was related to the age of the grafted tissue only;

that is, transplants from young animals survived longer than did those from old animals. In mice of four other genotypes, however, old and young skin grafts behaved the same way after transplantation. The growth rate of normal mammary gland transplanted into gland-free mammary fat pads of young isogenic female mice has been found to decline during serial transplantation and is dependent upon the number of cell divisions previously undergone. Other workers have found this to be essentially true for bone marrow cells and for single antibody-forming cell clones.

Other studies of this general type are described in more detail in Chapter 3. Although there are a number of reservations that must be voiced concerning the experimental design used in these studies, the overall thrust of the evidence suggests that the cells do age intrinsically. It is important to point out, however, that it is not the senescence of one particular tissue that limits the life-span of the animal. Indeed, in many cases, transplant studies show that the life-span of the transplant is longer than the maximal life-span of the organism.

Perhaps the most striking evidence that aging in cell culture reflects aging in the whole organism has come from the observation that the life-span of cells in culture is inversely proportional to the age of the donor. Initially, Hayflick showed, for example, that only 14 to 29 doublings occur in lung fibroblasts derived from adult human donors, whereas 35 to 63 doublings occur in cells from human embryos. George Martin and co-workers at the University of Washington in Seattle confirmed Hayflick's observations in 1970 by culturing fibroblasts from upper-arm tissue from humans ranging in age from embryonic to 90 years. They found a significant correlation between life-span and donor age; the older the donor, the less time the derived cells lived in culture. The average number of divisions was calculated to decline by 0.20 ± 0.05 standard deviation for each year of the donor's life. Other researchers have made similar observations. Most recently, Edward Schneider at the Gerontological Research Center in Baltimore, Maryland, compared a number of parameters of growth in cells cultured from healthy individuals 21 to 36 years old with those from healthy individuals 63 to 92 years old. In terms of population doublings, the average life-span for the cells derived from the younger individuals was significantly higher statistically than that from the older group (45 vs. 30). If one follows these cells through their life-span *in vitro*, one finds that the cells from the older individuals show age changes much earlier than the ones from the younger individuals.

Finally, Martin and other workers have studied cells from individuals with progeria (Hutchinson–Gilford syndrome) and Werner's syndrome, both diseases of accelerated aging. Victims of progeria begin to show growth deceleration, graying and loss of hair, atherosclerosis, and calcification of the blood vessels as early as 9 years of age. Persons with Werner's syndrome, a condition resulting from an autosomal recessive mutation, show similar symptoms but not until the third or fourth decade of life. These investigators found that fibroblasts from patients with these disorders have a shorter life-span in culture than fibroblasts from normal persons of the same age; only 2 to 10 doublings occurred in comparison to the normal 20–40 doublings. Other workers have reported decreased mitotic activity, DNA synthesis, and cloning efficiency for cells from these persons. In addition, cells from diabetic individuals, who also show an early onset of some of the degenerative changes associated with senescence, have a reduced ability to grow and survive in culture. Thus, it is clear that age-related changes *in vivo* are reflected in the growth of cells in culture and can be studied and evaluated at the cellular level.

It is also clear, however, that people and animals do not die because their cells "run out." We cannot use these cell cultures to ask in any direct way why people die. We can, however, use cell cultures as models to learn a variety of things about the cellular aspects of senescence.

We can, for example, learn something about the general characteristics of aging cells. We can determine whether differentiation takes place in these aging cells or not. We can examine whether aging in these cells results from the accumulation of random errors, one of the major hypotheses of aging. We can study functional failures in proliferative capacity as a model for functional failures in general. We can also study other functional changes that accompany changes in proliferative capacity of cells, using proliferative capacity as a precise measure of the age of the population.

Cell cultures can be used to study the cellular effects of such environmental agents known to be associated with aging *in vivo* as radiation, oxygen, pollutants, and carcinogens.

Finally, the two principal age-associated diseases, cancer and atherosclerosis, basically involve changes in the control of proliferation. We can use normal cell cultures to study the factors controlling the initiation and progression of cell proliferation as it is related to these diseases.

Scientists have taken many of these approaches in using cell cultures to study aging. However, basic to an understanding of the results that are obtained is a thorough analysis of the characteristics of cell behavior in culture.

POPULATION DYNAMICS

A fundamental question about the *in vitro* cell model system has been whether the decline in the proliferative capacity of aging cultures reflects a uniform decline in all the cells, a decrease in the fraction of cells in the population that are proliferating, or a combination of these factors. Two approaches have been taken to answer this question. One approach has utilized autoradiography, a technique which allows one to visually determine the location of a radioactive substance taken up by cells. If cells are labeled with radioactive (tritiated) thymidine, a precursor for DNA synthesis, one may determine the percentage of cells traversing the S phase of the cell cycle. (The cell cycle is commonly divided into G_1, the pre-DNA synthetic phase; S, the period of DNA synthesis; G_2, the post-DNA synthetic phase; and M, the mitotic phase, the period of division itself.) Analysis of such cultures has revealed that nearly 100% of the cells in young cultures synthesize DNA while only about 20% of the cells do so in old cultures. As the population ages, there is an exponential decline in the fraction of cells incorporating the radioactive precursor (Figure 2). The vast majority of cells that are not, or appear not to be, cycling have been shown by cytophotometry, a technique which measures the DNA content of individual cells, to be arrested in the pre-DNA synthesis or G_1 phase of the cell cycle.

The second approach to the population dynamics of these cultures has utilized time-lapse cinematography to measure interdivision times

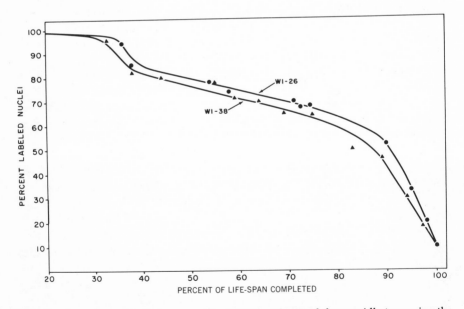

Figure 2. Percentage of cells incorporating [^{3}H]thymidine and thus rapidly traversing the cell cycle during the life-span of two human diploid cell populations.

(the time between mitoses) of individual cells. A great deal of heterogeneity, with much overlap in the interdivision time of cells from both young and old cultures, has been found. On the average, however, cells from old populations have longer interdivision times than cells from young populations.

The population dynamics of cells in culture, therefore, can be summarized by saying that individual cells asynchronously proceed through a series of proliferative states, ranging from rapidly proliferating to more and more slowly proliferating until, finally, what appears to be a total arrest of DNA synthesis and cell division occurs. These arrested cells then undergo changes in appearance and behavior which will be described below.

As a larger and larger proportion of the cells reach this state of arrest, the population appears more and more senescent until, finally, the few remaining cells in the population able to proliferate are not sufficient to repopulate the culture vessel, and the culture is lost.

Typically, the age of a culture is expressed in terms of the number of doublings undergone. However, a question of fundamental importance is whether the cells measure age by the passage of time or by the number of division events. To answer this question, Robert Dell'Orco of the Samuel Roberts Noble Foundation in Oklahoma and his co-workers kept cells in a nondividing state for various periods of time, ranging from 21 to 177 days, by reducing the serum in the medium. Then the serum was replaced and the cells were allowed to resume dividing until they reached senescence. The controls, which had been kept in a continuously proliferating state, became senescent after about 32 doublings. Those maintained 177 days became senescent after about 36 doublings. The total number of population doublings achieved was essentially the same, yet one group of cultures, those kept in maintenance, had lived 287 days and the other only 77. Chronological time, then, seems to be significantly less important than division events.

Many physiological changes have been found to accompany the decrease in proliferative capacity in aging populations of human diploid cells in culture. For one thing, as the cultures age, the cell density reached in the stationary or plateau phase decreases. Similarly, cells of microorganisms in culture pass through lag, logarithmic, stationary, and declining growth phases. The signal to enter a stationary phase in cell culture is related, in part, to cell-to-cell proximity or contact. Cells seem to become more "sensitive" to one another's presence as they get older. Furthermore, growth and division appear to become uncoupled. On the average, cells from old populations are larger than those from young populations. Nuclear size also increases, although DNA content remains approximately the same. Evidence is accumulating now that DNA repair may be deficient in aging cells and that cells deficient in DNA synthesis may also be deficient in repair.

Many studies have indicated that RNA content increases in aging cells, although RNA synthesis decreases. Protein content also increases with age while the rate of protein synthesis declines. Cellular lipids also increase with age. These observations are consistent with the view mentioned above that growth and cell division become uncoupled or unbalanced, thus allowing for the accumulation of many kinds of macromolecules.

The age-associated accumulation of the autofluorescent "age pigments" or lipofuscin granules, so prominent in aging cardiac muscle cells and neurons *in vivo,* is also found in aging cells in culture. Typically, an old cell appears to be engorged with secondary lysosomes, cell organelles associated with the digestion or breakdown of cell products. Biochemical studies have shown an increase in lysosomal enzyme activity in senescent cells. Possibly related to this is the finding that there may be alterations in the rates of protein turnover in older cultures. One other related set of observations is that altered or aberrant proteins may accumulate in older cells. There are conflicting reports about whether these aberrant proteins accumulate or not in older cells. If they do, they could result from faulty protein disposal, so that partially degraded proteins are formed in old cells. Perhaps the accumulation of lysosomes is due to faulty lysosome function, which results in deficient protein disposal processes. Alternatively, there may be a failure in the fidelity of the protein-synthesizing machinery, so that aberrant proteins are produced.

The study of cellular aging has involved attempts to modulate life-span in the hope that understanding the mechanism of modulation would elucidate the fundamental regulatory mechanisms involved in the biology of aging. Arthur Balin in our laboratory at the Wistar Institute has shown that reducing oxygen tension from an atmospheric Po_2 of approximately 137 mm Hg to 49 mm Hg or to 24 mm Hg produced no significant difference in the life-span of cells. High (twice-atmospheric) oxygen tension was extremely toxic, however, and shortened life-span drastically. Other workers have suggested that vitamin E prolongs life by trapping the by-products of oxidative action within cells. We have attempted to prolong life-span and to protect the cells against high oxygen tension with this agent; in our laboratory, the presence of vitamin E at two different concentrations and at four different oxygen tensions had no effect on the life-span of the cultures.

The only agent that has been consistently shown to extend cellular life-span is hydrocortisone, a hormone produced by the cortex, or outer

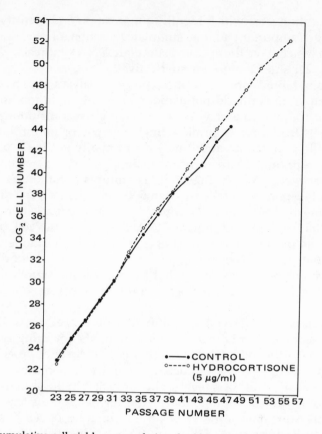

Figure 3. Cumulative cell yield vs. population doublings in the presence and absence of exogenously added hydrocortisone.

layer, of the adrenal gland. Workers in our laboratory have found that hydrocortisone at a concentration of 5μg/ml (14 mM), a concentration which is higher than the concentration found in circulating human blood, extends the life-span of cultures of human fetal lung fibroblasts 30 to 40% (Figure 3). DNA synthesis, measured autoradiographically, is increased significantly by exposure to hydrocortisone. Interestingly, there is no rescue with the hormone, for once a culture is senescent, hydrocortisone does not reverse the condition. However, if hydrocortisone is added at different points in the life-span of the population, the magnitude of the extension of the life-span is in direct proportion to the amount of time the culture is grown in its presence. It is clear that the life-span effect is not a general effect but is highly specific with regard to the molecular configuration of the hormone. The cell appears to have precise recognition sites which identify the position and identity of the atoms that are important to the stimulatory action of the hormone. Compounds which have these atoms stimulate DNA synthesis, whereas those which lack one or more are ineffective.

Autoradiographic and cytophotometric analyses of hydrocortisone-treated young and old populations of fetal lung fibroblasts have revealed that the hormone may act by delaying the transition of more and more cells from rapidly proliferating to slowly or nonproliferating fractions of the population, and it may act primarily in the G_1 phase of the cell cycle. Although hydrocortisone maintains a higher percentage of cells in the actively proliferating pool, as cultures age there is still an overall loss of responsiveness to the signal(s) that initiate(s) cell division. Tissue culture medium normally contains a serum supplement; since hydrocortisone can only stimulate DNA synthesis and cell division when serum is present, we have suggested that the hormone acts by amplifying the primary serum signal for division. This hypothesis is supported by our recent discovery that hydrocortisone significantly increases the uptake of ^{125}I-labeled serum protein by young, but not old, cells.

In attempting to understand cellular senescence, a question of major interest involves whether the "determinants" of aging reside in the nucleus or in the cytoplasm, or in both. The direct approach — to transplant by microsurgery the nucleus from a young cell into an old cell and vice versa — has been technically difficult, although some workers are reporting success in preliminary experiments. A more feasible approach to this same question involves the techniques of cell hybridization. In this procedure, cells of different types are incubated together in the presence of Sendai virus, which enhances the fusion of cells. The fused cells are heterokaryons, cells containing two nuclei. In some cases, nuclear fusion also occurs to form a synkaryon or hybrid. Neither heterokaryons nor hybrids prepared from mixtures of old and young cells acquire the characteristics of the young component, indicating that senescence is a "dominant" characteristic. However, Martin and his co-workers have shown that if senescent human cells are fused with HeLa cells, which are a tumor-derived cell line, or with virus-transformed cells, DNA synthesis is reinitiated in the senescent normal nuclei.

In another series of experiments, Hayflick and his associates at Stanford University enucleated cells using a substance called cytochalasin B, which loosens the attachment of the nucleus to the cytoplasm. The anucleate cells or cytoplasts were then fused with whole cells of various ages. The data obtained in these experiments suggest that it is the nucleus that is primarily involved in the phenomenon of senescence. These kinds of experiments should continue to produce exciting insights into the control of cellular aging.

In conclusion, it is clear that cells grown in culture age. The rate of aging is precise and reflects the previous age *in situ* and the aging of that cell type. In addition, a well-defined array of morphological and biochemical changes accompany cellular aging *in vitro*. The techniques are now available to dissect the factors regulating the aging of cells; by understanding these factors, we are brought closer to an understanding of aging in the intact organism.

ACKNOWLEDGMENT. This work was supported in part by U.S. Public Health Service Research Grant AG-00378 from the National Aging Institute.

BIBLIOGRAPHY

Balin, A., D. B. P. Goodman, H. Rasmussen, and V. J. Cristofalo. 1977. The effect of oxygen and vitamin E on the lifespan of human diploid cells *in vitro*. *Journal of Cell Biology* 74: 58–67.

Cristofalo, V. J. 1972. Animal cell cultures as a model system for the study of aging. *Advances in Gerontological Research* 4: 45–79.

Cristofalo, V. J. 1975. Hydrocortisone as a modulator of cell division and population life span. Pages 57–79 in V. J. Cristofalo, J. Roberts, and R. C. Adelman, eds. *Explorations in aging*. Plenum Press, New York.

Dell'Orco, R. T., J. G. Mertens, and P. F. Kruse, Jr., 1973. Doubling potential, calendar time, and senescence of human diploid cells in culture. *Experimental Cell Research* 77: 356–360.

Hayflick, L. 1965. The limited *in vitro* lifetime of human diploid cell strains. *Experimental Cell Research* 37: 614–636.

Hayflick, L., and P. S. Moorhead. 1961. The serial cultivation of human diploid cell strains. *Experimental Cell Research* 25: 585–621.

Martin, G. M., C. Sprague, and C. J. Epstein. 1970. Replicative life-span of cultivated human cells. *Laboratory Investigation* 23: 86–92.

Norwood, T. H., W. R. Pendergrass, and G. M. Martin. 1975. Reinitiation of DNA synthesis in senescent human fibroblasts upon fusion with cells of unlimited growth potential. *Journal of Cell Biology* 64: 551–556.

Schneider, E. L., and Y. Mitsui. 1976. The relationship between *in vitro* cellular aging and *in vivo* human age. *Proceedings of the National Academy of Sciences of the USA* 73: 3584–3588.

Wright, W. E., and L. Hayflick. 1975. Nuclear control of cellular aging demonstrated by hybridization of anucleate and whole cultured normal human fibroblasts. *Experimental Cell Research* 96: 113–121.

3

Is Limited Cell Proliferation the Clock That Times Aging?

DAVID E. HARRISON

Senility may not be pleasant to look forward to, but consider the alternative. If we didn't lose our health and vigor (not to mention our jobs) as we grew old, aging would be a pleasure. We often don't know enough to take full advantage of opportunities when they first appear; with time, we accumulate wisdom and perspective. Especially for creative and valuable people, life becomes a race between increasing wisdom and ability vs. declining health and capability — a race that mankind usually loses. These facts define a major goal for researchers in biological aging: to maintain optimal health, even though people get old.

To do this we need to locate the aging clock. We need to discover how losses in the multitude of biological functions that decline with age are timed. Human beings are kept healthy by an enormously complex set of interacting cells and tissues, each performing particular tasks. A greatly simplified analogy may be made to a bicycle, kept running well by interaction between its wheels, chain, pedals, and frame.

HOW AGING IS TIMED

The bicycle in Figure 1A represents one way in which aging could be timed. All its parts age, losing the ability to do their jobs, at the same rate. The entire machine is ready to collapse at once, like "The Deacon's Wonderful One-Horse Shay" in the poem by Oliver Wendell Holmes. The bicycle in Figure 1B represents a second possibility. None of its parts have aged except the front wheel. However, that wheel functions so poorly that the entire machine will be grossly shaken and banged about as it is used. It will end up in as bad condition as the one in Figure 1A.

DAVID E. HARRISON, Ph.D. ● The Jackson Laboratory, Bar Harbor, Maine 04609

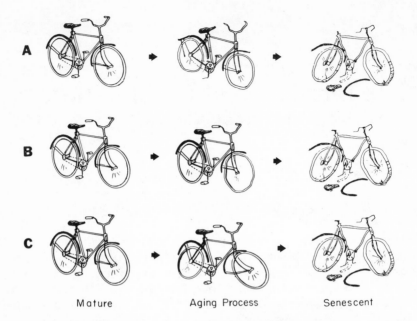

Figure 1. These bicycles illustrate three different possible ways aging could be timed on the cell tissue level: (A) All the parts are intrinsically timed to age at similar rates. (B) Only one key part, the front wheel, is intrinsically timed to age initially. Its malfunction damages the other parts, causing them to age also, although they could have continued to function normally if the front wheel had not aged. (C) No single part ages intrinsically, but their interaction becomes defective. The poorly aligned parts damage each other, causing them to age, although they could have continued to function normally if their interactions had remained correct.

The third way in which aging could be timed is illustrated in Figure 1C. No individual bicycle part becomes defective, but the parts no longer interact correctly. The wheels are not straight, the pedals and chain hit the frame and wheels, and these misalignments grow worse. Eventually the entire bicycle is destroyed.

Consider the individual tissues of a man or a mouse as functional units analogous to the parts of the bicycle. The same three possibilities exist:

    A. Aging is timed by an internal clock operating in every tissue. Thus, all losses in function with age are timed intrinsically within the defective tissue. Since most of our tissues age and become defective simultaneously, their clocks operate at similar rates.

B. One key tissue contains the clock. All other functions that decline with age are secondary to initial aging in this key tissue; thus, their deterioration is paced by the clock in the key tissue. Instead of a loss in function, the clock might be a normal function of the key tissue, for example production of an aging hormone.

C. The aging cells and tissues interact less effectively with each other, although they are individually healthy. However, they are so dependent on each other that defective interactions cause all to decline in functional ability.

Of course, the situation may be much more complicated. For example: all three possible timing mechanisms may occur at once, with intrinsic clocks operating at different rates. The first vital tissue to age (the one with the fastest clock) would be the key tissue; all tissues depending on it would become defective, as it did. Furthermore, the key tissue might be different in different people.

But no matter how complex the aging process is, we must start somewhere to unravel it. It is just possible that the cause of aging is simple and will be easily cured once we find it. Perhaps only the results of aging, not the causes, are hopelessly complex.

THE GERONTOLOGIST AS BICYCLE MECHANIC

The analogy with the bicycle illustrates this point, if you consider the researcher studying biological aging as analogous to a repairman assigned to fix the bicycle in Figure 1. To rejuvenate the bicycle in Figure 1A, each part must be repaired individually to counteract the aging process. However, the bicycles in Figure 1B or 1C could be rejuvenated by repairing only the front wheel or the alignments, respectively. These three possibilities would be difficult to distinguish because the initiating cause of aging would not be the sole defect. A careless repairman might miss the defective wheel or misalignment and concentrate on defects that were results, not causes. Of course, the repairs would be quickly undone because the cause of the problem would not have been corrected.

But how are we to distinguish causes from results? There is no lack of aging symptoms, but how, in a creature as complex as a man or a mouse, do we recognize the initiating causes of senescence? The system must be simplified, as illustrated in Figure 2. The aging mouse has many tissues whose function has become defective. To determine whether a

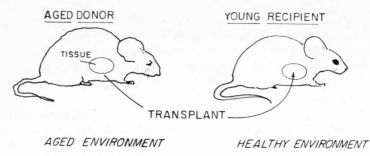

AGED DONOR   YOUNG RECIPIENT

TISSUE

TRANSPLANT

AGED ENVIRONMENT   HEALTHY ENVIRONMENT

	POSSIBILITY	RESULT
A.	AGING INTRINSIC IN ALL CELLS	DEFECT CONTINUES
B.	AGING TIMED BY ONE CRUCIAL TISSUE	DEFECT CURED
C.	AGING RESULTS FROM INTERACTION	DEFECT CURED

Figure 2. To determine whether or not aging in a particular tissue is intrinsically timed by an internal clock, that tissue is transplanted from the aged donor into a young recipient. If aging is intrinsic (possibility A), the tissue will become defective with age, just as if it had been left in the old donor. If aging is not intrinsic (possibility B except for the key tissue, possibility C for all), the old tissue will not become defective but will function as well as transplanted young tissue.

tissue contains its own aging clock, it must be taken out of the old mouse and studied in a young healthy environment. Its function will not improve if the loss is timed within the tissue removed. Instead, the old tissue in the young environment would continue to lose its functional ability as if it had remained in the old donor, and this would support possibility A. However, if possibilities B or C are true, the transplanted tissue's functional ability will be restored by moving it to a healthy environment. Of course, many different tissues must be studied to determine which of the three general possibilities in Figures 1 and 2 is most nearly correct.

As you might expect, actually doing these experiments is not so simple. There are many pitfalls that have only become apparent through years of work. I will illustrate with some specific examples.

TISSUE CULTURE STUDIES

From 1915 until about 1965, biologists often discounted possibility A, that aging was intrinsic within all or most tissues, because chicken embryo cell lines had been reported to live and multiply in tissue culture for indefinitely long times. Furthermore, many other cell lines, usually from tumors or from fetal mice, also grew steadily with no sign of slow-ing down.

When cells are grown in tissue culture, a bit of tissue is removed from the animal and placed in a sterile flask. Solutions containing nu-trients are given to the cells and periodically renewed. When the cells fill the bottom of a flask, a fraction of them is transferred to a new flask, where they have room and fresh nutrients for further growth (Figure 3). Thus, tissue culture seemed to be a way of removing cells from the limitations of the body in which they grow. Because, under these condi-tions, they appeared able to grow forever, this experiment suggested that they had no intrinsic clock, thereby disproving possibility A.

Modern tissue culturists could not make chicken cell lines live indefinitely. In the early work, fresh cells had accidentally been added with fresh culture media. Furthermore, the immortal cell lines from other species often had abnormal numbers of chromosomes compared to the species from which they came. Cells from these lines often had come from tumors, and even when they had not, often caused tumors when injected into animals like their donor.

On the other hand, there were cells that would grow in culture for a time, but then stop. These cells had normal chromosomes and did not cause tumors. Sometimes the normal cells would change their growth pattern and acquire the potential for unlimited growth. This transforma-tion was usually accompanied by the appearance of abnormal chromo-somes and the ability to form tumors. At present, researchers in tissue culture recognize two general classes of cells: normal cells with unal-tered chromosomes that eventually will stop growing, and transformed cells with abnormal chromosomes that often grow without limit.

Leonard Hayflick and his colleagues at the Wistar Institute and later at Stanford University made this point most clearly. They showed that normal human cells placed in their cultures could double only about 50 times if from embryos and 20 times if from adults.

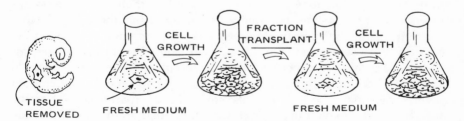

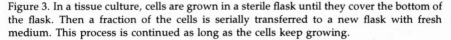

Figure 3. In a tissue culture, cells are grown in a sterile flask until they cover the bottom of the flask. Then a fraction of the cells is serially transferred to a new flask with fresh medium. This process is continued as long as the cells keep growing.

In a piece of deduction that electrified researchers on aging, Hayflick proposed that the limited proliferative capacity of normal cells in tissue culture offered a good model of senescence. He even referred to the number of times a culture had doubled as its age. Because a single cell type in tissue culture is relatively straightforward to study compared to an intact animal, Hayflick's model system is extremely attractive to many scientists. But is it a valid system in which to study aging?

We do not know the answer to this question. If aging in an intact individual is caused by the same thing that limits cell replication in tissue culture, then the model system is valid. In any case, it produces valuable information that would be difficult or impossible to obtain using more complex systems, as Chapter 2 demonstrates.

Whether or not gerontologists accept Hayflick's cultured cells, or any other system, as a model for aging is left to their intuition and their favorite theories and approaches. The decision cannot be based on clear-cut facts because the facts are ambiguous. This kind of situation holds true in most areas of science that are on the cutting edge of knowledge between the known and the unknown. While such areas are tremendously exciting and the most fun to work in, they are confusing and frustrating to people outside the field who don't know which scientist to believe. To learn who believes what, and why they do, I recommend Albert Rosenfeld's recent book, *Pro-longevity*.

UNCERTAINTIES IN THE CULTURE SYSTEM

Hayflick's results suggested that aging might result from cells running out of proliferative ability. However, although cells from old people tended to have reduced proliferative capacities, they had enough left to keep producing cells for much longer than the donor's remaining life expectancy.

Even the reduction observed may not have been intrinsic. Cells from elderly people may have been damaged by their residence on unhealthy old bodies. The possibility of damage from the internal environment of the aging individual besets all experiments with aging tissues. In the system used (skin fibroblasts), animal experiments are possible to determine whether such damage occurred. Skin grafts from old and young donors could be carried on young and on old recipients for several months. Then their cells could be compared to determine whether their behavior in tissue culture depended on the age of the skin and its cells or on the age of the recipient upon which they had been growing.

These kinds of experiments have their own problems — for example, making sure the cultured cells came from the original skin donors. Later in this chapter, I will discuss how this problem was solved for stem cell transplants by use of the T6 chromosome marker. The same marker could be used to identify skin fibroblasts. I believe that the combination of tissue culture and grafting experiments has great potential value in research on aging.

The final uncertainty stems from the simplified nature of the tissue culture system. One cannot prove beyond any doubt that the chemicals constituting the medium, methods for transferring cells, vessel surfaces, gas mixtures, cell interactions, or other facets of the culture system are optimal. Some workers have reported increased proliferative capacity after improvements in the culture system. Perhaps under ideal (as yet undiscovered) conditions, cells from old and young adults would not differ, or cells would not run out of proliferative capacity; perhaps old cells require a different culture system than young cells for optimum performance.

When dealing with as complex a creature as a man or a mouse, one can never be certain that these uncertainities have been eliminated. Therefore, the predictions from tissue culture are tested in living animals.

PROLIFERATIVE CAPACITY IN ANIMALS

If there is a loss of normal function with age resulting from limited proliferative capacities, it should show up in tissues whose cells must multiply to function. A number of different epidermal and epithelial tissues in mice have shown decreased proliferative activities and renewal rates with age. Unfortunately, these findings alone do not tell us whether the decreased proliferative rates are intrinsic to the old tissues (possibility A, Figure 1) or result from their residence in an aging body (possibilities B or C, Figure 1). This same reservation applies to liver cells, whose ability to regenerate is slightly delayed in older rats.

TRANSPLANTED TISSUES

To measure accurately how well old tissues are able to function, they should be transplanted into a healthy environment as described in Figure 2. Normal tissue interactions only occur when tissues are studied in an intact animal. This is an old idea, but only in the past 30 years have we understood how to do the long-term transplantations required for such experiments. The donor and recipient must be histocompatible;

that is, the recipient's immune system must not react against the transplanted cells.

Erich Geiringer of the Royal Hospital for Sick Children, Glasgow, Scotland, was one of the first modern investigators to understand the importance of distinguishing intrinsically timed aging changes in an organ from changes induced by the aging animal's internal environment. He transplanted adrenocortical tissue and showed that this tissue could function for at least 3 years, supporting normal growth rates and reproductive performances. This was 6 months longer than his rats lived, but was not long enough to prove conclusively that adrenocortical cells did not age intrinsically. Although his animals were diseased and not fully inbred, Geiringer took this into account in interpreting his results. He did the best he could and described what technical improvements were required.

Ovarian and skin grafts were transplanted by Peter Krohn of the University of Birmingham, who also saw clearly the importance of determining whether tissues aged intrinsically. When transplanting ovaries, Krohn had an obvious quantitative measure of normal function, the number of normal offspring. He showed convincingly that young fertilized ovaries or eggs function much less well in old than in young recipients, indicating that other factors besides the ovary affect reproductive ability with age. Furthermore, fertilized eggs from old donors were as viable as those from young donors in young recipients. Clear-cut results were not obtained with old ovaries because only a few functioned as well as young ones. This may have been due to increased sensitivity of old ovaries to damage during grafting, or intrinsically timed aging, or both.

Krohn's skin grafts from certain types of old mice grew as well as did grafts from young controls. With successive transplantations, the grafts became progressively smaller and many were lost, but this problem occurred with both old and young grafts. In 1966, the longest-lived graft was 6.7 years old. No clear functional tests were reported to show that the old skin was rejuvenated by residence on a young recipient. Furthermore, there was no way to prove that the functioning cells had in fact come from the original donor and had not migrated in from the successive recipients.

The problem of identifying the donor cells is less crucial in transplants of more complex organs. In these, it seems less likely that essential cells could migrate in from the recipients. Carel Hollander and colleagues at the Institute for Experimental Gerontology in Rijswijk, the Netherlands, transplanted kidneys from rats of different ages into young compatible rats whose own kidneys had both been removed. This was indeed a rigorous test of functional capacity, because unless the transplanted kidney quickly began to function, the recipient died. Comparing 3-, 18-, and 26- to 34-month-old donors, half the recipients survived for 16, 13, and 7 months, respectively. Some old kidneys functioned as well as the young ones. The maximum kidney life-span was 46 months, while the longest-lived rat of the strain used lived for 39 months. This suggests that old kidneys might function longer than rats can live. Old kidneys were much more likely than young ones to be damaged by lack of oxygen during transplantations. This increased vulnerability may have resulted from residence in an old body rather than from intrinsically timed aging. The difficulty in separating these factors prevents a clear-cut interpretation of these experiments.

When interpreting experiments of this sort, there is an important fact to remember. If even a few old tissues of a particular type remain fully functional for significantly longer than the maximum life-span of the donor species, they constitute strong evidence against intrinsically timed aging in that tissue (possibility A, Figure 1). The evidence would only be slightly improved if most tissues tested had remained fully functional, because residence in an old animal's body, or other extrinsic factors, could easily have permanently damaged many of the old tissue tested, causing failure to function.

Other intact tissues transplanted in experiments on aging include pituitary gland and thymus. Ming-Tsung Peng and Hive-Ho Huang in Taiwan reported that some 23-month-old pituitary transplants supported vaginal cycles and ovarian activity in recipients that had their pituitary glands removed, but 4-month-old transplants were more likely to succeed. Of course, these experiments were not adequate to fully test the functional capacity of the complex pituitary gland.

Katsuiku Hirokawa and Takashi Makinodan of the Gerontology Research Center transplanted old thymus grafts into recipients whose thymuses had been removed and found that they recovered some thymic functions nearly to the level allowed by young grafts. However, recipients of old grafts recovered more slowly than recipients of young grafts, and too few old grafts were given enough time for clear-cut results. It would be exciting if the thymus, an organ that shrinks to a shriveled fragment of its former self during the adult life-span, could be rejuvenated by residence in a young recipient. These experiments leave open the possibility that aging in even the thymus might be caused by factors outside the organ.

Charles Daniel and Kenneth DeOme of the University of California at Santa Cruz and Berkeley and colleagues have studied the proliferative capacity of mammary gland epithelial tissue. This tissue grows rapidly when a piece is transplanted onto its natural environment, the mammary fat pad previously cleared of all host epithelial tissue. After out-

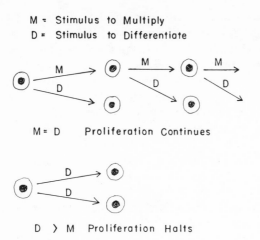

M = Stimulus to Multiply
D = Stimulus to Differentiate

M = D   Proliferation Continues

D > M   Proliferation Halts

Figure 4. A stem cell may multiply to produce more stem cells, or differentiate to produce cells for particular jobs. (The differentiated cells may multiply a great deal as they further differentiate.) If the pool of stem cells is to be preserved, the stimulus to multiply and produce more stem cells (M) must be balanced with the stimulus to produce differentiated cells (D), as illustrated in M = D. If the stimulus to differentiate is too great, as in D > M, too many stem cells will differentiate without producing more stem cells. Eventually, the stem cell pool will be exhausted.

growths from an initial transplant fill the fat pad, a piece of the freshly grown epithelial tissue is transplanted to another fat pad. After a few successive transplants, the outgrowth rate slows. The percentage of fat pad covered in 8 to 12 weeks declined from 100% to 10 to 25% by the fourth successive transplantation. This was interpreted as showing a "relatively short life-span even under optimum conditions." In fact, this interpretation is not entirely supported by the evidence available. Removing a tiny piece of tissue and transplanting it onto a damaged bed where it is stimulated to grow as fast as possible may not provide optimal conditions. The unusually strong stimulus to populate the bed resulting from the experimental procedure may have hastened exhaustion of proliferative capacity.

This is a general criticism resulting from the techniques required to transplant rapidly multiplying cells. The amount of proliferative capacity such cell types have depends on how much the earliest precursor or stem cells can replicate to produce more stem cells. Excessive stimulus to populate the host may channel too many stem cells into producing differentiated progeny, leaving too few to produce new stem cells. This effect is illustrated in Figure 4.

Assuming that proliferative ability is a good measure of mammary epithelial cell function, the important question is whether it declines with age, not whether it is damaged by successive transplantations. Researchers attempting to answer this question compared mammary transplants from young and old donors. These showed no difference in proliferative ability when serially transplanted to cleared fat pads in either old or young hosts. Unfortunately, virgin females were used as donors in both the old and young groups. Their mammary tissues had been deprived of the normal stimulation to proliferate, so these experiments did not completely answer the basic question. After a lifetime of normal functioning, is the old tissue intrinsically defective?

MARROW CELL TRANSPLANTS

The question of whether the tissue has really functioned normally throughout the life-span does not arise with marrow cells. As outlined in Figure 5, the essential cells in a marrow graft, the stem cells, must constantly proliferate to maintain their own numbers and to produce the huge numbers of short-lived blood cells and immune responsive cells required by the body.

Marrow cells are easily transplanted by injection as a suspension into the recipient's bloodstream (Figure 6). The recipients must be lethally irradiated to kill all their own stem cells. They are populated and their lives are saved by the marrow grafts. In such experiments, it is essential to have an unambiguous method for identifying donor cells, because if even small numbers of recipient cells survive the irradiation, they are able to multiply and repopulate the recipient. Then they may function normally, giving the false impression that the donor cells are still functional.

H. S. Micklem and John Loutit of the Radiobiological Research Unit, Harwell, England, used an abnormal chromosome to identify donor stem cell lines in the first well-designed experiment for studying the

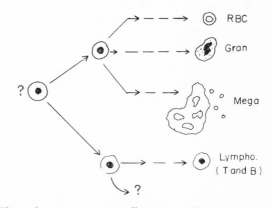

Figure 5. Stem cells in the marrow differentiate to form CFU (colony-forming units in spleens of irradiated mice) that are precursors of RBC (red blood cells,), Gran. (granulocytes, a type of white blood cell), and Mega. (megakarocytes, which produce platelets). Other stem cells differentiate to form the precursor cells that populate the immune system, differentiating into T and B lymphocytes. Possibly the same stem cell line can produce both kinds of precursors. Most of the cells produced by these processes are short-lived. Thus, the marrow stem cells must continue to produce red blood cells, granulocytes, megakarocytes, and T and B lymphocytes constantly throughout the life of a human being or a mouse.

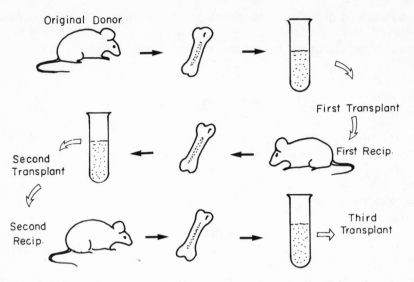

Figure 6. To transplant marrow cells from a mouse, they are rinsed from long bones (femurs and, possibly, tibias) into a single cell suspension in tissue culture media. Then the cell suspension is injected (usually intravenously for best implantation of stem cells) into the recipient, whose own stem cells have been destroyed or are genetically defective. Once the first recipient's marrow has been repopulated by the graft, the marrow cell line may be serially retransplanted into successive recipients.

proliferative capacity of marrow stem cells. After 1 year, marrow cells from the initial recipient were transplanted into a second irradiated recipient as in Figure 6. Some recipients were checked to be sure that they contained cells with abnormal chromosomes, proving that they came from the original donor. This procedure was repeated every year, and after about 40 months and three or four successive transplantations, the stem cell lines were no longer very good at populating lethally irradiated recipients and saving their lives. One recipient still had a few cells containing the abnormal chromosome after five successive transplantations over 60 months, but that was the maximum observed.

Micklem and Loutit believed that most marrow cell lines had functioned poorly after about 40 months, the maximum mouse lifespan, because they were intrinsically timed to age. Stem cell lines must have a high proliferative capacity and be able to produce many cells rapidly in emergencies to do the jobs outlined in Figure 5. If aging in even such versatile cells is intrinsically timed, it strongly supports possibility A, Figure 1. But there is an alternative interpretation of this experiment.

Aging may not have anything to do with the loss of proliferative capacity in repeatedly transplanted stem cells. The transplantation procedure itself may have damaged the cells. This possibility can be readily tested by transplanting old and young stem cells in parallel experiments. If the number of transplantations, not the age, is important, the old cells will function as well as the young ones.

CRITERIA THAT SHOULD BE SATISFIED BY TRANSPLANTATIONS ON AGING

After critically reviewing large numbers of transplantation experiments on aging, we see that many have flaws preventing clear-cut interpretations. To avoid the most obvious flaws, I devised three criteria that transplantation experiments on aging should meet and designed an experimental system which met them. These experiments, reported in 1973 and 1975, obviously build on the ideas and experience of previous investigators, especially Krohn, and Micklem and Loutit. The criteria are:

1. *Function.* Intrinsically timed aging should be measured by loss of the ability to function normally.
2. *Identification.* The transplanted tissues should be unambiguously identified.
3. *Control.* If aging is intrinsically timed in old tissues, they will lose functional ability compared to identically treated younger controls.

To meet criterion 3, in our experiment we compared marrow cells from old and young animals in parallel as outlined in Figure 7. This comparison provided controls for damage due to transplantation rather than aging. Instead of lethally irradiated recipients, we used mice with a hereditary anemia causing stem cell defects. With no irradiation, these mice were populated by donor cells that take over red blood cell production (Figure 8). This was demonstrated using donors with a different kind of hemoglobin from the recipients. The hereditarily anemic recipients produced only donor-type hemoglobin after they were cured, proving that their red blood cells were now being produced by stem cells from the donor. The hereditary anemia was never cured spontaneously, so both the donor hemoglobin type and the cure of the anemia identified the donor cells, meeting criterion 2. The function of red blood cell production can be rigorously measured in the recipients. This tests the functional ability of the transplanted old stem cells, meeting criterion 1.

The basic functional test is whether the anemic recipient is cured and normal levels of donor hemoglobin are maintained in the blood. These measurements indicated normal stem cell functions, as did the fact that cured anemic mice responded normally to erythropoietin, the hormone that stimulates red blood cell production. However, the most dramatic functional test is to severely bleed the cured mice and determine how rapidly they recover. This stresses the stem cell lines to produce red blood cells at their maximum rate. Any intrinsic defect should become apparent in these experiments. Figure 9 shows a summary of many such experiments. All the old and young stem cell lines tested supported normal recovery rates, even though some had functioned for 70 to 73 months and four serial transplantations. There was no evidence for intrinsic aging of the red-cell-producing stem cell lines.

However, only anemic mice that had been well cured were chosen for this experiment. Cell lines that had been retransplanted several times

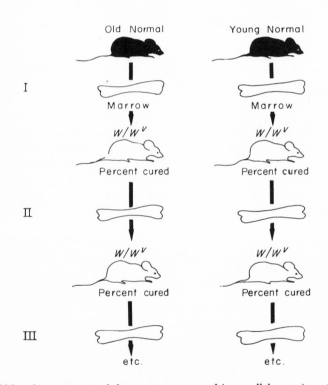

Figure 7. Old and young normal donors are compared in parallel experiments to distinguish the effects of aging from effects of transplantation.

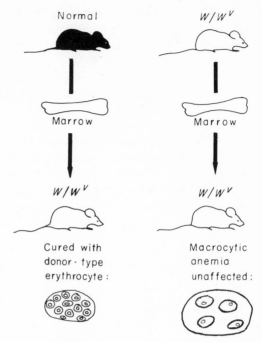

Figure 8. Genetically anemic W/W^V mice have defective stem cells. Normal stem cells populate their marrow, producing donor-type red blood cells (erythrocytes) and curing their anemia. The defect is qualitative, not quantitative, because giving a W/W^V anemic mouse more W/W^V marrow cells does not help it at all.

were less likely to cure anemic mice, regardless of their age. Figure 10 shows that the percentage of anemic mice cured by both old and young stem cell lines declined in a parallel fashion with repeated serial transplantations. After three successive transplantations, percentages began to drop significantly. Most transplantations were done at annual intervals; when done at 3-month intervals, percentages of mice cured dropped at least as rapidly, confirming that the loss in proliferative capacity depended on the number of serial transplantations, not on age.

Other evidence that transplantation damages stem cells was reported by Dane Boggs and his colleagues at the University of Utah. Using irradiation, they killed almost all of an animal's stem cells, then measured how rapidly the cells regenerated. When this procedure was repeated several times, the measured stem cell proliferative capacities were much higher than those reported by workers who transplanted stem cells into successive irradiated recipients. Apparently, the operation of transplanting the stem cells reduced their proliferative capacity.

Transplantation into an irradiated or anemic recipient may damage stem cells because it overstimulates them to differentiate and produce the specialized cells necessary to save or cure the recipient. Too many stem cells differentiate and too few multiply and make more stem cells, as illustrated when $D > M$ in Figure 4, and as previously discussed for mammary gland transplants. Eventually, the supply of stem cells will be exhausted.

IMMUNOLOGICAL STEM CELLS

Besides precursors of red blood cells, marrow contains the stem cell lines that populate our immune systems. Immune responses decline and become less specific with age; many of the diseases of senility may result from these defects. Are the immunological defects of aging intrinsic in immunological stem cell lines? Do these cells have internal clocks timing

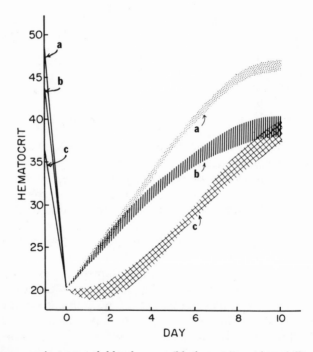

Figure 9. After removing as much blood as possible from mice without killing them, their hematocrits (percentages of blood that is packed red blood cells) fall to 20 from a normal level of 45 to 50. Normal mice show recovery pattern a; old mice show recovery pattern b, and W/W^V anemic mice that have not been cured show pattern c. These patterns enclose the 95% confidence limits, and they are very significantly different. Included in pattern a are not only normal mice but W/W^V mice cured by normal cells. Some cured anemic mice with normal recovery patterns had been cured by cell lines that had functioned normally for up to 73 months and through four serial transplantations.

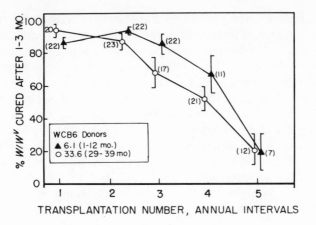

Figure 10. On the average, the ability of transplanted cell lines to cure anemic mice declined after three or four serial transplantations at annual intervals. (It declined even more after three or four serial transplantations at 3-month intervals). The numbers represent the number of donors used, with an average of 5.1 young W/W^v recipients used per donor. Mean values are plotted and the bars give the standard errors. Although the cell lines from the old donors (33.6 months old) had functioned for an extra 27.5 months of life, their ability to cure anemic recipients was equal to that of the cell lines from the young donors (6.1 months old).

their aging? This is a very controversial question now. To study it, we have found it necessary to include a fourth criterion for aging transplantation experiments: *health*. The old donors should be healthy enough so that the tested tissue has not been permanently damaged. In experiments meeting all four criteria, we found that old immunological stem cell lines did not appear defective. After 2 months or more in the young recipient, old marrow grafts supported the immune response against sheep erythrocytes as well as young ones. However, other workers reported that old stem cells were defective when tested within a week or less after transplantation. One group found defects in old stem cell function even after 4 to 6 weeks in young recipients.

An important difference in the conflicting experiments is the amount of time the stem cells spend in young recipients before being tested. I maintain that this time should be long (at least 3 months) because stem cell lines that were intrinsically capable of functioning normally might require this time in the young recipient in order to recover completely from their residence in the deleterious internal environment of the aging donor.

A problem arises because transplanting stem cell lines damages them. This damage might mask the effects of aging. In preliminary experiments, we have compared the effects of aging and transplantation. We found no difference between the ability of equal numbers of old or young marrow cells to populate a lethally irradiated recipient in competition with the same third marrow donor. However, after one transplantation, the stem cell lines are much less able to compete, no matter what their age. Thus, a single transplantation damages or exhausts stem cell lines more than a lifetime of normal functioning. Since stem cell lines are able to populate and save lethally irradiated recipients even after several successive transplantations, such experiments suggest that they have the capacity to function normally for much longer than the normal life-span. Red-cell-producing, and probably immunologic, stem cell lines provide examples of tissues in which aging is not intrinsically timed, at least not nearly as rapidly as their donors' age.

LIMITATIONS ON PROLIFERATIVE CAPACITIES

As predicted from Hayflick's tissue-culture studies, normally functioning tissues transplanted in animals cannot proliferate without limit. However, there is little or no difference in the performance of cells from old or young mature donors using the best-defined tissues. These include red-blood-cell-producing or immunologic stem cell lines and mammary gland cells. Even the limited proliferative capacities of such tissues appear sufficient for them to function far beyond their donor's maximum life expectancy. Thus, such limitations on proliferative capacity do not seem to cause aging.

There is one exceptional tissue that can be made to function normally and yet has a limitless ability to multiply. Beatrice Mintz of the Institute for Cancer Research in Philadelphia recently showed that cells from a cancer of the reproductive organs of a mouse are capable of developing into completely normal tissues when transplanted into very early mouse embryos. Yet, in culture or in transplants, the same cells have an unlimited proliferative capacity. These cells may be cancer cells whose transformation to cancer is reversed, or they may be normal undifferentiated mammalian cells that resemble the undifferentiated cells from fruit flies described by Ernst Hadorn of the University of Zurich that have a limitless proliferative capacity.

INTRINSICALLY TIMED AGING

Most tissues fail to show clear-cut intrinsic aging. This is strong evidence against possibility A in Figure 1. In rigorous experiments using old and young adult tissues compared in identical environments, at least some (and often most) of the old tissues function as well as young controls. These include fibroblast cells in tissue cultures, and the following tissues grafted in recipient animals: adrenal cortex, skin, pituitary glands, mammary glands, red-blood-cell-producing stem cell lines, and immunological stem cell lines. Even old kidneys function beyond rat life-spans if they avoid damage from the transplantation procedure, and old thymus glands show some normal functions in young recipients. Such findings provide a warning to researchers comparing tissues from old and young animals. Do not assume that the changes measured are intrinsically timed within the tissue. So far, no tissue has been clearly shown to age if removed from the aging animal. Differences between old and young tissues, even on the molecular level, are likely to be effects of aging, not causes.

RESEARCH ON AGING: COMPLEXITY AND HOPE

I have not discussed experiments that differentiate between possibilities B and C in Figure 1 because none have been reported. Indeed, knowledgeable people still disagree about possibility A. The complexity of the experiments leads to differences in interpretation that seem likely to persist. The tentative nature of even the simplest conclusions leads to a hard question about research on aging. Is it worth doing?

I suggest that it is. In fact, I believe that it is one of the best research efforts on which to place our bets. Basic studies, such as research on aging, may be considered to be wagers of research effort and money against social gains if the research succeeds. Although the odds that aging research will succeed in adding, for example, 10 healthy years to our adult lives may not be overwhelming, the payoff in increased health would be enormous. A large proportion of the ill health in a developed country like ours is a result of aging. Our chances of getting heart disease or cancer — indeed, almost all major diseases — increases exponentially with age. Even a small decrease in the aging rate would greatly improve our nation's health. On a personal level, how much would 10 extra healthy years be worth to you? To your parents? To your grandparents?

But what about overpopulation, at present the greatest threat to the human race? Wouldn't extended life-spans aggravate that problem? In fact, there would be surprisingly little effect, as long as people did not use their increased healthy adult years to have more children, and why should they? Most couples are physiologically capable of producing at least 10 children. They choose not to produce anywhere near their maximum already; increasing the maximum should not affect that choice.

But suppose that the population problem would be slightly aggravated. Should that matter? No one worried about overpopulation when introducing sanitary procedures, immunization technology, and antibiotics. These were used because they relieved the suffering and saved the lives of children. No objections were raised, although it was obvious that these children would grow up and have more children. Yet now when we propose to study how to keep people healthy longer as adults, people who won't have more children anyway, voices are raised in protest. Why? Is it somehow more just that we should suffer ill health when 50 than when 15? I would do anything in my power to protect the health and lives of my children. Should I feel less strongly about the health and lives of my parents?

AN EXPERIMENT

But no matter how much we want to, is there anything we can do about aging? In this final section, I will describe a simple experiment that might give hopeful results. There is a high probability that it will not succeed, and there may be other experiments that could give even more hopeful results. Nevertheless, I offer this as an example of what we are doing now and what it may mean.

Two mice that will not reject each other's tissues can be joined together surgically as illustrated in Figure 11. Their blood supplies are connected, so anything in one animal's circulation can get into the other's. At least some defects of aging may develop because of deleterious "aging factors" released into the circulation of old animals. Such aging factors may come from the pituitary gland since its removal has been reported to retard aging rates of tail tendon collagen, ovaries, and minimal oxygen consumption levels. Circulating aging factors would be able to enter and act on a young partner surgically joined to an old mouse.

We will join old and young mice as illustrated in Figure 11 and test parameters and functions that change with age. These include collagen solubility, immune responses, and urine-concentrating ability. If circulating aging factors are important in such changes, the young partner should appear increasingly older than its chronological age, perhaps approaching the age of the old partner. Once separated from the old partner, the young animal might return to its chronological age.

There are other possibilities. The old animal might be rejuvenated by its young partner, or a better circulatory connection made by joining

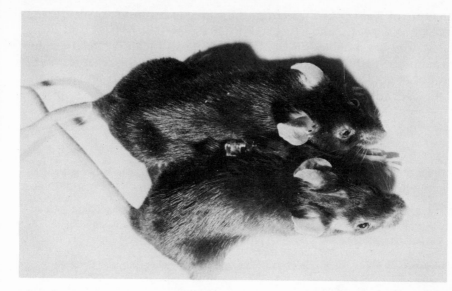

Figure 11. These two mice are artificial Siamese twins. Their skin has been surgically joined from shoulder to hip, and as it healed, their small blood vessels or capillaries meshed. The rate of blood mixing in the circulation between two such animals is about 1% of that within a single mouse.

major blood vessels might be necessary to give positive results. It might be necessary to remove the old animal's pituitary gland so that it can be rejuvenated. But consider the implications if the experiment worked and showed that circulatory aging factors were important.

Researchers could concentrate on identifying those factors and finding out where they came from and how they acted. With something positive to look for, it would be reasonable to hope that the production of aging factors could be retarded, that they could be neutralized, or that harmless compounds could be bound to their receptors.

Anticipation that things will get better in the future is one of mankind's great comforts. Luck and hard work in aging research may allow us to realistically look forward to good health late in life. It is important not to be overly optimistic; negative results must be interpreted realistically. Yet, there are hopeful possibilities that should be pursued. Despite its complexity, the field of research on aging remains a young and exciting area with the greatest potential.

ACKNOWLEDGMENTS. These studies were supported in part by National Institutes of Health Research Grants AG-00250 from the National

Institute of Aging, AI-11757 from the National Institute of Allergy and Infectious Disease, and HL-16119 from the National Heart and Lung Institute. I am indebted to Clinton M. Astle and Joan D. DeLaittre for dependable technical assistance. The drawings in this chapter were produced by Ruth Soper of The Jackson Laboratory Art and Photography Department.

BIBLIOGRAPHY

Albright, J. W., and T. Makinodan. 1976. Decline in the growth potential of spleen-colonizing bone marrow stem cells of long-lived aging mice. *Journal of Experimental Medicine* 144: 1204–1213.

Boggs, D. R., J. C. Marsh, P. A. Chervenick, G. E. Cartwright, and M. M. Wintrobe. 1967. Factors influencing hematopoietic spleen colony formation in irradiated mice. III. The effect of repetitive irradiation upon proliferative ability of colony-forming cells. *Journal of Experimental Medicine* 126: 871–880.

Cameron, I. L. 1972. Cell proliferation and renewal in aging mice. *Journal of Gerontology* 27: 162–172.

Daniel, C. W., K. B. DeOme, J. T. Young, P. B. Blair, and L. J. Faulkin. 1968. The *in vivo* lifespan of normal and preneoplastic mouse mammary glands: a serial transplantation study. *Proceedings of the National Academy of Sciences of the USA* 61: 53–60. (Also see the article by Young et al., below.)

Everitt, A., and J. A. Burgess. 1976. *Hypothalamus pituitary and aging.* Charles C Thomas, Publisher, Springfield, Ill.

Geiringer, E. 1954. Homotransplantation as a method of gerontologic research. *Journal of Gerontology* 9: 142–149.

Hadorn, E. 1968. Transdetermination in cells. *Scientific American* (Nov.) 219: 110–120.

Harrison, D. E. 1973. Normal production of erythrocytes by mouse marrow continuous for 73 months. *Proceedings of the National Academy of Sciences of the USA* 70: 3184–3188.

Harrison, D. E. 1975. Normal function of transplanted marrow cell lines from aged mice. *Journal of Gerontology* 30: 279–285.

Harrison, D. E., and J. W. Doubleday. 1975. Normal function of immunologic stem cells from aged mice. *Journal of Immunology* 114: 1314–1317.

Harrison, D. E., C. M. Astle, and J. W. Doubleday. 1977. Stem cell lines from old immunodeficient donors give normal responses in young recipients. *Journal of Immunology* 118: 1223–1227.

Hayflick, L. 1965. The limited *in vitro* lifetime of human diploid cell stains. *Experimental Cell Research* 37: 614–636.

Hayflick, L. 1968. Human cells and aging. *Scientific American* (Mar.) 218: 32–37.

Hirokawa, K., and T. Makinodan. 1975. Thymic involution: effect on T cell differentiation. *Journal of Immunology* 114: 1659–1664.

Hollander, C. F. 1971. Age limit for the use of syngeneic donor kidneys in the rat. *Transplantation Proceedings* 3: 594–597. (Also see the article by Van Bezooijen et al., below.)

Krohn, P. L. 1962. Heterochronic transplantation in the study of aging. *Proceedings of the Royal Society of London* B157: 128–147.

Krohn, P. L. 1966. Transplantation and aging. Pages 125–148 in P. L. Krohn, ed. *Topics in the biology of aging.* J. Wiley & Sons, Inc. — Interscience Division, New York.

Makinodan, T. 1978. Mechanism, prevention and restoration of immunologic aging. In

D. Bergsma and D. E. Harrison, eds. *Genetic effects on aging* (National Foundation–March of Dimes Original Article Series). Alan R. Liss, Inc., New York.

Micklem, H. S., and J. F. Loutit. 1966. Pages 171–173 in *Tissue grafting and radiation.* Academic Press, Inc., New York.

Mintz, B., and K. Illmensee. 1975. Normal genetically mosaic mice produced from malignant teratocarcinoma cells. *Proceedings of the National Academy of Sciences of the USA* 72: 3585–3589.

Peng, M.-T., and H.-H. Huang. 1972. Aging of hypothalamic–pituitary–ovarian function in the rat. *Fertility and Sterility* 23: 535–542.

Rosenfeld, A. 1976. *Prolongevity. A report on the scientific discoveries now being made about aging and dying, and their promise of an extended human lifespan—without old age.* Alfred A. Knopf, Inc., New York.

Strehler, B. L. 1977. *Time, cells and aging,* 2nd ed. Academic Press, Inc., New York.

Van Bezooijen, K. F. A., F. R. deLeeuw-Israel, and C. F Hollander. 1974. Long-term functional aspects of syngeneic orthotopic rat kidney grafts of different ages. *Journal of Gerontology* 29: 11–19.

Young, L. J. T., D. Medina, K. B. DeOme, and C. W. Daniel. 1971. The influence of host and tissue age on lifespan and growth rate of serially transplanted mouse mammary gland. *Experimental Gerontology* 6: 49–56.

4

Molecular Biology of Aging

DOUGLAS E. BRASH and RONALD W. HART

In higher animals and plants, senescence and death occur at each level of biological organization. While the organism ages, the immune system begins to attack its host, functioning of the brain deteriorates, tissues cease transporting metabolites and are more likely to become diseased with cancer or arteriosclerosis, elastic molecules of the skin and lung become rigid, and many cellular enzymes turn heat-sensitive.

Since molecular structure and function underlie all other levels of biological organization, and since there are strong, precise research tools at this level, investigation of the molecular biology of aging offers the prospect of both understanding and control.

For the purpose of this chapter, mammalian aging can best be characterized by 10 general phenomena. In the subsequent sections, we will attempt to identify and organize the possible molecular mechanisms for these phenomena of aging.

PHENOMENA OF MAMMALIAN AGING

1. Each species has a characteristic maximum life-span (MLS), thus indicating that aging rate is genetically determined. In recent times, as health care has improved, the mean life-span (mLS) of man has doubled. However, the MLS for man (about 110 years) has not changed in the last 50,000 years.

2. With increasing age there is a corresponding decrease in the physiological function of almost all organ systems, organs, tissues, and cells. Even in systems which maintain, with age, approximately normal

DOUGLAS E. BRASH, Ph.D. and RONALD W. HART, Ph.D. • Departments of Biophysics and Radiology, The Ohio State University, Columbus, Ohio 43210

levels of function, deterioration is often revealed under stresses, such as heat, emotional stress, and infection. The system may fail, or normal levels of function may be restored only sluggishly.

3. Increasing age is accompanied by age-dependent disease of two types: (a) Disease in which the incidence increases with age — such as arteriosclerosis ("hardening of the arteries") and cancer. Cancer has specific peak ages for different varieties of the disease, whereas arteriosclerosis or autoimmune disease may occur with increasing severity due to increasing incidences of arteriosclerotic plaques or aberrant immune cells. (b) Diseases in which the consequences are more serious with age — such as pneumonia and influenza, as well as minor accidents. This indicates an increasing susceptibility to the stress caused by these phenomena.

4. The basic aging process is not the age-dependent increase in the incidence of a given disease, but the increase in susceptibility to the stress produced by the disease process. The cure of arteriosclerosis and cancer has been mimicked by subtracting graphs of deaths vs. age for these diseases from the curve for death due to all causes. While the mLS increased, MLS increased by only 2 to 10 years. Only the second class of consequences of aging, indicating stress susceptibility, significantly alter the MLS. This means that senescence is the gradual incline in susceptibility to stress in general rather than an incline in any particular disease state. For example, it has been shown that the extent of cold shock or blood loss required for death of laboratory rats decreases with age (Figure 1). In this chapter, we will attempt to find the origin of both classes of aging phenomena.

5. The extent of tissue damage required to cause death in the elderly is small compared to that required to cause death in the young. This finding is the pathological correlate of the stress-susceptibility basis of aging.

6. Difference in MLS between species, in the few cases examined, corresponds with susceptibility of the species to a number of stresses. (This does not preclude the possibility that differences in mLS may relate to stress susceptibility.)

7. The percentage of a population dying is constant with age until its members reach maturity, at which time it rises exponentially.

8. Starvation of rats prior to maturity, but not afterward, delays development and prolongs life-span.

9. Senescence and death can occur at each level of biological organization and can be independent of the general appearance or age of the animal. For example, the cell may survive in tissue culture long after the animal has died; skin tissue transplanted serially from one mouse to another lives longer than any of its hosts; functioning of an organ can

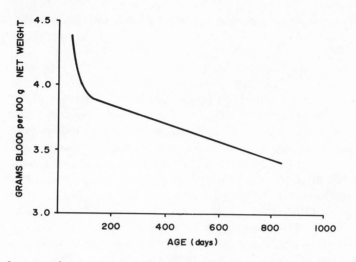

Figure 1. Increase of stress susceptibility with age: blood loss required to cause death of rats, as a function of age. (Redrawn from Simms, *Journal of General Physiology* 26: 169, 1942.)

decline markedly without widespread change in the enzyme levels of its cells; and the enzyme system for melanin synthesis senescences at different rates in pigment cells of skin, liver, and brain.

10. Other phenomena are: radiation can accelerate certain events resembling aging in mouse and man; a species' MLS is predictable from its brain weight and body weight; cells grown in tissue culture divide only a limited number of times; and there are human syndromes which prematurely exhibit certain facets of aging.

How are these phenomena related to the process of aging at the molecular level? This is the question we will attempt to answer in the following pages; however, before proceeding it may be useful to briefly review the molecular architecture of the cell.

THE MOLECULES OF THE CELL

The cell is dominated by large molecules containing thousands of atoms. Yet, there are only four varieties and they are each made by stringing small molecules (which we will call "subunits") into a long chain. These four varieties are: *proteins, carbohydrates, lipids,* and *nucleic acids.* For example, sugar and starch are carbohydrates, whereas fat is a lipid. Nucleic acids are probably not familar to the reader; however, we will elaborate on their role shortly. Each subunit of a nucleic acid chain contains an important side group called a base. Less prominent are small molecules, of only a few dozen atoms, which often act to control the large molecules.

Tissues are arrangements of cells held together by fibrous proteins such as collagen. The outer boundary of the cell, the cell membrane, is made up of various kinds of lipids. The many activities of the cell are controlled by chemical reactions. These events would proceed very slowly except for the presence of certain proteins called enzymes, which accelerate (catalyze) chemical reactions. Enzymes are generally globular, owing to coiling up of the protein chain. Many enzymes are also located on the cell membrane, more or less floating in the lipid, or on similar membranes within the cell.

In the middle of the cell is the nucleus, a membrane-surrounded compartment containing the enormous nucleic acid, deoxyribonucleic acid (DNA). DNA is a double helix consisting of two nucleic acid chains intertwined; it carries in the sequence of its bases the information for synthesis of all the other components of the cell. Thus, DNA has two functions: (1) it serves as the template for synthesis of the cell's molecules, and (2) it replicates itself, so that upon cell division each daughter cell will contain all the cellular information.

Synthesis of the cells' molecules occurs as follows. The long DNA molecule is used to make working copies of those genes which will be used. The copies are made of ribonucleic acid (RNA) and are called messenger RNA (mRNA); the reaction is catalyzed by an enzyme which matches up DNA bases to analogous RNA bases. (A similar process occurs when DNA replicates to make DNA copies for daughter cells.) The mRNA moves outside the nucleus and is used as a template for making proteins. An enzyme lines up subunits of protein in accordance with the sequence of bases on the mRNA. The protein that is synthesized may be a structural protein, such as collagen, or it may be an enzyme. Enzymes then catalyze many reactions, such as burning sugar for energy; digesting food; or synthesizing lipids, carbohydrates, proteins, nucleic acids, their subunits, and small molecules, such as hormones.

In the body there are many types of cells, each containing different chemical reactions and therefore different enzymes. Consequently, any one type of cell uses only a fraction of its DNA. The rest is turned off by a coating of proteins. The on–off status of the genes is controlled by a second set of proteins which can selectively remove the first set. The action of one type of hormone is to attach (bind) to a receptor protein in the cells, which then binds to the second set of proteins mentioned above, changing the state of differentiation of the cell. The other type of hormone binds at the cell membrane to a receptor protein which controls the synthesis of a small molecule called cyclic adenosine monophosphate (cAMP); this molecule and other small molecules have widespread control effects on cell differentiation, metabolism, and division.

A SYSTEMS VIEW OF AGING

The mechanisms and alterations underlying senescence can be better understood when organisms are viewed as systems.

A system simply consists of interconnected components achieving a function they could not perform separately. For example, whereas the circulatory system supplies the body's tissues with oxygen, the heart, blood, or lungs alone could not. Some components are composed of subcomponents; for example, lungs consist of connective tissue and several types of cells. The lungs can thus be considered a component from one point of view and a system from another. Since cells, in turn, consist of molecules, living organisms form a hierarchy of levels of systems and components: atom, molecule, cell, tissue, organ, organ system, organism.

The three principles of systems to be discussed in this section are particularly relevant to aging. The conclusions we will draw from these principles are that any change with age is probably senescent and will lead to increased stress susceptibility; that senescence can propagate throughout the organism, both within a level and up and down the hierarchy; and that some of these propagations are more important than others.

Principle of Operating Ranges

A senescent event is by definition a deteriorative change. On the molecular level enzyme activities may increase, decrease, or remain constant with age, or one type of molecular bond may be replaced by another. The significance of these changes relates to the *principle of operating ranges:* a system functions correctly only within a fixed operating range. For example, if the intracellular potassium ion concentration of a neuron is too high, the neuron will not fire, whereas if it is too low, it will fire spontaneously. These special conditions constitute an operating range for the neuron within which it will perform a definite function and off which function declines (Figure 2). The center of the operating range is the point at which optimal function is attained.

Corollary I: Any age-dependent change (in mature individuals) away from the center of the operating range is senescent. For example, a change in potassium concentration with age is not itself obviously de-

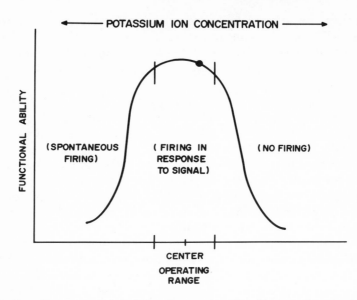

Figure 2. Operating range and deviation from center. (Specific case of neuron in parentheses.)

teriorative; yet such a change results in decreased neural function and is senescent. Similarly, addition of a carbon atom to DNA is not obviously senescent, yet it may impair DNA function.

When a stress is put on a system, such as damage to DNA or loss of blood, the system deviates slightly from the center of its operating range. However, while within the limits of its operating range the system will still function, but not as well. A large stress may run the system off its operating range completely, and the system will cease to function; a small stress can have the same effect if the system is already near the edge of its operating range as a result of normal fluctuations or previously incurred damage (Figure 2).

Corollary II follows from the last consideration: a displacement from the center of the operating range increases stress susceptibility. Age-dependent increase in susceptibility to minor stresses is a simple manifestation of a cumulative age-dependent drift away from the centers of operating ranges as a result of accumulated impairments. *It is this accumulation of damage which may be best equated with senescence.*

In living organisms, the progression of senescence is opposed by two life-span-extending mechanisms: redundancy and restoration.

An organ with many redundant components (cells) will show no senescence until a large number of cells are altered. Evolutionarily, redundancy is the most primitive way to extend life; once a component has evolved, it is merely necessary to make many of them. Redundancy is one possible explanation for the relation of MLS to brain and body weight. In some cases, such as the circulatory system, there is little redundancy. There is only one heart and, indeed, heart failure commonly leads to death. It should be kept in mind that, even in an organ with many cells, a minimal cell alteration is really still senescent, for the organ is nearer the edge of its operating range and is more susceptible to stress.

Restoration systems attempt to return a stressed system to its normal state before it leaves the operating range (see Figure 3, upper left). The immune system, body-temperature regulation, cell turnover (destruction of old cells and replacement by new ones), molecular turnover, and DNA repair are a few examples of these systems.

It is important to recognize that no restoration system is totally error-free or complete, and that the extent of restoration varies widely between species, organ systems, and even from one cell type to another. For example, carbon and hydrogen atoms added to DNA by a compound called ethyl nitrosourea (ENU) damage DNA but are removed quickly in most organs except for the brain. These differences are major in that ENU causes only brain tumors.

Principle of Propagation

The *principle of propagation* states that displacement of a component from the center of its operating range will put varying degrees of stress on components which depend upon the first component for their own proper functioning. From a systems standpoint, senescence would then propagate throughout the organism, both within levels as well as between levels of biological organization. These propagations will form *senescence pathways*. For example, ovarian senescence, resulting in failures of pregnancy, is contributed to by alterations in the hypothalamic region of the brain.

Corollary I: Many of the alterations seen with age in a given system are indirect and are due to prior alterations in other systems. Corollary II: Senescence at the organ system level may have arisen from senescence begun at molecular levels.

Senescence pathways are delineated in two ways. The *cause* of senescence in an organ, for example, may be found by replacing the suspected causal organ with one transplanted from a young animal, or by supplementation with extra amounts of implicated hormones or chemicals and then observing whether senescence is halted or reversed. This approach was used in the investigation of ovarian senescence mentioned above. The *effects* of the senescence of a given parameter may be found by specifically altering that parameter only and observing the subsequent effect on other parameters known to be age-dependent.

Principle of Rate-Determining Steps

If aging is initiated by damage to the organism — and in the next section we will discuss evidence that this may be the case — then each damaged cell, tissue, or organ will be the beginning of a senescence pathway. The multiplicity of these senescence pathways means that *there is no single cause of aging.* In fact, even if normal aging were initiated by a programmed event, elimination of that event would still result in aging due to damage. However, it is impossible to halt all processes that cause damage, and thereby achieve immortality. So the goal of research on aging must be to search not for the cause of aging, but for control points. Thus arises the *principle of rate-determining steps:* Some senescence pathways are far more important than others. These may be possible sites for control.

There are several senescence pathways which may be expected to be major: the immune system, the nervous system, collagen, and DNA damage. Each pathway should be delineated, its relative significance estimated, and the question asked whether any of these potentially major senescence pathways are interconnected.

Three points may be conveniently noted here: (1) Since various pathways have different relative significances, one can expect no agent — for instance, radiation — to mimic exactly the natural aging process, or any agent — such as free-radical inhibitors — to exactly reverse it. (2) Even if aging were programmed, other pathways, such as graying of hair or wrinkling of skin, might be proceeding entirely independently of the program-initiated pathway or of other pathways. (3) Each of the numerous "theories of aging" unconsciously postulates itself to be the rate-determining step.

Since senescence seems to be an accumulation of impairments which can propagate their effects, the two questions which remain are: What initiates these impairments?" and "At what biological level and in which organ system do deviations from optimization begin?"

INITIATION OF SENESCENCE

The onset of senescence in a healthy organism requires an initiating event, in which at least one system is nudged away from the center of its operating range. The initiating stress for this event may originate from either internal or external sources; however, the relative significance of any particular form of stress in aging is not known. The fact that stress susceptibility increases with age implies either the repeated appearance of initiating events or the steady progression of the sequels to such events. Two classes of theories of the initiating event have been proposed — damage theories and programmed aging.

Damage theories are based on the inadequacy of restoration systems. The initiating event is an incompletely restored alteration following a stress. Errors accumulate with age until the resulting senescence leads to death. Originally, it was proposed that the stresses were internal obligatory stresses, such as free radicals (reactive molecules which can initiate random chemical reactions) formed as by-products of metabolism, altered hormone levels from development, or the massive thymus cell death which accompanies infections. Recently, damage theories have come to include exogenous stresses, such as radiation and toxic chemicals, and endogenous molecular errors of protein or DNA synthesis.

In programmed aging, the initiating event is not due to an inadequate design but to appearance of an optional destructive agent which overwhelms an adequate design. Since the destructive agent is optional, its appearance is presumably governed by a clock. There are three elements involved: a clock which varies with time but does not senesce, a trigger which is sensitive to some aspect of the clock, and the stress mechanism. The clock and trigger are usually postulated to be at the organ level in the neural network or at the molecular level in "the developmental program." Two stress mechanisms which have been proposed are a "death hormone" and the release of degradative enzymes into the cell. The destruction caused by the stress mechanism would then propagate further impairments — just as in damage theories — leading eventually to the death of the organism. Often, any sign of temporal regularity in aging is taken to indicate a clock and programmed aging.

Clearly, damage to system components does occur, and clearly both internally timed events such as maturation, and internal clocks such as the biological rhythms, do exist. Indeed, both of these processes may reasonably be involved in aging; however, the question is: Are they?

Programmed aging does exist in plants, in embryos, and during metamorphosis. Since plants will be discussed elsewhere in this book, we will focus our attention only on the latter events. In chick-limb morphogenesis, it has been found that at a particular time, certain cells undergo a differentiation which will later lead to morbidity and engulfment by white blood cells. The morbid cells have decreased DNA synthesis and also symptoms of decreased protein synthesis. The differentiation may be reversed by diffusible molecules from other tissues. Such behavior is typical in embryology, where, for example, primitive skin tissue differentiates to primitive nervous system tissue upon stimulation by molecules diffusing from primitive backbone and muscle. Metamorphosis also appears to be a programmed event. For example, frog metamorphosis is triggered by thyroid hormone, which causes not only the development of the limbs and middle ear, but also resorption of the intestinal epithelium and the tail. Cells and tissues are thus capable of responding to inducing molecules by dying in the space of a few hours, just as easily as they are of differentiating.

Evidence for a role of programmed events in initiating the basic aging process, however, is slim. It has been proposed that the limited doubling potential of cells in tissue culture is due to a developmentally programmed differentiation. The hormone hydrocortisone can extend cell life-span, but this has not been shown to be due to dedifferentiation. Even if it were, differentiations resulting in aging might arise from damage-induced repression or derepression of cellular DNA. Asymmetric mitoses, which are seen in differentiation, have been observed in culture, with the larger daughter showing a limited number of subsequent divisions. An alternative explanation to a programmed event might be that these cells arise as a result of DNA damage. Indeed, such cells are often seen in cultures following application of agents known to produce DNA damage. In addition, recent work has shown that aging of cells *in vitro* in many ways is not analogous to aging of cells taken from biopsies of patients of different ages.

Programmed aging was originally proposed to explain several data which suggested a timed destructive agent: (1) the existence of a genetically determined and well-defined species MLS; (2) the sudden rise of mortality rate after maturity, or the involution of the thymus at that time; (3) the rodent starvation experiments; and (4) the existence, for mammals, of a correlation between life-span and gestation period, age at puberty, heart rate, and metabolic rate, as if all were directed by a single clock.

Damage theories can account for this and other data as follows: *Damage accumulates, reaches a threshold of tolerance, and death occurs. During damage accumulation, senescence results.*

Both the tolerance threshold and the rate of accumulation of damage may be genetically determined, thus timing senescence and death. Specification of the rate of the accumulation of damage can be achieved by specifying the efficiency of restoration systems. The resulting differences in stress susceptibility between species enables their life-spans to be different despite exposure to similar rates of damage. With regard to the second and third points above, we will see later that the changes in hormones and in cell division rates which accompany maturity may drastically alter the amount of genetic damage to the cell. The final correlation is a consequence expected from an increase in body size between species, together with the increased embyrological differentiation associated with the greater number of cells in the embryo. The phenomena of programmed aging can thus be accounted for within the damage theory. With these points in mind, one can begin to examine the biology of aging, and in our case, the molecular biology of aging.

We will now examine the molecules of the cell with the intent of finding *rate-determining, damage-initiated senescence pathways which can propagate a deviation from optimization on the molecular level up to higher levels of the organism, thereby increasing their stress susceptibility.*

INITIATION OF SENESCENCE AT THE MOLECULAR LEVEL: DNA AS THE PRIMARY MOLECULAR TARGET OF AGING

DNA, RNA, protein, and lipid membranes are each possible targets for significant molecular senescence. Each is a major component of the cell, and each shows alterations with age. Owing to the fact that RNA, intracellular protein, and lipids turn over, it is reasonable that these cellular constituents are probably not important determinants of senescence. However, DNA may be expected to be a major physiological target in the aging process because of its (1) large size, (2) presence in unique copy, and (3) function as the initial template for all DNA, RNA, protein, carbohydrate, and lipid syntheses. It has been shown that DNA damage can result in altered cell growth, division, and transcription, and in cell death, mutation, and malignant transformation. Since all these phenomena are characteristic of the aging process, it is a reasonable hypothesis that alterations in DNA structure and function may play a large role in the aging process.

This relation has been shown to hold in *Paramecium*. Their life-span can be shortened with UV-induced DNA damage and restored by specific removal of this damage. *In higher organisms, there is no direct evidence that genetic damage has any role in aging.* However, DNA alterations have been indirectly implicated in aging.

1. Transplantation of young nuclei into old cells rejuvenates the old cells, whereas cells with old nuclei and young cytoplasm act old. Thus, determination of senescence lies in the nucleus rather than the cytoplasm. This fact militates against cytoplasmic proteins, lysozomes (cytoplasmic compartments containing degradative enzymes), aging pigments, and the cell's outer membrane as determinants of senescence, and argues in favor of DNA.

2. In rats, chromosomal aberrations increase with age at a more

rapid rate than in the longer-lived dogs. Correlations involving chromosome aberrations are not perfect, probably because they represent only the gross forms of DNA damage.

3. Individuals with Down's syndrome or Klinefelter's syndrome, characterized by chromosomal anomalies, prematurely exhibit facets of senescence.

4. Ionizing radiation, the primary target of which appears to be DNA, induces in many respects accelerated aging of *Drosophila*, rats, and chick cells in tissue culture. These studies must be interpreted with caution, as the effect actually varies widely depending on the strain of mouse or species of cell used.

5. Individuals who die of the early onset of one type of autoimmune disease are deficient in repair of DNA damage caused by ionizing radiation.

6. Ultraviolet (UV) radiation, the primary target of which is DNA, causes accelerated aging of skin. In addition, the life-span in culture of cells from one type of patient deficient in excision repair of UV-induced DNA damage is shorter than that of normal cells. Life-span of a number of mammalian species correlates well with their capacity to repair UV damage to their DNA. (The prevalence of this correlation is curious, for if life-span were determined by a single type of repair, one would expect syndromes and chemical or physical agents *exactly* mimicking normal aging to occur.)

7. Cells from individuals exhibiting the "premature aging" syndromes show an increased mutation frequency.

8. Extraneous proteins attach to DNA with age, and such DNA–protein cross-links accumulate more rapidly in short-lived rodents than in long-lived rodents.

Membranes and extracellular proteins, such as collagen, remain potentially important molecular paths of senescence *in vivo*. Their alteration can plausibly be treated as a sequel to alteration of DNA; however, their direct damage will also be considered below.

SENESCENCE PATHWAYS AT THE MOLECULAR LEVEL: CONSEQUENCES OF DNA DAMAGE

Figure 3 outlines the possible consequences of DNA damage in somatic cells. At first glance it is a formidable figure, but it is really simply a tree which cancels out some of the potential consequences of DNA damage. The goal of the diagram is to attempt to present some of the data on the molecular biology of aging in an organized form.

To guide the reader, the italic numbers in the text correspond to the numbers on the diagram. The data will be clearest if the reader follows the text, occasionally referring to the diagram. The continuity of the ideas will be made clearest by following the diagram and referring to the text.

We will first describe DNA damage and its repair. Notice that DNA damage has two main classes of consequences: mutation and gene derepression or repression. Mutations will be discussed at some length, since many of the effects of mutations on aging may be the same as those resulting from gene derepression or repression. Finally, we will describe the role of gene derepressions and repressions in aging.

DNA Damage and Repair

DNA can be damaged by physical, chemical, or biological agents generated internally or externally (1). These cause DNA *breaks* (breakage of one or both strands), *distortions* (kinking or opening up of the intertwined strands), and *adducts* (addition of extra molecules to the DNA).

Examples of internal factors producing DNA damage are: (1) normal cellular metabolism, which generates free radicals and reactive metabolites capable of cross-linking DNA to proteins, RNA, and itself; (2) metabolism of certain externally originating molecules, producing excited molecules which form DNA adducts; (3) heat from body temperature, which causes a loss of bases and subsequently single-strand breaks and single-stranded regions in cellular DNA; and (4) the action of enzymes which can degrade DNA.

Examples of external factors producing DNA damage include: (1) ultraviolet light, which joins some adjacent bases to form a bulky product which distorts the helix; (2) gamma rays and x-rays, which generate free radicals and thereby alter or remove bases; (3) chemical mutagens and carcinogens, which either bind to DNA and form adducts, produce DNA strand breaks, or slip between bases to form nonbound intercalations; (4) ultraviolet light used in conjunction with certain chemicals causing chemically mediated DNA–DNA, DNA–RNA, and DNA–protein cross-links; and (5) viral DNA can insert into the host DNA, altering the information content.

The potential effect of these lesions relates roughly to the size of the distortion of the DNA helix. Small adducts interfere with DNA replication only slightly, whereas bulky adducts, joined bases, intercalations, cross-links, and strand breaks are the source of many cellular dysfunctions. Nevertheless, small nondistorting damage will have a substantial effect if it alters the information content of the DNA, as do base modifications which cause the wrong RNA bases to be matched with DNA bases.

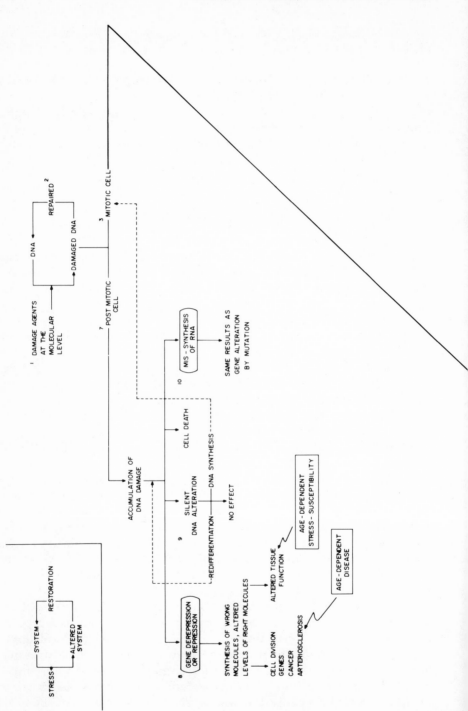

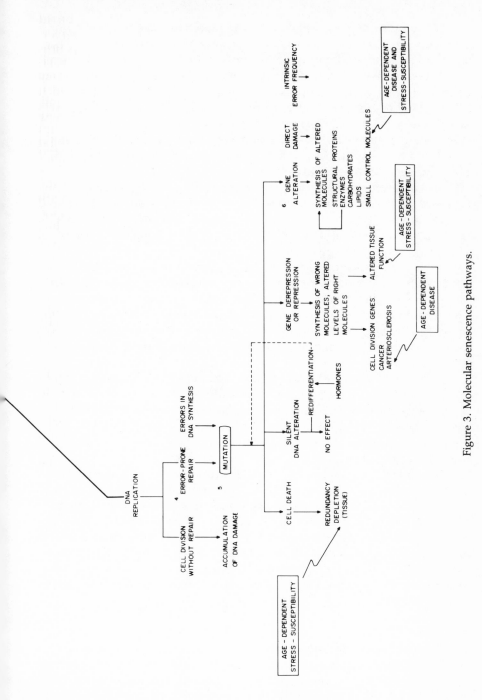

Figure 3. Molecular senescence pathways.

Many of these types of DNA damage have been shown to accumulate with age: DNA–protein cross-links, single-strand breaks, single-stranded regions, and chromosomal aberrations.

There are many DNA repair systems (2). These fall into three broad classes: strand-break repair, excision repair, and postreplication repair. The last will be discussed later. Strand-break repair rapidly rejoins broken single strands and, in some organisms, broken double strands. Excision repair systems remove the damage from the DNA; an enzyme called endonuclease nicks the DNA near the damaged site, after which the damaged region is removed and replaced. There are several separate endonucleases for recognition of DNA distortions. A given cell type may be more proficient in one of these types of repair than in others, and several disease syndromes, including some which prematurely exhibit facets of aging, have a deficiency of one repair system while the other DNA repair systems appear to be normal.

If strand-break or excision repair is complete, the integrity of the DNA is restored; however, the fact that DNA damage accumulates with age in postmitotic tissue indicates that repair is not always complete. This accumulation of DNA damage does not seem to be due to a decline in the function of repair systems with age, since most investigators have found a decline occurring in these repair systems only in very old cultures.

The consequences of damage and repair are likely to be very different in dividing (mitotic) and nondividing (postmitotic) cells, due to the consequences of entering DNA replication or cell division with unrepaired damage still present. We will consider mitotic cells first (3). Cells which divide only occasionally, such as liver and kidney, are equivalent to postmitotic cells until they begin DNA replication.

Minor DNA damage, such as small adducts and viral incorporation, often have no effect on replication or division and will be passed on to the new cells during cell division. Minor damage can thus accumulate in the DNA. We will consider the effects of accumulated damage later, when we consider postmitotic cells. On the other hand, if strand breaks or DNA–DNA cross-links are present during cell division, the chromosomes will be unequally distributed. The result of such major alterations in the integrity of cellular DNA is usually reproductive death of the cell.

Mutations

When DNA replication reaches one of the bulky helix distortions, it passes over the damaged region and then continues along the strand. Of the two resulting DNA duplexes, one is perfect, and the other contains the damage in the parental strand and a gap in the newly synthesized strand. The gap, presumably opposite the damaged site, is repaired by the postreplication repair system, which in mammalian cells apparently fills in the gap by synthesizing new DNA. While most enzymes which synthesize DNA (DNA polymerases) make few errors, evidence is rapidly accumulating for existence of an inducible error-prone polymerase (4) which inserts incorrect bases. The resulting incorrect base sequence is called a mutation and can result in the synthesis of altered proteins by that gene.

While the new base sequence may be incorrect as far as the organism is concerned, it is simply a new sequence of DNA and thus cannot be detected as incorrect by any repair system. These mutations thus become permanent changes in the DNA and are inheritable. Mutations can have enormous effects on the organism; for instance, sickle-cell anemia arises from a single mutation. Mutations are also presumed to form the basis for evolution. If, however, the gap is repaired by an error-free polymerase, there is no mutation and therefore presumably no biological effect.

One possible explanation for the effect of starvation on rodents is that starvation during the growth phase of the rat's life slows down cell division, enabling excision repair of damage *before* replication, thereby reducing accumulated mutations early in life. Starvation after maturity would not have the same degree of effect, since nearly all division stops when growth halts at adolescence.

Mutation can also arise from errors in DNA replication, both from mispairing of bases at damaged sites and from the inherent frequency of error of the enzymes that synthesize DNA. The latter is a senescence pathway which does not have its origin in DNA damage. The relative importance of these pathways in generating mutations is unknown.

Mutation is shown on the diagram as occurring only in mitotic cells, since there is evidence that a round of replication is required for mutation. The various events above can all occur simultaneously in the same cell if it has several types of damage.

As we mentioned, a mutation can lead to various physiological changes at all levels of biological organization, including many occurring at the molecular level (5). Immediate cell death can result from mutation in a critical gene, such as one responsible for synthesizing proteins required in energy metabolism. Since cell death is really loss of cell function, even very subtle mutations can ultimately lead to cell lethality. For example, if a cell's function is to divide, as in the cells which give rise to blood cells or in certain types of cells in tissue culture, then loss of reproductive capacity is called cell death, even though many aspects of cell metabolism may be normal. Additionally, loss of the capacity of a neuron to fire would in many respects constitute neuron death.

Cell death is a stress at the tissue level, for it depletes the redundancy of the tissue and thus results in one of the phenomena of aging — increased stress susceptibility. Notice how senescence is propagated from the molecular level (DNA damage) up through the cell level (cell death due to mutation) to the tissue level (depletion of redundancy). Propagation also occurs in other parts of Figure 3 as well and may move vertically as well as horizontally, thus affecting function at various levels of organization. Rather than describe in detail the upward propagation of the consequences of stresses in a chapter on molecular aging, we will view all arrows at the bottom of Figure 3 as summing to give the "composite ability of the organism to handle stress." Also not shown are many restoration mechanisms at each level; for instance, cell death may be partially compensated for by proliferation or enlargement of the cells which remain.

Incidentally, a dead cell might be less of a stress on the tissue than an aberrant cell. In such a case, it would be advantageous for a cell to have a low tolerance threshold for damage, so that it would die rather than cause the tissue to malfunction. In a tightly operating system such as the nervous system, this stratagem might be particularly prevalent where redundant cells are available, and may be one explanation for the relation of MLS to brain weight.

Most mutations are probably not at lethal sites. Since in a differentiated cell most of the DNA is not expressed, there is a significant probability that a mutation may be located in a nontranscribable region (a region which has been turned off), as well as in a transcribable region (a region which has been turned on) or in a (transcribable) gene-control region.

A mutation in a nontranscribable region will have no effect on the cell and thus will be "silent." A silent mutation may become transcribable if redifferentiation occurs. Such redifferentiation can occur during regeneration of injured lens tissue or in response to hormones. Widespread hormonal changes occur at puberty and at menopause. These may expose mutations which have accumulated silently for years. The increase in the hormone prolactin at menopause seems to be associated with breast cancer incidence, and one promoter of carcinogenic transformation has been shown to alter gene expression. The replacement of a constant death rate by a rising one at maturity may be due to massive exposure of silent mutations by developmental hormones. Since the rodent-starvation experiments delay development, the life-span extension may represent a delaying of the unmasking of accumulated mutations during development. Starvation also decreases hormone levels and may contribute to life-span extension in this manner. Redifferentiation of silent mutations can also explain how a wasp in which all mutations are dominant (the haploid *Habrobracon*) can have the same life-span as a normal wasp.

The involvement of redifferentiation events here and, we will see, in postmitotic cells is reminiscent of the two-step theory of cancer. In that theory, an initiation event is followed, perhaps years later, by a promotion event which triggers the cancer. A two-step theory of aging may be applicable to arteriosclerosis, autoimmune disease, and stress susceptibility, with mutation or DNA damage as initiator and redifferentiation as promoter.

Mutations in gene-control regions can probably cause gene derepression or repression, resulting in the synthesis of the wrong molecules — perhaps the synthesis of hemoglobin in a collagen-synthesizing cell. The effect might be less extreme, with synthesis of correct molecules in abnormal amounts. Assays of innumerable enzymes of energy production, DNA synthesis, and cellular metabolism have actually shown increases and decreases with age. Synthesis of structural proteins, such as collagen, also declines with age, as does the presence of some hormone receptors on the cell membrane, and also the activity of enzymes capable of inactivating free radicals. The overtones of these changes are ones of deterioration. Since synthesis of many of these proteins is subject to physiological regulation, it is usually not clear whether these variations are simply physiological responses to alterations in other parts of the system. In old tissue culture cells, the division apparatus declines while metabolism is still functioning, resulting in enlarged cells with age. Elevated levels, with age, of one enzyme have been linked to mutation; this is important, because we will see later that gene repression and derepression can also be caused by direct DNA damage. Clearly, such changes may drastically alter the functioning and stress susceptibility of tissues.

Mutation of genes which control cell division may lead to greater proliferation of cells. There is evidence that many age-related diseases are proliferative. For example, tumors arise from a single cell, and most carcinogens are mutagens. Arteriosclerotic plaques also originate from a single cell. Cells of the immune system, and possibly the immune-system cells which attack the host in autoimmune disease, arise by proliferation of a precursor cell when stimulated by an invader. Cell proliferation can also arise from direct action on a nonmutated cell by cell-division initiators, such as cholesterol, platelet-aggregating factor, and hypertension.

The appearance of these age-dependent diseases in the DNA-damage senescence pathway makes it extremely probable that both age-dependent disease and age-dependent stress susceptibility are fundamental aging processes. Age-dependent stress susceptibility, however, still retains the distinction of being the primary factor in controlling MLS. Supporting this view is the fact that one age-dependent disease is tightly coupled to species MLS: cancers in short-lived species occur earlier than in long-lived species, in correspondence to their shorter life-span.

Mutations in transcribable regions are immediately expressed as altered genes (6). These lead directly to synthesis of altered structural proteins and altered enzymes.

Elastin, an extracellular protein responsible for the elasticity of tissues such as blood vessel and lung tissues, forms cross-links between its own molecules with age and becomes less flexible, and thus may contribute to the increasing rigidity of these tissues and to wrinkling of skin. Calcium precipitates onto old elastin, a phenomenon reminiscent of arteriosclerosis.

Since very few collagen molecules turn over, damaged collagen molecules will not be replaced. Collagen becomes stiffer with age as a result of cross-links of several types. Stiffness of collagen-containing tissues may restrict movable tissues such as lung, and may also interfere with tissue perfusion. While these phenomena do occur with age, the role of collagen in causing them is not clear. Bone collagen seems to have a role in calcium resorption with age, and this appears to result in brittle bones. Loss of teeth may also be collagen-related in aging, as it appears to be in scurvy.

Cross-links in collagen most likely arise not from mutation but from direct postsynthetic modification by reactive molecules such as free radicals and aldehydes. Thus, collagen aging and elastin aging, and the pathologies that may result, may be a significant senescence pathway which is separate from the DNA-damage pathway. If so, it is difficult to see a control point for averting skin wrinkles and brittle bones. The rate of cross-linking is sensitive to temperature and to thyroid hormone; these might be exploited, although cross-linking also occurs *in vitro*. Thyroid hormone changes could result from the DNA-damage pathway.

Collagen-synthesizing cells can make various forms of collagen. It may be that the greater number of cross-links with age is actually due to switching to synthesis of a different variety of collagen and thus may be under DNA control. However, the low turnover rate of collagen makes this an unlikely explanation.

Altered enzymes, as measured by their increased sensitivity to heat or to protein-digesting enzymes, accumulate with age in nematode worms, mice, and human tissue culture cells. Incorporation of the wrong protein subunits also increases with age. These alterations may be due to mutation; they also may be due to errors during protein synthesis or to postsynthetic modifications. At this time it is difficult to distinguish which of these explanations is correct.

An intrinsic frequency of errors in protein synthesis, leading to alterations in proteins, would include errors in enzymes that synthesize proteins; this vicious circle could lead to more errors and an "error catastrophe" of altered proteins, mutations, and cell dysfunction or death. Although these phenomena do occur during aging, they may have other origins. Early onset of these events has been produced artificially with protein subunit analogs and DNA base analogs and may occur naturally in short-lived strains of bread mold. However, replication and protein synthesis by viruses proceeds unaltered in infected old cells, and there is no enhanced mutation or heat sensitivity, indicating that the DNA synthesis and protein synthesis machineries are not invariably defective in normal aging.

Subsequent to enzyme alteration, innumerable cellular molecules synthesized by these enzymes would be altered, including, for example, membrane lipids, carbohydrates, energy systems, hormones, and cAMP. However, it appears that the altered enzymes discovered to date are not conducting erroneous syntheses, but rather are completely inactive. The result would be altered levels of the molecules mentioned, and not erroneous molecules. Furthermore, synthesis of an enzyme is often stepped up to compensate for the fraction that are inactive, and there may be no net functional change.

There may also be direct damage to these nonprotein components, for instance oxidation of the membrane lipids. Since membranes turn over rapidly, this may not be too detrimental, although there is evidence that oxidized lipids in the nuclear membrane may damage DNA by free-radical reactions. Membrane stabilizers and antioxidants have been used to extend mLS and, in a single case, MLS.

There is no general age-related trend in the actual data on nonprotein molecular changes. For example, membrane composition changes slightly with age, activity of the apparatus on which proteins are synthesized decreases; lipid synthesis increases; and cAMP levels are altered differently depending on the species and cell and on whether the experiment is done *in vivo* or in tissue culture.

From these changes, however, numerous subsequent effects can be imagined. For instance, cAMP exerts widespread cellular control, and loss of histocompatibility antigens (proteins with carbohydrate groups on the cell's surface which mark the cell as the host's rather than an invader's) could result in autoimmune diseases. From either of these events, age-dependent disease and age-dependent stress susceptibility could follow.

Mutations are very rare, and it has been proposed that mutation rates, usually measured in tissue culture, are too low for mutation to occur in a significant number of cells and thus cannot have a major role in aging. However, there are additional factors involved, such as tissue culture conditions, cell type, and conditions required for the induction of error proneness in DNA repair. A crucial point is that proliferation of a single mutated cell can result in substantial numbers of mutated cells. It is interesting that the age-dependent diseases do seem to be proliferative. Perhaps the major senescence pathway for age-dependent disease is mutation in mitotic cells (and derepression/repression of division-control genes in postmitotic cells). The major senescence pathway for stress susceptibility might, on the other hand, be gene derepression/repression in both cell types. Finally, it may be that somatic mutation is actually not important in aging.

Gene Derepression and Repression

In a postmitotic cell (7), there is either no cell division (nerve cells) or an extremely slow rate of division (liver, heart, and smooth muscle cells), and damage can accumulate without being converted to mutation. This situation might also occur by accumulation of minor DNA distortion even in a mitotic cell. Between 20 and 50% of the DNA damage induced by chemical agents appears to be very slowly repaired; thus, damage may accumulate with age as a result of the occurrence of damage outracing its repair. The reason for slow repair seems to be that regions of the DNA which are turned off by the coating of proteins are also not as accessible to the endonucleases and polymerases required for repair. Such accumulated damage can have several consequences, the first two of which are familiar from our discussion of mutations: repression of a transcribable region or derepression of a nontranscribable region; silent accumulation in a nontranscribable region; and pairing of transcribable DNA bases with incorrect RNA bases. Any of these (except silent damage) may damage a lethal target causing loss of cell function.

Gene derepression and repression events (8) due to mutation of control genes have been mentioned before as occurring in mitotic cells. The same effect may occur in postmitotic cells by more direct mechanisms: for example, DNA–DNA cross-links, DNA–protein cross-links, and the joined bases caused by radiation, have been correlated with decreased ability of the chromosome to be used for synthesizing messenger RNA. Interestingly, in many cases, transcription of individual genes appears to be terminated prematurely rather than never begun. All the types of damage just mentioned have been suggested or shown to accumulate with age.

The two sets of chromosomal proteins which govern differentiation are reported to undergo age changes in proportions and in state of chemical modification, but other reports see opposite changes or no change.

Immunological methods suggest that proteins present only in fetal life begin to reappear with age, indicating erroneous derepression. Single-stranded regions of DNA increase with age and may serve as transcribable sites which should not be transcribable.

Derepression of genes which control cell division may be the cause of the age-related proliferative diseases, cancer and arteriosclerosis, when they occur in postmitotic tissues. The likelihood that cancer may involve such nonmutational derepressions or repressions is shown in an elegant experiment in which tumor cells were injected into an early mouse embryo; the tumor cells took part in the development of an entirely normal mouse.

Derepression or repression of other genes leads to erroneous presence or absence of proteins. Derepression of histocompatibility antigens could result in autoimmune disease. Changes also occur postsynthetically — as in chemical degradation of lens crystallins, possibly related to formation of cataracts — and are independent of DNA damage. In either case, the result is age-dependent stress susceptibility in postmitotic tissues.

Damage in nontranscribable regions will accumulate silently (9) unless the state of differentiation changes or the cell resumes division. For example, damage accumulating silently in the ova until fertilization may be the reason for the dramatic increase in birth defects and spontaneous abortions in pregnancies of older women.

Upon stimulation by cell-division initiators, such as hypertension, certain hormones, or cholesterol, or by derepression of a gene controlling division, or by regeneration as in the case of lens, liver, or skin wounds, the cell becomes mitotic, with the corresponding mutational consequences already outlined. Liver regeneration is accompanied by massive proportions of chromosomal abnormalities and cell death. Hormones, viral infection, and regeneration can, as we mentioned before, cause the cell to redifferentiate, masking damage or unmasking silent damage.

Damaged DNA, especially when involving minor damage such as adducts, is known to cause mispairing of DNA bases during replication, resulting in mutation. A similar phenomenon may be expected for pairing of RNA and DNA during mRNA synthesis, resulting in altered RNA and proteins in postmitotic cells (10). There are no experimental data bearing on this point. The same phenomenon would not be as important in mitotic cells, since DNA damage is diluted upon cell division. Mistranscription of mRNA could also arise from the intrinsic frequency of error of transcription enzymes, and would be a second way of generating an error catastrophe.

RATE-DETERMINING STEPS AND CONTROL POINTS

The principal senescence pathways at the molecular level seem likely to be: (1) DNA damage, with two branches — somatic mutation and gene derepression or repression; (2) collagen and elastin cross-linking; and perhaps (3) membrane lipid damage.

Evidence for the relative significance of these pathways is meager: (1) Capacity to repair UV damage is correlated with the MLS of several mammalian species. However, the only definitive correlation as to the role of DNA damage in aging would involve an analysis of all repair systems in all tissues and cell types of each species. The only direct evidence for the role of any particular form of DNA damage or repair system would require its selective manipulation and subsequent examination of the effect of this manipulation on age-related parameters. Until studies such as these are performed, any correlations or exceptions to correlations must be taken cautiously. (2) The collagen hypothesis rests on plausibility arguments. (3) The MLS of mice has been extended by feeding of membrane stabilizers. A more common result of membrane stabilizers is to extend mLS.

The principal senescence pathways at higher biological levels have been plausibly argued to be the immune system and the nervous system. It is possible to connect these to molecular pathways by proposing that cells of these systems are especially sensitive to deleterious molecular effects such as DNA damage, membrane damage, or decreased tissue perfusion, and thus senesce rapidly. Studies in which skin tissue or bone marrow cells are transplanted from one host to another over several life-spans, provide good evidence that these are not the sensitive cells. Cells which do not senesce particularly rapidly, but which propagate their effects widely, could also act as critical cell types. Evidence for either variety of critical cell type, however, is nonexistent, and the higher-level senescence pathways may originate as a function of the higher levels of biological organization.

The most prevalent phrase in discussing senescence pathways was "may lead to." The reason is that although many age changes have been described, very few of the arrows connecting them have been examined. There are two reasons for this: (1) Much gerontological data has been obtained by researchers in other fields who, in examining the phenomenon of their interest, also examined it with age; (2) in biology, it requires considerable thought and work to vary a single parameter and no other, in order to look at effects of unambiguous causes.

Control points for membrane-damage and collagen-cross-linking pathways are not evident. One hope for forestalling skin wrinkling and more serious collagen-related pathologies is that they are traceable to neural control of thyroid hormone level. Control points for mLS diseases, such as cancer, arteriosclerosis, and autoimmune diseases, have been sought by drug and transplant methods and may relieve the suffering that often accompanies the end of life-span. The consequences of DNA damage, which include both mLS and MLS senescene phenomena, have an obvious control point in DNA repair. The requisite repair capacity may even be present without augmentation by genetic engineering, since it seems that there are high levels of repair in embryonic cells but that these levels in shorter-lived species decline sharply with subsequent differentiation.

ACKNOWLEDGMENT. These studies were supported in part by NCI Grant CA-17917 and EPA Grant R-804201.

BIBLIOGRAPHY

General

Goldstein, S. 1971. The biology of aging. *New England Journal of Medicine* 285: 1120.

Kohn, R. R. 1976. *Principles of mammalian aging,* 2nd ed. Prentice-Hall, Inc., Englewood Cliffs, N.J.

Watson, J. D. 1970. *Molecular biology of the gene,* 2nd ed. W. A. Benjamin, Inc., Menlo Park, Calif.

Technical

Cutler, R. G. 1976. Cross-linkage hypothesis of aging: DNA adducts in chromatin as a primary aging process. In K. C. Smith, ed. *Aging, carcinogenesis, and radiation biology.* Plenum Press, New York.

Hart, R. W. 1976. Role of DNA repair in aging. In K. C. Smith, ed. *Aging, carcinogenesis, and radiation biology.* Plenum Press, New York.

Sacher, G. A. 1975. Maturation and longevity in relation to cranial capacity in hominid evolution. Pages 417–441 in R. Tuttle, ed. *Antecedents of man and after. I. Primates: functional morphology and evolution.* Mouton Publishers, The Hague.

5

Cellular and Metabolic Aspects of Senescence in Higher Plants

HAROLD W. WOOLHOUSE

Senescence is a collective term for the deteriorative changes in living organisms that lead to death. Senescence may be caused by pathogens, environmental stresses, or by inherent physiological changes in the organism; it may affect the whole organism or some of the organs, tissues, or cells.

Figure 1 depicts the life cycle of an angiosperm in which the succeeding stages are marked by unbroken arrows. With each of these phases of the life cycle, there are associated manifestations of senescence; in Figure 1, these are shown by broken arrows pointing outward from the cycle. The majority of these manifestations of senescence represent normal developmental processes that are inimical to the achievement of specific events in the programmed cycle of development of the plant.

In this chapter, I shall proceed around this cycle, through the successive stages of plant development, and consider briefly some of the salient features in the pattern of regulatory activity and the nature of the changes that occur during the course of this spectrum of processes of senescence.

SENESCENCE IN DRY SEEDS

There is evidence from many species that seeds stored in the dry state, under normal conditions of temperature, humidity, and oxygen concentration, lose viability much more rapidly than seeds from the same source stored in an imbibed (dampened) state. This is not to deny

HAROLD W. WOOLHOUSE, Ph.D. ● Department of Plant Sciences, University of Leeds, Leeds LS2 9JT, England

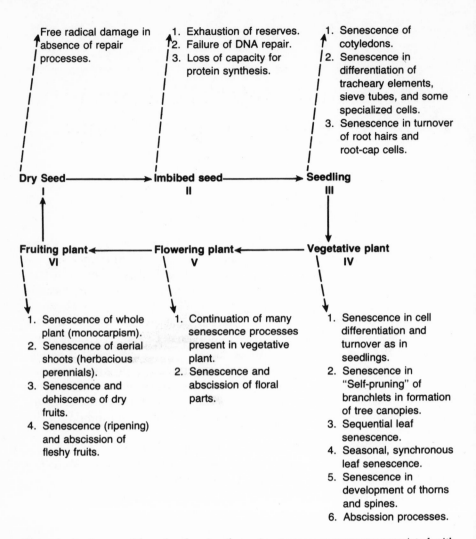

Figure 1. Angiosperm life cycle, showing the major senescence processes associated with each phase of the cycle. Unbroken arrows denote the successive stages of development; broken arrows point to lists of the senescence processes at each stage.

that under exceptionally dry conditions at subzero temperatures, the dried seeds of many species may be stored without damage for very long periods. Such conditions are, however, relatively rare in the natural state and often expensive to maintain for practical purposes. The causes of loss of viability of dried seeds are not fully understood; examination with the electron microscope suggests the possibility of some membrane damage, and the mitochondria appear shrunken, but there are technical problems involved in fixing and sectioning of dried seeds. Moreover, it is not clear whether these observed deformations are related to impaired cellular functions. Extracts from aged dry seeds show a reduced capacity for protein synthesis compared to extracts of fresh seeds, and this could be a factor in the impaired viability.

Seedlings germinated from old dried seed stocks often show abnormalities; similarly, a greater proportion of cytological defects are apparent in dividing cells of seedlings grown from old seed. This has led to the suggestion that a cardinal factor in the aging of dried seeds under ordinary atmospheric conditions may be the accumulation of damaged macromolecules, including deoxyribonucleic acid (DNA).

SENESCENCE OF IMBIBED SEEDS

Research on senescence of imbibed seeds is most conveniently carried out using species or cultivars that have a specific dormancy mechanism that can be released when required. This permits the seeds to be maintained imbibed for defined periods without germination; for example, certain cultivars of lettuce that have a light requirement for germination are suitable. Loss of viability in dried seeds is greatly accelerated by treatment with high-energy radiation, and there is a parallel increase in the numbers of observable cytological abnormalities in germinated seedlings following such treatments. Imbibed seeds show a much greater resistance to radiation damage. If a selected radiation dose is administered intermittently to dried seeds and the seeds are imbibed for 48 hours and then redried between doses, the adverse effects of the radiation on the viability of the dried seeds is much reduced. It is argued that the periodic wetting of the seeds permits a recovery from the radiation damage.

Preliminary studies indicate that the DNA of aging seeds is more fragmented than of fresh seeds. The cause and significance of this damage is unknown. It may arise from the progressive activation of a DNA-degrading enzyme in the aging seeds, and it is conceivable that these breaks in the strands of chromosomal DNA are crucial primary events in the loss of viability of aging seeds.

SENESCENCE PROCESSES IN CELLS, TISSUES, AND ORGANS OF DEVELOPING SEEDLINGS

The nonbotanical reader may find it helpful to refer to Figure 2 and its caption for orientation to the major parts of a seed plant.

The cotyledons are the first part of the developing seedlings to show

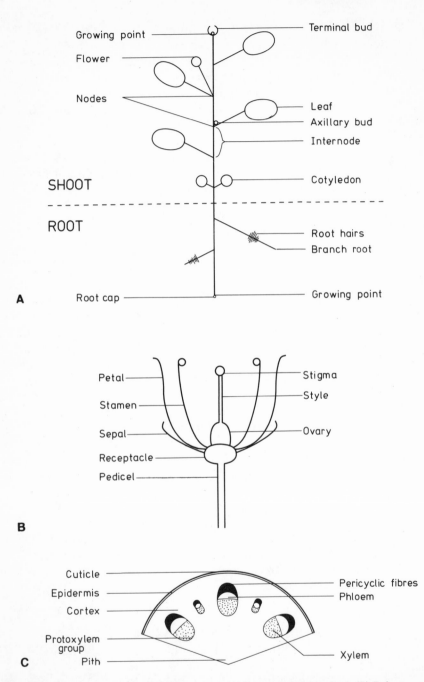

Figure 2. A typical plant, showing the organs. (A) Parts of a whole plant. (B) Enlargement of a longitudinal section through the center of a flower. (C) Transverse section of a segment of a young stem.

senescence, but equally important (if not so immediately obvious) is the death of differentiating cells of the xylem, the loss of nuclei from the sieve tubes, and turnover of cells as root hairs and root cap cells come and go.

In the discussion which follows, it will be necessary to make frequent reference to plant hormones. These are chemical messenger substances produced in one part of the plant and then transmitted to another region where they evoke a specific response. There are five groups of hormones known from higher plants — auxins, giberellins, cytokinins, abscisic acid (ABA), and ethylene — and all have been implicated in one way or another in regulating plant senescence. Broadly speaking, the auxin indolylacetic acid (IAA) and the giberellins are involved in regulating aspects of cell enlargement, the cytokinins regulate cell division, ethylene influences the shapes to which growing cells develop, and ABA is concerned, as its name suggests, with the abscission or shedding of organs and with processes by which many plants stop growing and become dormant. These plant hormones each possess a number of other distinctive functions in the plant, which will be referred to in more detail later, as necessary.

Senescence of cotyledons has been extensively studied in species of the gourd family, in which they develop as large photosynthetic structures, convenient for analytical and experimental purposes. The cotyledons of many plants serve only a storage function; they remain below ground and never develop chloroplasts or become active in photosynthesis. The pattern of events in the senescence of storage cotyledons has been little studied. Even within a single plant family, the legumes, one finds a wide range of cotyledon structures, showing varying degrees of differentiation between storage and photosynthetic functions. The patterns of withdrawal of reserves differ greatly in these different structures, and it is probable that there are many variations in the mechanisms of timing of senescence at the cellular level.

If we adhere to the definition of senescence provided at the beginning of this chapter, the differentiation of cells, such as xylem vessels, tracheids, and fibers, which in the mature state contain no living cells, must fall within this discussion. This is a specialized and difficult subject in which progress has been slow. There are essentially three problems: First, what are the factors that cause a cell to embark on the pathway of specialization, leading to its death? Second, through what cellular mechanisms do these factors operate? And third, what are the details of the actual differentiation processes as they occur? Space does not permit a consideration of these questions in each of the various cases of cell differentiation listed in Figure 1, but a brief consideration of xylem differentiation will serve to illustrate the problems.

It is now clear from detailed anatomical work that xylem develop-

ment always occurs in actively dividing tissues, usually the apical meristem or the cambium, and work with excised root segments and tissue
cultures provides compelling evidence that specific localized cytokinin
concentrations are one of the essential factors determining the onset of
xylem differentiation. There is evidence that growth hormones, particularly cytokinins, regulate and promote the cell divisions of the meristems, but an additional independent function for cytokinins in the determination of xylem differentiation is indicated by the following factors.
When 1-mm cuttings (explants) of the outside layer of seedling pea
shoots were placed in a tissue culture medium, the *proportion* of cells
differentiating as xylem elements increased as cytokinin concentration
was increased (i.e., the effect on differentiation was greater than on cell
division). Finally, it can be shown that transient treatment of explants
with cytokinin will increase the rate of cell division without inducing
differentiation (i.e., inducing cell division is not, of itself, sufficient to
cause xylem differentiation). The current hypothesis is that cells must
undergo cytokinin-regulated division to a stage in the mitotic cycle
when they are capable of responding to a cytokinin-stimulated differentiation.

It is beyond the scope of this review to provide a detailed account of
the subsequent events as the cells differentiate and die, but a few salient
points must be noted. First, characteristic patterns of thickening of the
walls of different types of xylem are determined by the deposition of
cellulose. In those regions of the wall that become thickened, there is a
dense development of microtubules at the periphery of the cytoplasm;
these structures are aligned parallel to the orientation of the cellulose
microfibrils, and they are generally credited with some role in controlling the orientation of microfibril deposition. The hormone auxin is almost certainly involved in the regulation of the growth of the primary
wall, and it has been suggested that other regulators may be necessary
for the determination of the characteristic secondary wall structure. The
deposition of lignin, the aromatic-alcohol polymer typical of xylem tissues, commences with the onset of secondary wall development, but the
prominent phase of lignin synthesis is a relatively late event in the life of
the cell, the lignin being impregnated in the regions of the wall already
thickened with a cellulose framework. The activity of a particular enzyme involved in synthesis of lignin precursors in differentiating xylem
correlates strongly with the lignification phase; the enzyme has been
much studied in a variety of tissues and exhibits complex regulatory
properties, but as yet nothing is known concerning which of these are
involved in the specific case of activation of lignin biosynthesis. In any
event, it is probable that it is the rise in this enzyme that represents the
"point of no return" beyond which the inevitable developmental con-

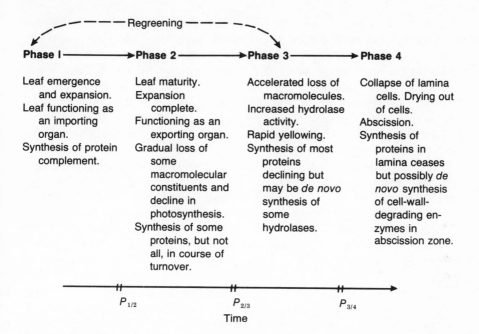

Figure 3. A generalized scheme in which four phases in the life of a leaf are distinguished, and some of the salient attributes of each phase are listed.

sequence is the senescence and death of the cell. The final critical event in the dying xylem cell is the formation of openings between adjacent cells to form the conducting tubes.

SENESCENCE PROCESSES IN THE MATURE VEGETATIVE PLANT

Foliar Senescence

Figure 3 is a generalized diagram, summarizing the phases in the life of a leaf and some of the cardinal events associated with each. Caution is required in generalizations of this kind, because it is well known that the pattern of events can vary greatly with species, with the conditions under which the leaf originally developed, the position of the leaf on the plant, whether it is on a vegetative or flowering plant, the extent of environmental stresses, and whether it is attached to or detached from the plant.

In the bristlecone pine *(Pinus aristata)*, individual leaves may persist in a functional state for up to 30 years, and there are now extensive catalogs of life-spans of leaves for other species, showing a range from a few days to months or years. In many herbaceous plants, leaf senescence occurs sequentially from the base toward the shoot apices, although in some deciduous trees, the seasonal senescence may affect all the leaves simultaneously. Many herbaceous plants can be maintained for long periods in the vegetative condition by subjecting them to an unfavorable photoperiod or by removal of the young flowers. When flowering occurs, the whole plant may die. However, normal leaf senescence from the base of the plant is often arrested during flower and fruit development, so that although the whole plant may die prematurely, the actual life-span of individual leaves at a given position on the stem may be longer than that of the corresponding leaf on a vegetative plant. Periodic water stress and nutrient depletion, particularly of the nitrogen supply, are factors that often drastically modify the rate of senescence of leaves under field conditions. The detailed study of how these various factors operate in relation to the mechanism of senescence has been under investigation for about 100 years, and throughout this period, many studies of the stages of the processes involved have made use of leaves excised from the plant or of discs cut from the leaves. Such systems have two advantages: (1) the isolated tissues usually deteriorate at a greater rate than when attached to the plant, permitting more rapid experiments, and (2) the cut surfaces are convenient for feeding labeled metabolites and regulatory compounds. Against this must be set the uncertainty of the extent to which the changes in isolated leaves resemble the events in intact leaves, bearing in mind that the import and export of materials is prevented, and wound responses are evoked in the cells damaged by cutting.

Studies of the senescence of leaves and petals and the senescence (i.e., ripening) of fruits have generated a heterogeneous assortment of facts concerning both genetically determined and environmentally modified variations in these processes. An apparently common characteristic, however, is in the pattern of the sequence of events, a period of slow change marked by gradual decreases in certain cell constituents and physiological processes, followed by a phase of much more rapid change. In Figure 3, which depicts in a generalized form the sequence of events in the life of a leaf, these two phases are labeled P_2 and P_3. The events surrounding P_2 and P_3 are sufficiently distinct to encourage the view that the separation of these two phases is significant; yet, it is equally clear that the gradual changes occurring in P_2 must give rise ultimately to the more rapidly occurring changes in P_3.

Thus, in the case of leaves, there is generally a period after the completion of expansion during which the rate of photosynthesis steadily declines before the final, more rapid, drop. This phase may last a few days or for many months, depending on the species and other factors.

This decline in the rate of photosynthesis during phase 2 is partly caused by a decrease in the gross photosynthetic rate but is also, in part, accounted for by a progressively increasing rate of photorespiration — that is, the rate of light-dependent efflux of CO_2 that proceeds simultaneously with the net influx of CO_2 in photosynthesis.

It has been known since the end of the last century that there is a gradual loss of soluble protein from senescing leaves during P_2; subsequently, it was shown that the ribonucleic acid (RNA) content also decreases. Further analysis has shown that, in terms of bulk, the greatest losses of RNA and protein are from the chloroplast.

Experiments on the effects of permitting root development on the petioles of excised leaves led to the demonstration that factors synthesized in the roots and carried in the transpiration stream are essential for the maintainance of protein synthesis in leaves. Cytokinins have been shown to be important components of the transpiration stream in this respect, but other factors may also be involved. If the young leaves of a plant such as dwarf bean, *Phaseolus vulgaris,* are greased or contained within a plastic bag or are removed altogether, so that the transpiration stream is diverted exclusively through the older senescing leaves, the senescence process can, in a sense, be reversed. There is a resumption of chloroplast protein synthesis; the leaf regreens, and its photosynthetic activity is restored. In other species, such as *Xanthium pensylvanicum,* the cocklebur, a regreening process cannot be induced, and it has been suggested that in such species the chloroplast genome may be degraded or irreversibly repressed at the completion of leaf expansion (P_1/P_2). Excised leaves of many species lose the capacity for chloroplast protein synthesis when kept in darkness. Reillumination of the leaves after a few hours of dark treatment leads to a resumption of chloroplast protein synthesis, but after about 24 hours of dark treatment, this ability to recover on returning to the light is lost. In species such as tobacco, in which cytokinins will delay the senescence of darkened excised leaves, the capacity for chloroplast–protein synthesis is not sustained by cytokinin treatment, nor does cytokinin treatment during the dark period greatly affect the ability to recover on reillumination of the leaves. In certain *Prunus* species (e.g., cherries), auxin is required to delay the senescence of darkened excised leaves, while in dandelions, docks, and other species, gibberellic acid (GA) treatment or combinations of GA and cytokinin are most effective in retarding senescence. There is some evidence of a lower endogenous level of the particular hormones that delay senescence in different species, suggesting that all three hormones may be needed, acting in concert to maintain leaf metabolism; those which are most effective on leaves of a particular species are the ones that are in limiting supply.

Excised leaves maintained in the light senesce more slowly than darkened leaves, and the effects of applied hormones are less pronounced. If a part of an excised leaf is darkened, that area will senesce while the illuminated portion remains green, suggesting that if a vital light-generated component is required, it cannot be transported to the darkened zone. There is a specific chemical agent that blocks the water-splitting reactions of photosynthesis so that CO_2 fixation cannot occur and does not counteract the effect of light in delaying senescence. It has been suggested that there may be a requirement for a high-energy intermediate process not inhibited by the chemical agent.

Hormone treatments of attached leaves are notably less effective in delaying senescence; this has led to some doubts concerning the significance of studies with excised leaves. There are, however, a number of studies suggesting that a central problem is effective penetration of the hormones, positive results having been obtained with attached leaves by using more frequent application at higher concentrations.

Treatment of excised leaves of many species with abscisic acid (ABA) causes acceleration of senescence, and it is well known that exposure of leaves to low water potentials, a stress that causes a rapid rise in the ABA content of leaves, also brings about accelerated senescence. However, analysis of leaves of *Phaseolus vulgaris* (bean) undergoing senescence while attached to the plant has failed to reveal any increase in the endogenous ABA content. It should be noted that a host of less specific compounds and a variety of inhibitors also delay senescence in excised leaves, emphasizing further the need for caution in the interpretation of results obtained using this technique. Because of our lack of knowledge concerning the mode of action of plant hormones, it is impossible at present to relate these complex hormonal effects to the fundamental metabolic changes accompanying leaf senescence.

When attached leaves are examined in the rapid phase (P_3) of senescence, or when senescence is accelerated by excission or stress treatment, it is generally found that there is a rise in the activity of certain hydrolytic enzymes, as measured *in vitro*. Much of the work on this aspect of the subject has been concerned with relatively crude extracts of leaves, and there is at present no good evidence as to whether this represents a late burst of *de novo* synthesis of a specific group of terminal degradative enzymes or the release of active enzymes from previously inactive complexes. Indeed, although it is reasonable to suppose that this increased hydrolytic activity is involved in the final degradation of macromolecular constituents of the leaf to form soluble substituents which can then be exported to other parts of the plant, it has not been critically demonstrated that this extra enzyme complement is necessary for, or involved in, such a process. It could be that the hydrolases mediating the normal processes of turnover of macromolecules are sufficient to achieve this task if, as we have seen in the case of the chloroplast, the synthetic side of the turnover process is arrested.

Metabolic and fine structural studies have shown that the respiratory activity and the integrity of the mitochondria and plasma membranes of leaf cells are retained relatively intact until a late stage in phase 3 of leaf senescence. Teleologically, this is understandable since an energy source and a functional membrane system will be required for the active loading of degradation products into the export channels of the phloem. There is as yet no clear indication of the differential mechanism by which these aspects of the fine structure and metabolism of the leaf cells are maintained intact while the photosynthetic apparatus is being dismantled.

Abscission

Abscission is the natural process that leads to the detachment of a part of a plant. Leaf fall, at the conclusion of senescence, involves cell separation along a morphologically well defined plane across the base of the petiole. It has been pointed out that this is probably one example of a more general phenomenon that is involved also in the abscission of petals, the rupture of anthers for the release of pollen, the dehiscence (splitting open) of fruits for the release of seeds, and probably also the separation of cells associated with the development of air spaces during the ripening of fleshy fruits such as apples. The implication of hormones in the abscission process is shown by the fact that application of IAA to a petiole stump delays abscission; it is suggested that IAA is acting as a substitute for the young leaf blade. The mechanism of this regulation is complicated. The site of application of IAA can alter the direction of the abscission response. Applied to the distal side of the abscission zone, it can inhibit abscission, but applied to the proximal side, it can accelerate it. The amount of IAA applied can alter the response. High concentrations of IAA inhibit abscission, low concentrations accelerate it. The time at which IAA is applied to the severed petiole can alter the response. Applied immediately after the removal of the leaf blade, it inhibits abscission, but applied 18 to 24 hours after removal of the blade, IAA accelerates abscission. These responses gave rise to the general impression of a correlation between more rapid abscission and decreasing IAA levels in the tissues distal to the abscission zone.

Exposure of leaves to the gaseous hormone ethylene accelerates abscission, and ABA has also been shown to accelerate abscission when applied to petiole stumps or to intact plants of cotton. There is a correlation between increasing amounts of diffusible ABA from coleus leaves and the onset of abscission. Interpretation of the experimental work on ethylene effects on abscission is particularly uncertain in experiments with parts removed from the plant, because ethylene production is stimulated at the cut cell surfaces. One feasible explanation of the situation is that as a leaf senesces, there is a rise in the proportion of ABA to IAA in the leaf lamina and petiole distal to the abscission zone, which serves as the trigger for an increase in ethylene production. Ethylene inhibits IAA transport down the petiole leading to a reduced IAA level in the abscission zone; this aspect of ethylene action in abscission can be simulated by application of specific inhibitors of auxin transport. Reduction of the IAA level in the abscission zone potentiates the cells to accept a direct stimulus from ethylene diffusing through the intercellular space of the petiole, which induces them to synthesize hydrolytic enzymes. In the abscission zone (as in other tissues such as pea epicotyls), this action of ethylene in activating hydrolytic enzymes, some of which are involved in cell wall metabolism, brings about a reorientation of the direction of cell expansion. This causes the cells on the proximal side of the abscission line to expand radially, generating stress forces across its cell walls. At the moment of rupture, when the leaf falls, there is a sudden increase in the width of the remaining base of the petiole as this tension is released.

SENESCENCE PROCESSES AT THE FLOWERING STAGE

In ephemeral, annual, biennial, and longer-lived but monocarpic (one that produces flowers and fruits only once and then dies) species, the whole plant dies after flowering or fruiting. This propensity is to be found widely scattered in many families of plants in which the majority of species are long-lived perennials. This observation, along with comparative taxonomic studies, suggests that the monocarpic habit represents a derived condition that has evolved independently along many lines. In some species, senescence may result from conversion of all the vegetative apices of the shoots to floral tissues and an inability to initiate new vegetative meristems. Cambial activity is also observed to stop at the time of flowering in these plants, but the cause of this generalized loss of meristematic activity is not understood. It has been known since the turn of the century that the life-span of many monocarpic species can be greatly extended by removal of the flowers as soon as they are formed. Earlier studies suggested that the plant died following flowering and fruiting as a consequence of a preferential removal of essential metabolites to the developing seeds. Subsequent work has discounted this hypothesis, and inquiry has shifted to suggestions ranging from irreversible effects on the growth of the root systems to the concept of positive lethal signals arising from the developing seeds. A great deal more physiological work will be needed on these phenomena before a start can be made on any analysis of these effects in metabolic terms.

Senescence of Floral Parts

Following the early observations of Goethe, comparative morphologists came to interpret the structures of floral parts in terms of variously modified leaves. With respect to the senescence of the sepals and petals, this view is supported by several features analogous to foliar senescence, but it must be emphasized that the pattern of events is probably almost as varied as is the morphology of the angiosperm flower. In many species — the Orchidaceae (orchids) are a prime example — the sepals and petals may be long-lived, but accelerated senescence of these parts and a simultaneous growth of the pistils follow rapidly once pollination is achieved. In some families of plants, the petals may senesce, but the sepals persist and may contribute in photosynthesis; in other taxa, all sepals and petals senesce and are shed as the fruits develop, while in others, both persist to form encapsulating structures.

Interest in petal senescence has increased greatly in recent years as a result of transport problems in the cut-flower trade. Crops such as carnations, roses, and gladioli are receiving particular attention. Cultivars differ greatly in the vase life of the cut flowers, which affords a valuable tool in comparative physiological studies; it should be borne in mind, however, that many cultivars are sterile forms in which the collection of stamens and often the pistils may be replaced by petaloid structures, so that aspects of petal senescence that relate to interactions between the floral parts may be much modified in these forms. Petals lack stomata, a fact that complicates many studies of foliar senescence, and chloroplasts are absent in the majority of flowers, although they may be developed in a modified form as chromoplasts.

There is a loss of RNA and soluble proteins from petal cells in the course of senescence, distinguishable in many species as a gradual, followed by a rapid, phase (P_2 and P_3), the latter accompanied by an increase in activity of hydrolytic enzymes, as in leaves.

Senescence of the petals in cut flowers of many species can be delayed by treatment with cytokinins and accelerated by supplying

ABA. If leaves are left attached to the flowering stems, however, these effects may be reversed; ABA, by causing closure of stomata on leaves, reduces water loss from the shoot, and so delays senescence in the petals, while cytokinins, by delaying leaf senescence, increase water stress in the shoot which accelerates senescence of the petals. Rose varieties with long-lived flowers have higher endogenous levels of cytokinins than do short-lived varieties.

Placing cut flowers in a 4% sucrose solution will feed the flowers through the shoot and will double the vase life in many varieties. A possible explanation is that the solution acts as a substrate for respiration; however, the observation that sucrose feeding will counteract the action of ABA in accelerating petal senescence suggests that the mechanism of action may be more subtle.

Recent work suggests that some of these effects of hormones on the senescence of petals may involve the metabolism of ethylene in the flowers; this aspect of the problem is now being studied.

SENESCENCE IN THE FRUITING STAGE

The Maturation and Ripening of Fruits and Seeds

Commercial interest in the regulation of fruit ripening has stimulated an abundance of experimental work during the past 40 years. The accumulation of isolated facts concerning the process is impressive, but progress toward any real understanding of the ordering of events is painfully slow.

It is convenient to approach the subject from three standpoints: (1) respiration and fruit ripening, (2) hormonal regulation of ripening, and (3) metabolic events in the ripening process. Before progressing to these matters, it is important to be clear about what one means by the terms "fruit" and "ripening." For the purposes of this discussion, we shall confine ourselves to the fleshy tissues that enclose or support the ovules. We should bear in mind, however, that there are many dry fruits, in which a regulated senescence process leads to the development of a mature, desiccated structure.

Anatomical studies of different families show that almost any part of the inflorescence may contribute to the swollen cells of a fleshy fruit. Despite this diversity of morphological origins of the tissues, the developmental events involved in ripening share a number of common features.

"Ripening" is a blanket term encompassing a host of changes that are not necessarily closely interrelated. Thus, in many fruits, "ripening" may include a color change in the outer layer of tissues in which chloroplasts are transformed into chromoplasts, and the timing of these events can vary greatly relative to the production of soluble sugars and aromatic compounds and the softening process that characterizes the ripening of the internal tissues.

Respiration. It has been known for many years that the rate of respiration of many fruits may decline gradually as the fruit matures and then increase several fold over a relatively short period during the later stages of ripening. This phenomenon is referred to as the climacteric rise in respiration; both the mechanism and the significance of the process have been the subject of much research and speculation. With hindsight, the climacteric rise could perhaps be best described as the "great red herring" of postharvest physiology.

Although a climacteric rise is observed in many fruits, there are some — grapes are a notable example — in which it does not occur. Further evidence that the climacteric is not a *sine qua non* of ripening comes from several sources. In pears, the softening and allied aspects of ripening may precede the onset of the climacteric; also, tomatoes, pears, and bananas will ripen in an atmosphere containing 2.5% oxygen, but at this oxygen tension, the climacteric rise in respiration does not occur.

Recent studies of the rate of glycolysis in maturing fruits have provided clear evidence that the complement of glycolytic enzymes present in the tissues is entirely adequate to catalyze the observed rise in climacteric respiration without the need for synthesis of more glycolytic enzymes. This is supported by the finding that the climacteric rise can still take place in fruits injected with cycloheximide, a potent inhibitor of protein synthesis in cells of higher organisms.

Hormonal Regulation. Ethylene has been accorded a preeminence in hormonal studies on ripening, arising from the early observations that ethylene can induce ripening in many fruits and may be produced in copious quantities in the course of ripening in some fruits.

In fully developed green pears, the characteristic softening of the tissues and formation of sugars can be induced by ethylene at a low internal concentration, but for the induction of the climacteric rise in respiration, a high internal concentration is required. We may conclude that different receptor mechanisms are involved in the softening processes and in the induction of the alternative respiratory pathway. The bulk production of ethylene by some ripening fruits occurs after the onset of the softening process; there is substantial evidence that in fruits, such as apples and pears, methionine is the precursor for the low levels of ethylene that initiate the early stages of ripening. The evidence concerning the subsequent bulk production of ethylene, associated with the climacteric rise, is conflicting, but it probably arises from secondary reactions and in the ripening cells and is a consequence, rather than a cause, of the ripening process. In support of this view, it is noteworthy that in some species, such as mango, no climacteric production of ethylene is observed.

It is now well established that the early stages of fruit development are regulated by hormones produced in the seeds. Production of auxins, gibberellins, cytokinins, ABA, and ethylene are found at various stages in fruit development and often in a specific sequence. It is increasingly clear that it is the relative proportions of different hormones that control fruit development and regulate translocation of materials into the fruit. It is not, however, possible at the present state of knowledge to pinpoint the hormonal changes that are critical in the release of the ripening process. In tomatoes and other fruits, correlations have been demonstrated between high cytokinin content and delayed ripening. We thus have an analogous situation to that in leaves, where high cytokinin content is also associated with a slower rate of senescence. Substantial further progress in understanding the hormonal control of fruit maturation and senescence will probably depend, as with so many other aspects of plant senescence, on progress in elucidating the basic mechanisms of hormone action.

CONCLUSIONS

In this discussion, we have seen that at all stages in the life of a plant, from the seedling to the mature fruiting stage, specific groups of cells or whole organs are undergoing sequences of developmental changes which lead directly to their death. One may find many comparable situations in animals, as the cells of the skin and of various organs die and are renewed.

We have referred at length to the involvement of hormones in regulating these diverse senescence processes, but little has been said concerning how the hormones actually operate at the molecular level. This is, quite simply, because we do not know. What is quite certain is that it is important that we should delve deeply into these problems, for when they are understood, it will almost certainly follow that the senescence can be controlled. When we think of the importance of fresh green vegetables for market, of cut flowers for the winter, or of a controlled supply of nicely ripened fruits, it becomes clear that an understanding of plant senescence in all its aspects is of potential value to us all.

BIBLIOGRAPHY

Abeles, F. B. 1973. *Ethylene in plant biology*. Academic Press, New York.

Baumgartner, B., H. Kende, and P. Matile. 1975. Ribonuclease in senescing morning glory. *Plant Physiology* 55: 734.

Biale, J. B. 1964. Growth maturation and senescence in fruits. *Science* 146: 880–888.

Butler, R. D., and E. W. Simon. 1972. Ultrastructural aspects of senescence in plants. *Advances in Aging Research* 4: 157.

CNRS International Colloquium. 1975. *Facteurs et la maturation des fruits*. No. 238, Paris.

Coombe, B. G. 1976. The development of fleshy fruits. *Annual Review of Plant Physiology* 27: 507–528.

Hulme, A. C., ed. 1970. *The biochemistry of fruits and their products*, Vol. 1. Academic Press, London.

Hulme, A. C., ed. 1971. *The biochemistry of fruits and their products*, Vol. 2. Academic Press, London.

Kende, H., and A. D. Hanson. 1976. Relationship between ethylene evolution and senescence in morning glory flower tissue. *Plant Physiology* 57: 523–527.

Leopold, A. C. 1961. Senescence in plant development. *Science* 134: 1727–1732.

Molisch, H. 1938. *The longevity of plants (Die Lebensdauer der Pflanze)*. Trans. E. H. Fulling, New York.

Roberts, E. H. 1972. *Viability of seeds*. Chapman and Hall, London.

Simon, E. W. 1974. Phospholipids and membrane permeability, *New Phytologist* 73: 377–420.

Tetley, R. M., and K. V. Thimann. 1975. The metabolism of oat leaves during senescence. *Plant Physiology* 56: 140–142.

Torrey, J., D. E. Jasket, and P. K. Hepler. 1971. Xylem differentiation: A paradigm of cytodifferentiation in higher plants. *American Scientist* 59: 338–352.

Villiers, T. A. 1972. Ageing and the longevity of seeds in field conditions. Page 265 in W. Heydecker, ed. *Seed ecology*. Butterworth, London.

Vonshak, A., and A. E. Richmond. 1975. Initial stages in the onset of senescence in tobacco leaves. *Plant Physiology* 55: 786–790.

Woolhouse, H. W., ed. 1967. *Aspects of the biology of ageing. Symposia of the Society for Experimental Biology*, No. 21. Academic Press, New York and London.

Woolhouse, H. W. 1974. Longevity and senescence in plants. *Science Progress* 61: 123–147.

Woolhouse, H. W., and T. Batt. 1975. The nature and regulation of senescence in plastids. In N. Sunderland, ed. *Perspectives in experimental biology*, Vol. 2. Pergamon Press, New York.

6

The Biological Significance of Death in Plants

A. CARL LEOPOLD

Life may be compared with a candle flame, a unit of matter which has persistent characteristics in time although its very composition is continually being exhausted and renewed. The simile can be extended to say that the flame of life of any individual is a revolving population of structures, including populations of cells, tissues, and sometimes even of organs, as well as of chemical components. A central feature of life is that it is composed of and built upon components which are continuously undergoing turnover.

Biological turnover involves not only nutrients, cells, and tissues or organs, but also a ubiquitous turnover of organisms or individuals. In the animal kingdom, individuals are systematically removed from populations by predation, stress, or aging processes; in the plant kingdom, individuals are ordinarily removed from populations through aging, or through stress, or in many species through a programmed internal lethal process, a process of senescence.

AGING AND SENESCENCE

For the purposes of this chapter, aging may be defined as the processes associated with the accrual of maturity in time, whereas senescence may be defined as the deteriorative processes which are natural causes of death. Ordinarily, aging involves many processes or changes which need not have any particular relevance to death in a causal sense, whereas senescence involves an amplification of specifically lethal con-

A. CARL LEOPOLD, Ph.D. ● Boyce Thompson Institute for Plant Research, Cornell University, Ithaca, New York 14853

ditions, and ordinarily senescence is a programmed or integrated set of lethal events carried on by the organism in a dynamic manner (presumably under programmed genetic control).

The differences between aging and senescence are not always clear, but can be illustrated in an idealized manner by a comparison of curves for the survivorship of populations of organisms through their life-expectancy time; for example, in Figure 1, the survivorship of a population which does experience aging but not senescence will ordinarily proceed in a logarithmically declining curve as illustrated in curve A. In a given unit of time, half of the population may fail to survive, and the repeated removal of half the population at successive time intervals results in such a tapering curve. Examples of this type of survivorship will be expected for most wild animals, especially those that are subject to continuous predation or stress conditions. The survivorship of a population which does experience senescence will ordinarily proceed as illustrated in curve C, where the individuals will essentially all survive over nearly the entire life-span for that species, and then in an abrupt gesture, all will die. Examples of this type of survivorship will be expected for most annual and biennial plants (e.g., soybean, corn, each of the annual grains, annual weeds, annual garden flowers, biennial beets, carrots, onions and for other species which flower only once in their lifetime (e.g., century plant, bamboos, bananas). In between the characteristics of survivorship by slow attrition and by sudden death, many organisms experience a tendency to die after reaching a given age as illustrated in curve B in Figure 1; a familiar example is that of man, who has an ability for high survivorship during the years of growth and maturity, followed

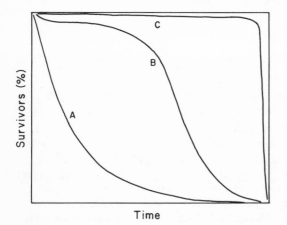

Figure 1. Generalized survivorship curves for (A) populations which do not show senescence; (B) populations, such as man, which are more likely to die in the later years of their life-expectancy period; and (C) populations, such as annual plants, which die en masse at a rather specific time. The maximum life expectancy is plotted as the extreme right limit of the graph. (From A. C. Leopold, Senescence in plant development, *Science* 134: 1727–1732, 1961.)

by a period of declining survivorship in the latter half of life. The gradual loss of survivorship in later life may be a natural expression of the accumulative changes with aging — the accumulation of changes with time which are not themselves naturally lethal but which gradually lower human viability.

In a general way, then, one could say that survivorship curves may decline in a logarithmic fashion due to predation or continuous (or often repeating) stress, or may be logarithmic after a given state of maturation due to aging; in species which experience senescence, survivorship may remain very high until the time of an abrupt termination of life.

In addition to the sudden-death type of behavior, plants also show another distinctive expression of senescence. In the normal development of most plants, there is a normal senescence and death of separate parts of the plant in an orderly manner. Ordinarily, there is an orderly dying off of the oldest leaves and a progressive death of successively higher leaves along the axis of the plant stem. In the case of perennial plants, there is often an annual senescence and death of all the leaves — the autumnal coloration and fall of leaves; in some perennials, this deciduous leaf senescence is associated with a senescence and shedding of branches, a natural pruning of the smaller branches from the tree. Such abscission of branches is conspicuous in the case of white oak trees, which drop large numbers of short branches each fall, each with a button-shaped abscission zone at its basal end. The cypress is another tree which sheds small branches in association with the senescence and abscission of its leaves. In the case of herbaceous perennials, such as the nonwoody garden perennials (hollyhock, lilies, tulips, peonies, chrysanthemums), the annual senescence occurs not only for leaves, but also for the stems, leaving the roots and basal buds to provide for new growth in the following year. Another striking example of normal senescence of plant parts is the senescence of flowers and fruits; in many species, the time of pollination of the flowers results in a sudden collapse of the flower petals and other parts, as illustrated by the pollination of orchid flowers and Easter lilies (which have a prolonged life if one removes the anthers as the flowers open). The senescence of fruits is another example of the orderly, programmed regulation of specific plant parts. Recognizing, then, that plants may have programmed capabilities of bringing about the senescence and death of localized plant parts, we can presume that programs for senescence must be a common type of genetic regulation.

Several physiological changes can be generally associated with senescence. In the case of leaves, the onset of senescence is characteristically associated with a decline in chlorophyll content; the loss of green color is often associated with increases in red or yellow pigments especially anthocyanins and carotenoids. Another type of physiological change is a deterioration of protein components; the loss is astonishingly large, as, for example, in the soybean leaf where over 80% of the protein is lost before the leaf enters senescence. A similar and parallel loss of nucleic acids also occurs, and over 90% of the total nucleic acids is lost before the leaf enters senescence. Another physiological change is a striking increase in leakiness of the cells of the senescing tissue; in the case of leaves, large amounts of inorganic and organic nutrients may actually leak out of the organs as they enter senescence; in the case of senescing flower petals, large amounts of water may leak out of the petals, leaving them collapsed and wilted.

LONGEVITY AND SURVIVAL

While turnover may be a central feature of life, the time involved for normal turnover of individuals varies enormously between species. Annuals are programmed to die after a single season's growth, whereas other species may persist for exceedingly long times. Famous for longevity are the sequoia trees, with life spans as long as 3000 years, and the bristlecone pine trees, which have been estimated to reach the age of 4000 years. Less readily recognized are the survival times for clonal species such as the grass clones of the prairies of the western United States; it has been estimated that these clones may have persisted since the time of the last glaciation — perhaps 15,000 years. A similar longevity has been suggested for the clones of aspen trees in the northcentral states; these, too, might have persisted from the time of the last glaciation.

As with the long persistence of some clones, there can be an extensive persistence or longevity of tissues or organs which are removed from the parent plant; thus, even leaves removed from a mature plant can root and not only escape the approaching senescence for which the parent plant is destined but can show a rejuvenation and commence another and extended life-span. When tissues are excised from plants, they, too, escape from the limits of the parent plant's longevity limits; tissues from tomato plants have been cultured continuously since about 1945. The normal limits of longevity, then, are characteristics of the whole plant, and can be avoided in the case of clonal growth, which may successively rejuvenate the plant, or in the case of isolated tissues, which can grow more or less indefinitely.

For many species of plants, longevity is specifically linked to flowering and fruiting; this is most evident in the case of the annuals which die at the completion of their reproductive cycle (e.g., soybeans, corn, the small grains, and many annual garden plants). The uniform death of millions of wheat plants in the month of July is clearly not attributable to the end of the frost-free growing season, but can only be accounted for as a consequence of the formation of some internal lethal signal within each plant — a communal hari-kari. The onset of senescence in the grains and soybeans is often the cause of serious losses of yields, especially when the crop was late in developing the grains or beans, and as yet the agronomist has no effective means of delaying the onset of senescence. Again, in the case of biennial plants, the completion of the flowering process and fruiting cycle is the time at which the entire plant dies, as illustrated by onions, beets, and carrots. In other species, the flowering and fruiting processes are not linked to a specific time period, but several years may pass before the plant enters the reproductive state, and after its completion, the entire plant dies. The century plant is a familiar example of this type of limited longevity.

The common correlation of senescence with the completion of reproduction presents one of the major clues to the nature of the process of senescence in plants; the interpretation was suggested by physiologists in the early part of this century that reproductive activity led to the death of plants through an exhaustion of nutrients from the vegetative parts into the fruits.

The wide diversities of survivorship between species of plants, from the short-lived annuals, with a predetermined senescence at the end of the reproductive cycle, to the long-lived conifers, with life-spans extending over millennia and no apparent predetermined terminus, indicate that the nature of the death processes is not at all uniform among plant species, some species having built-in programs for death and others not.

ECOLOGICAL ROLE OF SENESCENCE

In zoology, it is generally considered that the death of individuals has only negative values, such as the sloughing off of the older components of the populations, limiting the size of the populations, maintaining a high proportion of individuals in the reproductive state, and removing, as in the case of man, those individuals with the most learning. Analogies between the zoological and the botanical situations may be instructive, however, for one can readily see that in many plants the onset of senescence and death has distinct and positive values in terms of ecological adaptation, natural selection, and the efficiency of internal physiological functioning.

Many species of plants utilize senescence as an adaptation of the plant community to seasonal events. Many of the spring bulb species are adapted to the earliest part of the growing season, when there is minimal competition with other plants for light; the completion of fruiting occurs at about the time that competition for light becomes intense, and senescence then causes the plants to die back until the following early spring season. Another type of seasonal adaptation is found in corn, soybeans, and many annual weeds, which complete their reproductive activities shortly before the onset of frost in the autumn, and death of the plants just precedes the end of the frost-free season. Ragweed, blooming in the fall and dying at the end of the growing season, is a well-known case. Adaptations of species to seasons of sufficient moisture for growth are seen among desert species. These examples will suffice to illustrate that senescence may limit the growth of a plant species to a certain part of the growing season and provides seasonal adaptation to that limitation.

Another positive function of plant senescence may be through an enhancement of natural selection. Evolutionary rates of change may be presumed to be a product of two variables: the degree of variability between individuals of a species, and the rapidity of turnover of the individuals in the species. Thus, evolutionary change may occur most rapidly in species which have a high degree of variability and a short longevity. It is relevant to note that the species which show the greatest aggressiveness in adapting to new environmental situations — the weeds — are predominantly annuals with prominent senescence characteristics at the end of the reproductive cycle, which force an annual turnover of all individuals in the population. If an advantageous genetic change occurs in species of annual weeds, it is spread through the population at a maximal rate because of the complete turnover of the breeding population each year. In contrast to the annual weeds, the long-lived perennials, such as forest trees, continue to produce progeny — or at least produce seeds — with unchanging genetic constitution over long periods of time. Unlike the situation in many animal species, which terminate reproductive activity after a period of maturity, perennial plant species ordinarily continue to produce seeds over their entire life-span, and the retarding effect of this prolonged reproductive activity on evolutionary adaptive change seems evident.

Organized programs of senescence within the individual plant provide numerous physiological efficiencies. The progressive senescence of older leaves may serve not only to maintain a high photosynthetic efficiency through the elimination of leaves that have deteriorated in this capacity or have become shaded, but it also provides for a recycling of nutrients from the older to the younger leaves. The hydrolysis of chlorophyll, proteins, and nucleic acids in senescing leaves has already been mentioned; the extent to which nitrogenous substances may be mobilized from older leaves is enormous. The mobilization of inorganic nutrients from older leaves provides a major source of nutrition for the plants; in the case of oat plants, the root system ceases to take up nitrogen from the soil when the plant is about halfway through the life cycle, and essentially, all subsequent development is achieved through a redistribution of nitrogenous materials, mainly from the older leaves.

Another physiological efficiency achieved through the senescence of parts of the plant relates to flowering and fruiting. The senescence of flowers serves to remove pollinated flowers from the stage of insect attractions; the senescence of fruits plays an important role in the development of their edible characteristics, which make them attractive to animals in the process of seed dissemination. The occurrence of natural stem pruning in some species of trees has already been mentioned, and senescence also provides for a natural pruning of fruits to the point that the plant can support the maturation of its entire seed and fruit load.

THE UBIQUITY OF DEATH

Programmed death is a common feature in plant biology. All higher plants utilize dead cells in the form of xylem — chains of hollow dead cells which serve as conduits for the flow of sap; the orderly death of xylem cells is an ubiquitous feature of development in plants. In the case of trees, the accumulated xylem and associated fibers constituting wood are characteristic of older tissues, but the maturation of xylem cells to the dead, hollow state occurs in very young regions of the growing points.

The death of cells in meristematic tissues is a common event in the development of some plants. For example, in the development of leaves of some legumes, clusters of cells lose their nucleic acid content and die at regular intervals along the meristem surface, thus leading to the development of compound leaves by the subsequent growth of nondying portions of the meristem. An analogous system of embryonic cell death occurs in the development of the feet of birds, where localized clusters of dying cells in the embryonic foot lead to the differentiation of the several toes of the bird foot. Returning to leaf development, the holes which develop in the leaves of *Philodendron* or *Monstera* originate as clusters of cells which die in the early state of development of the leaf meristem.

Another type of meristematic senescence has been suggested from experiments with pea plants; as pea plants become senescent, the meristems in the terminal bud may lose their capability for growth, as evidenced by a failure to reinstate growth when grafted onto nonsenescent plants.

The general occurrence of organ death has already been mentioned, including the death of leaves as a normal expression of progressive senescence up the stem, of the senescence of aboveground parts in perennial forbes, of the autumnal or deciduous senescence in woody perennials. So, also, has the senescence of flowers and fruits been mentioned as normal developmental events.

The abrupt death of entire plant organisms is characteristic of most annuals and biennials, and of certain other species which die as a normal consequence of reproductive development. Those plants which do not have a built-in program of senescence may die through gradual deteriorative or aging changes, including a gradual decrease in growth rate and lowered viability.

The death of *species* of plants and animals from the biosphere was a major concern of Charles Darwin, as well as of subsequent students of evolution. It may not be reasonable to consider the disappearance of species as a normal component of a species life cycle, but the widespread occurrence of nonsurvival of biological species forms a cornerstone of the concept of evolutionary development. Students of plant geography can generally distinguish between young and old existing plant species on the basis of their relative aggressiveness or conservatism with respect to their ability to move into new environmental locations. It is common for "old species" of plants to be unaggressive about moving into new locations, and to be restricted to a few sites in which they have persisted over long periods of time; such "old species" are termed relic species by the plant geographer. An example of old or relic species of plants would be the yellow lady slipper, in contrast with the aggressive young species of cocklebur and dandelion.

An even wider scope of death, or the nonsurvival of biological entities, occurs in the aging and disappearance of *floras*, associations of plant species. An example is the flora of the Carboniferous period, which was the principal contributor to the formation of coal deposits of the world. The species which composed that flora are well known from fossil remains, and that association of species has disappeared. Again, as with species, plant geographers can distinguish between young and old existing floras on the basis of the relative aggressiveness with which a flora can move into new geographic locations. In North America, the grassland flora is considered to be young and aggressive in contrast to the Allegheny flora (the beech–maple forest of the New York–Virginia region), which is old and relatively immobile.

That death is ubiquitous in plant biology, from tiny cells to enormous floras, should not itself be a notable observation. The ubiquitousness of death becomes more impressive, however, when one perceives that programmed death has been so extensively exploited as a physiological benefit, and that organismal death has been such a basic necessity for evolutionary development and change.

MECHANISMS OF SENESCENCE

The most conspicuous fact about the onset of senescence is that so many species of plants experience overall senescence at the time of completion of the reproductive phase; a Viennese botanist, Hans Molisch, developed the theory that the fruits exhausted the nutrients from the other parts of the plant and thus killed it — *Erschöpfungstod*. Striking deferral of senescence can be achieved through the simple device of pruning the flowers off a plant, thus avoiding the lethal effects of the fruiting process, as illustrated in Figure 2. Following his theory,

Figure 2. Deferral of senescence in soybean plants by the removal of flowers and fruits during the growing season. (From A. C. Leopold, E. Niedergang-Kamien, and J. Janick, Experimental modification of plant senescence, *Plant Physiology* 34: 570–573, 1959.)

several lines of supporting evidence have been found, including, for example, an extensive mobilization of nutrients, both organic and inorganic, from the vegetative into the fruiting parts; the lethal effects accrue during the entire period of flowering and fruiting, and fruits exert a localized effect on surrounding leaves, but if nearby leaves are removed, the effect is transmitted promptly to more remote leaves and branches.

Some serious complications of Molisch's theory have been found; for one, it has been found that plants of spinach which have only male flowers, and hence are unable to develop fruits, show the same onset of senescence at the end of flowering, as do female plants at the end of fruiting; removal of the male flowers defers plant senescence in a manner comparable to the removal of female flowers. Another complicating piece of information is that cultivars of peas show striking differences in their tendencies toward senescence, some cultivars producing a single set of fruits after which the plant becomes senescent, and other very closely related cultivars producing enormously greater quantities of fruits without becoming senescent at all. The *Erschöpfungstod* concept does not appear to provide a comprehensive answer to mechanisms of senescence.

Studies of the senescence of older leaves have established that, in addition to the senescence-inducing effects of flowers and fruits, there are distinct senescence-inducing effects from the vegetative growing points of the plant. This can be demonstrated in a manner similar to the fruit-removal experiments: removal of the active growing points of plants can be shown to prevent the senescence of the older leaves. There is indirect evidence, too, that roots may provide a positive effect, deferring the onset of senescence in the older leaves. The senescence of the foliage of onion plants when the bulbs have been completely filled is known to be preceded by a termination of growth of the roots. Systemic effects of one part of the plant on the physiological functioning of another remote part are called correlation effects, and the effects of flowers, fruits, shoot tips, and roots on the senescence of other parts are clearly systemic or correlation effects.

How can correlation effects occur? How can physiological influences be transmitted from one part of the plant to another, and there cause regulatory changes? There are two main possibilities, one being that there can be a direct translocation of constituent substances from one part of the plant to another — a source-to-sink mobilization, which could have physiological effects on the organs which serve as the source. The other main possibility is that there may be chemical or hormonal signals transmitted from one part to another, there serving to regulate a physiological event. Each of these types of physiological mechanisms appears to be involved in the correlative effects of senescence.

Directed translocation does occur from the leaves and stems to the flowers and fruits; this can be demonstrated precisely through the utilization of radioactive tracers introduced into the leaves and followed as they move into the fruits. Likewise, there is a directed translocation from the leaves into active growing points, which also has been demonstrated through the use of radioactive tracers. Even within individual leaves, particularly the leaves of grasses, there is a directed translocation from the tips of the leaves to the base; this can be studied in detail in isolated individual leaves, and can be readily associated with the tendencies of leaves to show characteristics of senescence first at the tips, and then progressively down toward the base of the leaf. In sugar cane, this directional translocation plays an important role in the accumulation of sugars in the stems of the plants, from which it is commercially harvested. A generalization could be made that directed translocation occurs as an integral component of all types of senescence that have been studied, with the onset of senescence associated with a mobilization of nutrients out of the plant part.

Hormonal factors are the most readily understood carriers of correlation effects, and the entry of the hormonal concept into the physiology of senescence began with the discovery that the application of some hormones could prevent the senescence of leaves on cuttings of cocklebur. At the present time, there are five known plant hormones: auxins, cytokinins, gibberellins, ethylene, and abscisic acid; each of these is known to have marked influences upon the development of senescence, but the most impressive effects are from cytokinins. When a cytokinin is applied to a leaf on a cut stem, it has profound effects in deferring the leaf senescence. Some species are prevented from senescence, not by cytokinin, but by gibberellin; and others by the presence of both cytokinin and auxin. Ethylene and abscisic acid generally hasten senescence.

The possibility that hormones might be responsible for the correlative nature of senescence arose when it was found that roots provide a major source of cytokinins and gibberellins to the shoots of plants, and conditions under which the shoots tend to become senescent correlate rather well with conditions under which the roots provide lesser amounts of cytokinins and gibberellins. A further role of hormones relates to the involvement of directed translocation; experimentally, it can be shown that when fruits are removed from isolated cuttings of beans, for example, the tendencies toward senescence are lessened, but when treatments of auxins, cytokinins, and gibberellins in various combinations are applied, directed translocation of nutrients to the region of application occurs in a manner similar to the directed translocation to growing fruits.

Returning to the role of hormones in senescing leaves, it has been known for several decades that older leaves contain lowered levels of auxins, cytokinins, and gibberellins; the application of various combinations of these hormones to isolated leaves (depending upon the species of plant being studied) can defer or prevent the onset of leaf senescence. Another possibility is that there may be a lethal hormone which brings about organ death. In addition to the known effects of ethylene and abscisic acid in enhancing leaf senescence, there is some evidence that ripening fruits may produce a lethal substance which is translocated from the fruits to the stems and leaves.

The hormonal involvements in leaf senescence, then, can be seen to have several components: (1) The relative hormone levels in various parts of the plant can determine whether or not that part will become senescent; (2) the supply of hormones from the roots to the shoots can affect the amounts of hormones in the shoots and, hence, alter the tendencies to become senescent; (3) the localization of hormones in fruits or growing points can account for directed translocation of nutrients out of the leaves to these sites of localization; and (4) there is the possibility of a lethal hormone being formed in the ripening fruit which may move to the leaves and stems.

The most notable biochemical changes associated with senescence are the deterioration of chlorophyll, protein, and nucleic acids. It is well known that the synthesis of these various components is closely linked, for nucleic acids are the major directors of protein synthesis and of chlorophyll synthesis. The hormones which are known to defer the development of senescence in leaves also serve to maintain the leaf content of nucleic acids and, thereby, of proteins and chlorophyll. For example, the application of cytokinin to a leaf which would normally be about to enter senescence will result in the maintainance of a markedly higher level of nucleic acids in the leaf and, of course, markedly higher levels of proteins and chlorophyll. It is not altogether clear whether the hormonal stimulus to maintain higher levels of these components is due to a stimulation of synthesis of the nucleic acids by the hormone, or whether, alternatively, the hormonal effect may be due to a retardation of hydrolysis of the nucleic acids.

The possibility of explaining the senescence of leaves through the actions of hormones is somewhat dimmed, however, by the fact that the best experiments showing hormonal regulatory effects are with isolated leaves, and when one attempts to modify or prevent development of senescence in leaves which are intact on the plant through the application of hormones, the effects are much less impressive. Many of the hormonal experiments on senescence have been done with excised leaves or excised leaf pieces, held in darkness to hasten the development of senescence; thus, there are two artificial conditions imposed, the excised condition, and the artificial imposition of darkness. It is not at all certain that either of these experimental conditions allows a fair representation of the situation in the intact plant under normal light conditions.

A prominent physical change observed to be associated with the onset of senescence is the loss of integrity of plant membranes; old leaves become very leaky as they enter senescence, and large amounts of nutrients can leak out of them. Increased leakiness is known to be also associated with the senescence of fruits, flower petals, and other plant organs. The lowered integrity of membranes can be presumed to participate in the development of senescence in two ways. First, it may contribute to a lowering of the effectiveness of cellular anabolic activities, such as a lowered effectiveness of respiratory production of energy and of photosynthesis. A second possibility is that the lowered effectiveness of membranes as barriers in the cells could lead to the disorganization of tissues in such a way that enzymes may attack substrates that they would not be able to reach under nonsenescent conditions. Several hydrolytic and oxidative enzymes are known to become more active during senescence. It has also been suggested that the onset of senescence may result from the synthesis of an increased amount of hydrolytic enzymes. The release of harmful enzymes might also occur as the breakdown of membranes surrounding organelles in the cells which contain packets of hydrolytic enzymes.

The mechanisms involved in the development of senescence, then, appear to include especially a directed translocation of nutrients out of the organs which will become senescent, an unequal distribution of hormones between the senescing and the nonsenescing organs, and a deterioration of membranes in the cells or organs which are becoming senescent.

POSTMORTEM

The flame of life incorporates the continuous turnover of a panoramic spectrum of life units, from the smallest organelles within cells to the largest arrays of plant species. Examination of the range, role, and mechanisms of the processes of death allows three major suggestions. (1) Death of organisms is a central component of the process of evolution; it is a basic element of evolutionary selection. It is beyond the realm of imagination that evolutionary processes could give rise to an organism that could evade this requisite. (2) The death of parts of organisms and of organisms need not be passive; the development of programmed lethal stages has been utilized by higher plants for a wide array of physiological, ecological, and adaptive advantages. (3) Given the essentiality of organismal death in evolutionary processes, one may presume that, whereas senescence imposes death with associated advantages, predation and stress impose death in a more passive manner, and that aging provides the ultimate and final assurance of the turnover of organisms.

BIBLIOGRAPHY

Comfort, A. 1956. *The biology of senescence.* Holt, Rinehart and Winston, Inc., New York. 257 pp.

Leopold, A. C. 1961. Senescence in plant development. *Science* 134: 1727–1732.

Leopold, A. C. 1975. Aging, senescence, and turnover in plants. *BioScience* 25: 659–665.

Molisch, H. 1938. *The longevity of plants.* Science Press, Lancaster, Pa.

Woolhouse, H. W., ed. 1967. *Aspects of the biology of ageing. Symposia of the Society for Experimental Biology*, No. 21. Academic Press, New York.

7

Aging in Lower Animals

WILLIAM F. VAN HEUKELEM

Biologists have long recognized that the study of simpler animals often provides understanding of how our bodies work. Basic biological processes are similar in very diverse organisms. Most of what is known about the transmission of nerve impulses, for example, has come from studies of the giant nerves of squid. Ideas on the nature of hormone action and the role of hormones in turning on or off the activity of genes have come from studies of insect metamorphosis and crustacean moulting cycles. Fruit flies have been invaluable in the study of genetics because of their rapid generation time (about 2 weeks) and because giant chromosomes in the salivary glands of their larvae are easy to see. Many more examples could be given to illustrate the usefulness of lower animals in biological research, but these few should give a general idea of the extent to which invertebrate animals have served in the progress of biological thinking.

There are many good reasons for choosing lower animals for the study of aging. The relative simplicity of invertebrate systems is one. Rapid generation time, short life-span, and small size are a few other advantages offered by many lower animals. Consider, for instance, the space, labor, time and expense involved in maintaining 1000 fruit flies until they reach old age (about 2 months) as opposed to maintaining 1000 mice for 3 years to answer the same question. If the question could be answered equally well using fruit flies, it would be foolish to use mice in the experiment. (There are, of course, questions that require the use of mammals to obtain answers.)

Like skeletal muscle fibers and the central nervous system of mammals, adults of many invertebrates are composed almost entirely of fixed postmitotic cells. Such tissues have lost the power of self-renewal by cell

WILLIAM F. VAN HEUKELEM, Ph.D. • Pacific Biomedical Research Center, Kewalo Marine Laboratory, University of Hawaii, Honolulu, Hawaii 96813. Present address: Horn Point Environmental Laboratories, Cambridge, Maryland 21613

division and offer the advantage that one can be sure he is working with an old cell in an old organism rather than a young cell in an old organism. Adult rotifers, nematodes, and insects are examples of organisms composed of fixed postmitotic cells. The existence of widely varying life-spans in such forms suggests that "wear and tear" on non-renewable cells is not a basic mechanism in senescence.

That scientists have used lower animals extensively in studies of aging was illustrated by a recent bibliography published in the journal *Gerontology*. This bibliography listed 451 references dealing with aging, life-span, and parental age effects of fruit flies *(Drosophila)* alone. Other insects that have been used extensively in studies of aging include houseflies, blowflies, mosquitoes, and flour beetles. Simpler organisms that have played important roles in research on aging include protozoans, rotifers, nematodes, and water fleas.

LIFE-SPANS OF LOWER ANIMALS

Some of the longest-lived animals are found among the invertebrates. Sea anemones, for example, have been kept in aquaria for up to 90 years, and some freshwater mussels are thought to attain ages of 100 years or more. A fact that becomes evident in a survey of invertebrate life-spans is that degree of complexity does not correlate with life-span. The long-lived sea anemone has a very simple nervous system — only a nerve net, no brain at all, while invertebrates with the most highly developed brains, octopuses and their kin, have life-spans ranging from 6 months to a few years, depending on species. Closeness of biological relationship does not correlate with longevity either. Some hydroids, for example, which are close relatives of the long-lived sea anemone, have life-spans of only 1 week. Turban shells may live 30 years, whereas related mollusks, nudibranchs, may live only 90 days. Nor does size seem to contribute to life-span. The giant clams of the tropical Pacific grow to 1 meter (nearly their maximum length) in about 20 years while the freshwater mussel mentioned above reaches its maximum length of only 150 mm in 100 years. Interested readers should consult the 1964 edition of A. Comfort's book, *Ageing, the biology of senescence* (New York, Holt, Rinehart and Winston, Inc.), in which a six-page table lists known maximum longevities of a number of invertebrate animals.

PROGRAMMED LIFE-SPANS

That each species of animal has a characteristic life-span (some sea anemones may be an exception, in that they may have indeterminate life-spans) argues in favor of the notion that information determining duration of life of each species is contained in the genetic code, like information specifying other characteristics, such as structure, color, and orderly growth. The genetic code may be compared with a computer program that specifies an orderly sequence of events, such that as one is completed, a command is given to execute another sequence of events (subroutine), and so forth, until all the information on the program is used up (all the commands have been carried out). Programmed events in the life cycle of short-lived organisms obviously occur more rapidly than in long-lived species. The most important event that must be specified in the genetic program, to ensure survival from generation to generation, is reproduction of the organism. All other commands in the program (cell multiplication, structure formation, growth, etc.) prepare the organism for this command. Once this event has been achieved, the program need no longer specify maintenance, wound healing, regeneration, disease resistance, and so on, except when necessary for the care of young and maintenance of the adult population until the next generation is competent to reproduce. If juveniles require several seasons or years to attain sexual maturity, maintenance of a healthy reproductive adult population through this period is essential insurance against accidental destruction of the juveniles which would result in extinction of the species. The program may specify only a single spawning as in cephalopods (octopuses, cuttlefish, and squid), Pacific salmon, and European eels; multiple spawnings within one season as in nudibranchs, sea hares, and many insects; or many seasons of reproduction in succeeding years as in oysters, sea urchins, lobsters, and many fish. In any case, clear characteristics of senescence do not become evident until after the age of reproductive maturity. In organisms that spawn only once, senescence and death are not necessarily programmed, in the sense of being specified by a "death clock"; rather, it is possible that the program simply runs out of information (i.e., "reproduce" is the final command, and "continue maintenance" is not specified). Death, then, might occur from a variety of causes, depending on which unregulated system fails first — lack of enzymes, hormone imbalance, lack of protein synthesis, disease resistance, replacement of wornout cells, and so on.

GENETIC CONSTITUTION AND LIFE-SPAN

Direct evidence that the life-span of animals is controlled by their genetic constitution comes from selective breeding experiments with insects and mice. Different inbred strains of the same species have characteristic life-spans. One researcher produced five different strains of fruit flies which exhibited average life-spans ranging from 14 to 44 days, and each strain bred true for longevity. Inbred lines usually exhibit inbreeding depression and have shorter life-spans than wild stock. Crossing two inbred lines often results in hybrids with longer life-spans than either inbred strain. This phenomenon is referred to as hybrid vigor. For example, when two inbred lines of fruit flies were crossed, the progeny had mean life-spans of 47 days even though the average longevity of the two parental strains was only 37 days and 21 days. Hybrid vigor is most marked in the first (F_1) generation and declines with continued inbreeding. A clear-cut segregation of life-span characteristics was found in the second (F_2) generation in one fruit-fly experiment; some second generation flies exhibited life-spans characteristic of one original parental line while others exhibited longevity characteristic of the other parental line. Another example of the effect of genetic constitution on life-span is that, as in man, the male of many invertebrate species is shorter-lived than the female. Males of one housefly strain have mean life-spans of 17.5 days while females live an average of 29 days under the same conditions. Black widow spiders, flour beetles, water beetles, and fruit flies are other examples of species in which females normally outlive males.

PARENTAL AGE AND LIFE-SPAN

Offspring from old rotifer mothers have shorter life-spans than those from young mothers. The life-span of progeny from old mothers decreased in successive generations, but this phenomenon has not been found in several experiments using insects.

In a study of the evolution of senescence, artificial selection for shorter life-spans produced shorter average life-spans in most lines of flour beetles after 11 to 12 generations. Fitness traits early in life were selected for by discarding adults after 10 days in selected lines. Flour beetles produce eggs over a 200-day period and may live as long as 600 days under the environmental conditions used in these experiments. Mean longevity of males was reduced from 271 days to 231 days, and female longevity from 228 to 207 days. In another experiment, in which selection was continued through 40 generations and eggs used were only those laid during the first 3 days, average life-span of females was reduced from 30 to 20 weeks and that of males from 35 to 20 weeks. These experiments employed strains of flour beetles in which males outlive females. Experiments utilizing similar techniques have produced altered life-spans in fruit flies and houseflies, lending credibility to the idea that characteristic life-spans have evolved for each species.

EFFECTS OF ENVIRONMENT ON LIFE-SPAN

Temperature

In 1908, Jacques Loeb asked the question: Is there a temperature coefficient for duration of life? Loeb (1859–1924) was known as a "mechanist" and tried to demonstrate that various phenomena of life could be reduced to physical–chemical laws. Loeb and J. H. Northrop, working at the laboratories of the Rockefeller Institute for Medical Research, were the first scientists to manipulate the life-span of animals experimentally with different rearing temperatures. Working with fruit flies, they found that flies reared at 20°C lived 54 days, those at 25°C lived 39 days, while those at 30°C lived only 21 days. Similar results have since been obtained using rotifers, nematodes, water fleas, various insects, sea hares, cuttlefish, sea squirts, and fishes, and it seems likely that all "cold-blooded" animals live longer at the lower end of the temperature range within which they can complete their full life cycle. In general, the temperature coefficient for life-span has been found to lie between 2 and 3; that is, animals reared at a cool temperature live 2 to 3 times longer than individuals of the same species reared at a temperature that is warmer by 10°C. As these differences are similar to the different rates at which chemical reactions proceed with changing temperature, it was assumed that a basic mechanism controlling length of life was somehow related to the rates at which chemical reactions occurred within the body, and Raymond Pearl (1879–1940) at Johns Hopkins University published his "rate of living" theory in 1928 to explain such temperature-induced differences in life-span in terms of rate of energy expenditure throughout life. Metabolism is higher, eggs develop faster, juveniles grow at increased rates, mature, reproduce, senesce, and die earlier at elevated temperatures than at low temperatures (i.e., increasing the temperature speeds up the rate at which the genetic program runs). Thus, two groups of animals reared at different temperatures will be physiologically different in age, even though they are chronologically the same age. Animals at the lower temperature will be physiologically younger at any point in time than those reared at higher temperatures, even though the two groups would be the same physiological age; for example, at the time of first reproduction, which would occur later in time at the lower temperature. All this is exactly what one would predict if animals behaved like chemicals in a test tube.

The rate-of-living theory was adequate to explain the differing life-spans of animals kept at constant temperatures during experiments. In 1958, 1962, and 1963, however, J. Maynard-Smith of University College, London, published results of experiments which cast doubt on the simple rate of living theory. In these experiments, fruit flies were kept at different temperatures for part of their life-span, then switched to another temperature for the duration of their life. Flies maintained at 30°C for about half of their expected life and then transferred to 20°C, lived as long as flies kept continuously at the lower temperature. The rate-of-living theory predicted that the transferred flies should have lived longer than those kept continuously at the high temperature but not as long as those kept at the low temperature throughout their life. Clearly, the results did not agree with those predicted by the rate of living theory, and Maynard-Smith proposed the threshold theory to explain his results. According to this theory, the flies' life-span can be divided into an aging phase and a dying phase. The aging phase is unaffected by temperature and leads to a decline in vitality. When vitality falls below some threshold level, the flies begin to die, and the rate of dying is faster at a higher temperature than at a low temperature. Stated in another way, the threshold level of vitality necessary for survival is higher at a high temperature, and the difference in survival at different temperatures is due not to different rates of aging but to different thresholds of vitality necessary for survival. Later experiments by different workers, using different species of fruit flies, mosquitoes, rotifers, and tropical fish, indicate that neither the rate-of-living theory nor the threshold theory adequately explain results of temperature-transfer experiments. Some experiments with different species and strains of fruit flies gave results that agreed with the rate-of-living theory, whereas experiments with other strains yielded results that agreed with the threshold theory. The life-spans of males agreed with predictions of one theory while life-spans of females in the same experiment agreed with those predicted by the other theory. In other cases, results obtained did not agree well with predictions based on either theory. Furthermore, fish reared at a high temperature for about half their life and then transferred to a lower temperature lived 76% longer than those kept at the lower temperature continuously.

When the rate-of-living theory was formulated, it was thought that poikilothermous (cold-blooded) animals were strictly at the mercy of their environmental temperature. Biologists now regard animals as adaptive control systems. We now know that many animals can adjust their metabolism to a new temperature. Such forms of adaptations, perhaps by synthesis of new isoenzymes which operate more efficiently at the new temperature, could explain results of temperature-transfer experiments. For example, if new machinery for enzyme synthesis were developed to adjust to a new temperature, a new "clock" for this system, with time set to zero, might be created. Similar adaptations might occur in hormonal and immune systems. Clearly, temperature-transfer experiments offer some intriguing results which warrant further investigation.

Rotifers lived longer when reared under a daily cyclic temperature which changed from 15°C to 25°C (average temperature = 20°C) than control animals living at a constant temperature of 20°C. The same animals had progressively longer life-spans when reared under constant temperatures of 25, 20, and 15°C. The slight increase in life-span (over rotifers kept at 20°C) under the cyclic temperature may indicate that an internal circadian rhythm is involved in longevity as has been demonstrated using light with fruit flies (see below).

In hydroids, an annual rhythm of longevity has been demonstrated. Individual hydranths that make up the hydroid colony had different life-spans at different times of the year, even though the animals were kept under constant temperature and lighting conditions in the laboratory. At 10°C, mean life-spans in the winter and summer were about 10 days, while during late fall and spring average length of life was about 16 days. These animals, thus, have programmed changes in longevity at different times of the year which appear to anticipate environmental changes occurring in their natural habitat.

Light

Fruit flies and houseflies have been found to live longer when reared in total darkness than when reared under continuous light. Under varying conditions of light and dark, *Drosophila* live longer when the light and dark periods add up to 24 hours than when night and day periods add up to either greater or less than 24 hours or when they are kept under continuous light. These animals, like most, have internal "clocks" that run in cycles of about 24 hours (circadian rhythms) even in the absence of environmental clues (light–dark cycles or temperature cycles), and they seem to do best when reared under lighting conditions close to their own internal rhythms.

There is good reason to believe that cephalopods (octopuses, cuttlefish, and squid) would complete their life cycle faster if subjected to a long-night (or constant-darkness) rearing regime than to normal or long-day lighting conditions. These animals grow rapidly, reproduce only once, and then die. Light, acting through the eye–brain–optic gland system, inhibits secretion of the hormone responsible for ripening the gonads. Long days would, thus, delay spawning and prolong the life-span, whereas long nights or constant darkness would have the opposite effect. The hormone responsible for maturation of the gonads appears to shut off systems responsible for blood protein synthesis, digestion, feeding, growth, regeneration, and wound healing. Experiments need to be performed to elucidate the mechanisms involved. Assuming that expression of senescence depends on secretion of the optic gland hormone, optic gland removal before sexual maturity or keeping the animals in constant bright light should, in theory, prevent sexual maturity and the senescent changes that follow, resulting in animals that continue to grow and stay young indefinitely. There is no apparent feedback from the gonad in this simple system as there is in vertebrates, and it appears to offer the first clear opportunity to break the chain of events leading to senescence of any organism. The genetic program that specifies "develop, grow, mature, reproduce (and perhaps senesce and die)" might be arrested in the "grow" phase, and the later commands might never be triggered. The program code is undoubtedly contained in genes, and as hormones are known to activate the expression of genes, withholding the hormone could mean that the next step in the program might never be activated. At present, all this is largely conjecture. The ideas presented grew out of work designed to elucidate the control of reproduction in cephalopods by Martin Wells (Cambridge, England) and Alain Richard (Wimereux, France) and my own studies of growth and life-span in octopuses. The problem has yet to be approached as a system in which aging could be studied.[1]

Nutrition

Underfeeding early in the life cycle has proven to be one of the most successful ways of prolonging the life-span. This method has been used successfully on mammals (rats) as well as invertebrates. For example, rotifers transferred to fresh pond water and fed daily lived 34 days, whereas rotifers transferred daily but not fed lived 45 days. Increase in life-span appears to be gained by slowing the rate at which the program runs, as in lowering rearing temperature. The most dramatic results are

[1]Dr. Jerome Wodinski of Brandeis University has just published some results that partially support the hypothesis presented above (see *Science* 198: 948–951, Dec. 2, 1977). He removed optic glands from female *Octopus hummelincki* 4 to 17 days after they laid eggs. Wodinski reported that these animals ceased incubating their eggs, resumed feeding, and lived an average of 175 days after egg-laying as compared to normal females which only lived an average of 43 days after egg-laying.

Wodinski's results are in contrast to those I obtained when I performed the same experiments with *Octopus maya* in January of 1975 (unpublished). Female *Octopus maya* that had the optic glands removed after spawning continued egg-incubation behavior, refused food, and did not live any longer than normal animals. Thus, in the above hypothesis, I specified that the optic glands should be removed from young octopuses (i.e., before the optic glands have begun secretion) in order to see if life-span could be prolonged. The different results obtained may indicate a species difference in the details of optic gland effects on the whole organism.

obtained by underfeeding before the attainment of sexual maturity — a final stage in the program and one at which life-span has largely been determined. Underfeeding often delays the onset of sexual maturity and thus the expression of postreproductive senescence.

Dormant Periods

Many invertebrates enter stages of "suspended animation" (diapause, hibernation, anabiosis, encystment, etc.) that enable them to survive periods of unfavorable environmental conditions. Such stages may occur as resting eggs (brine shrimp eggs, for example), in a larval stage, or during adult life. Animals in these arrested stages of development often remain viable for many times the normal life-span of the animal. Rotifers, for example, normally live from 8 days to 5 months, depending on species, but desiccated individuals have been revived after as long as 27 years and encysted rotifers after 59 years. Resting eggs of nematodes have begun development after 20 years of dormancy. Tardigrades ("water bears") can be dried up and revived at least 10 times in the laboratory and can revive from the dried state after 7 years. These animals normally live only a few months if not dried. Wasps may live several years in larval diapause (a period of dormancy between stages of active life), although the adult life is only a fraction as long. Such dormant stages are often triggered by environmental stimuli, such as changing temperature, humidity, day length, and nutrition, and can extend the life-span of an organism many times. Changes in day length of only 10 to 14 minutes are known to result in diapause instead of continued development of insects. In freshwater organisms, drying up of ponds in the summer or freezing in the winter triggers the formation of dormant stages. In other species, they are a normal part of development, not dependent on environmental stimuli. The animal does not seem to age during such stages, and when reactivated by proper environmental conditions, the developmental program resumes where it left off. Larvae of some nematodes enter a semidormant larval stage in response to starvation. These larvae do not feed but continue to move and may live as long as 70 days. If the larvae are supplied with food by placing them in a bacterial suspension, development resumes, and the larvae become mature adults, reproduce, and have the same postlarval life-span (10 days) as adults that never passed through the semidormant state.

Metamorphosis of some insect larvae can be prevented by implanting larval endocrine glands that secrete juvenile hormone. Such larvae molt to become giant larvae, instead of turning into adults at a stage when they would normally do so. Some insects have been maintained in the larval condition for as many as six extra larval molts. Although no one has yet tried to keep an insect in a larval stage indefinitely by such means, an experiment has been performed which indicated that it would be possible to do so. In this experiment, imaginal discs from larvae were transplanted into adult female fruit flies. Imaginal discs are larval structures that normally differentiate into adult (imago) organs when exposed to the hormone for metamorphosis at metamorphosis. Ordinarily, these discs would then live the same length of time as the adult (i.e., about 1 month). The hormone for metamorphosis (ecdysone) is not present in the adult female, and the discs retain their larval form when cultured in adult female fruit flies. The imaginal discs were transplanted more than 160 times from one young adult female to another over a period of 6 years. Exposing some of the discs to the hormone for metamorphosis during the course of the experiment showed that they retained their original ability to change into adult structures at any time. Examples of dormant periods that prevented metamorphosis of whole insect larvae and the long-term culture of imaginal discs demonstrate clearly that life-span is intimately related to the developmental program and can be greatly extended by arresting the program at some point. In general, this point seems to be prior to the attainment of sexual maturity.

AGE-RELATED CHANGES IN LOWER ANIMALS

Changes that distinguish senescent from young or mature animals have been sought at many levels, including the whole organism, organ, tissue, cell, subcellular, and biochemical levels. Age-related differences often are surprisingly difficult to find.

At the level of the whole organism, there is generally a deterioration in motor ability, decline in metabolic rate, decrease in fecundity, and loss of weight during senescence.

In old rotifers, the cilia beat feebly, the animals crawl rather than swim, become sluggish, respond slowly to stimuli, and shrink in size. The normally transparent animals become opaque and granular in appearance. Calcium and lipofuscin (age pigment) accumulate, and there is a decrease in the number of ribosomes (sites of protein synthesis) in the epithelium of the stomach. Decreasing enzyme activity has also been reported in senescent rotifers.

Sea slugs or sea hares produce up to 50% of their body weight in eggs during the early part of the 6 months of their reproductive life, but egg production gradually declines and ceases altogether during the last week or two of life. The animals become sluggish — much more so than young sea slugs — lose weight, and the skin, normally heavily pigmented in young animals, becomes translucent. Velocity of nerve response in mature reproductive individuals of one species of sea hare is only one-half as fast as in prereproductive individuals.

Octopuses reduce their food intake in old age and eventually stop feeding altogether, lose weight, become lethargic and flabby, and lose most of their ability to change color. Young animals heal rapidly when wounded and regenerate lost arms, but in old postreproductive animals, the skin does not close over a severed arm tip even after several weeks, and regenerative powers are nil. The "liver" (an organ that functions in digestion and energy storage) shrinks to about one-third of its maximum size, and activity of digestive enzymes from the liver and salivary glands is only 8.1 and 3.8%, respectively, of the activity shown by extracts from young animals. Nuclei of nerve cells in the large nerve cord of the arms shrink, become round rather than rectangular, and lose structural detail, a sign of cell death. Blood protein synthesis is, apparently, shut off by the hormone that is responsible for both sexual maturation and rapid yolk protein synthesis in the ovary.

Insect Flight Muscle

The most thoroughly investigated system in studies of aging in invertebrates is the flight muscle of insects. This is a system with an extremely high energy demand in which wing-beat frequency may reach 300 per second in fruit flies. By comparison, maximum frequency of wing beat in hummingbirds is only about 70 per second. Decline in flight performance of insects, measured as duration of sustained flight till exhaustion, has been reported in aging fruit flies, houseflies, mosquitoes, and milkweed bugs. Seven-day-old fruit flies can sustain flight for 110 minutes and beat their wings over 2 million times during this period. By 33 days, they can only fly for 19 minutes and beat their wings only 170,000 times. Old male houseflies have frayed wings and can no longer fly after 2 weeks of adult life.

The ultrastructure of flight muscle of insects has been studied at a number of laboratories, but the results have been contradictory. Authors of the most recent studies have concluded that there are no major visible degenerative changes in flight muscle of senescent houseflies, blowflies, or fruit flies. However, an increase in whorl-like figures in mitochondria (sites of energy transformation in cells) of aging insects, and a lack of cytochrome C oxidase activity within these whorls indicated some loss of function in supplying energy for flight. A reduction of energy stored as glycogen, or the ability to use it, has also been found in the flight muscle of a variety of insects.

Decline in mitochondrial efficiency, measured by the ability of iso-lated mitochondria to transform energy, has been reported for some insects. Decreased activity of several enzymes associated with the energy supply for flight have also been found in aging insects. Since insect flight muscle is composed of fixed postmitotic cells, and wornout cells cannot, therefore, be replaced, it is surprising that more dramatic changes have not been found in old animals that are unable to fly. For far more detailed treatments of this subject, readers should consult two recent reviews on aging changes in insect flight muscle, one by R. S. Sohal of Southern Methodist University, Dallas, Texas, the other by G. T. Baker III of the University of Zurich, Zurich, Switzerland.

Sundry Age-Related Changes in Lower Animals

Loss of specific activity of enzymes with age has been demonstrated in nematodes, as have alterations in ribosomes, sites of protein assembly in cells. Correct ribosome activity is essential for accurate translation of messenger RNA, and defective ribosomal particles would likely impair protein synthesis. Decreases in protein synthesis and in overall biosyn-thesis have been demonstrated in insects, and a decrease in the amount of mitochondrial DNA, but not nuclear DNA, was found in old fruit flies.

An increase in cross-linking of biological macromolecules has been suggested as a mechanism leading to senescence. Increasing collagen cross-linking occurring in aging animals was thought to support this suggestion, but in a recent study, no increase in DNA cross-linking was found in senescent fruit flies. If increasing cross-linking of mac-romolecules is a general mechanism causing senescence, the increase should occur in the large DNA molecule as well as the collagen molecule.

Abnormal structures (nuclear inclusions) have been found in the cells of aging fruit flies as well as mammals. It is thought that these inclusions may decrease the volume of the nucleus and impair proper functioning of cells, thus contributing to senescence and death. Ac-cumulation of lipofuscin (age pigment) has been found in cells of old protozoans, nematodes, rotifers, leeches, bees, fruit flies, and house-flies. In fruit flies, such pigment accumulation has been associated with increased numbers of lysosomes, as it has in fixed postmitotic cells of mammalian brain and heart muscle. Lysosomes have been called "suicide bags" and, when cells are starved, the lysosomes are known to release enzymes that digest various cell parts as a source of energy for the cell.

A decrease in the number of brain cells occurs in old bees. About

35% fewer cells were found in old bees than in young bees. Loss of nerve cells in some parts of the brain of senescent rats and man have also been reported.

Dysfunctions of the immune system are also found in aging vertebrates, but there have been few studies on the immune system of invertebrates. In fact, we don't even know if major groups, such as mollusks and arthropods, possess an immune system. Starfish, earthworms, and sea squirts have well-developed immune responses, and capacity for at least some immunorecognition has been found in corals, sea anemones, hydroids, and even sponges. Whether or not progressive failure of immune response and increased autoimmune phenomena accompany aging in these groups is, as yet, unknown.

In short, a host of structural, functional, and biochemical changes characterize the aging invertebrate, and the changes are similar to those found in senescent vertebrates. At present, none of these changes has been clearly demonstrated as a cause of senescene, but all of them appear to be changes that occur as a result of senescence.

PROSPECTS FOR FUTURE RESEARCH

In spite of all the research that has been done to date and all the knowledge that has been gained from this research, a major breakthrough in understanding aging is yet to be achieved. Aging occurs in time, and animals have build-in biological clocks that measure the passage of time. Each organism is probably composed of a population of biological clocks which somehow control the programmed events in different tissues and cells of the body and are, therefore, involved in determining the life-span of the animal. We can alter the rate at which some programs run by changing temperature or light, but other programs are closely controlled by internal clocks that operate independently of environmental change. Is there a master clock in each organism that determines life-span? Probably so, but we do not know where it is, what it looks like, or exactly how it operates. In forms that have one, the central nervous system seems a likely site for such a master clock, but we do not know where it is in the central nervous system or how to control it. A good deal more research on the involvement of circadian clocks in aging seems warranted if we are to understand aging.

Although redwood trees and some sea anemones seem capable of indefinite life, life-span appears to be programmed in some way for most species. This suggests that if we can discover how to prevent expression of final steps in the program, we could extend the life of many organisms indefinitely. We know enough, in theory, to break the chain of programmed events leading to sexual maturity and subsequent death in two invertebrate groups, insects and cephalopods. Insect metamorphosis has been prevented by continuing to supply juvenile hormone, but we do not know how long this can be done. Cephalopods do not metamorphose, but it appears likely that sexual maturity and senescence could be prevented by withholding a hormone. In both insects and cephalopods, secretion of the hormones involved is controlled by the brain, which receives information on environmental variability, such as light and temperature. We know enough about the chain of events in these two systems so that we have been able to formulate clear, testable theories, but the theories have yet to be tested.

Lower animals will undoubtedly play increasingly more important roles in research on aging. They offer simplified systems that can be manipulated, and their short life-spans make it possible to obtain quick tests of theories. The speed at which we can reject theories that do not withstand the rigors of testing will be a major factor determining our rate of progress in research on aging.

I suspect that we will first define a chain of events in an invertebrate animal such as the following. After a measured amount of time at specified conditions of temperature, light, and nutrition, gland X is released from nervous inhibition, secretes a hormone in gradually increasing quantities that stimulates protein synthesis in the gonads, and switches off genes controlling protein synthesis in other areas of the body, resulting in cessation of repair functions, depletion of essential enzymes, and, eventually, death due to loss of function in some essential system. Only after defining such a sequence of events in detail will we be able to control, reverse, or prevent senescence in any organism.

ACKNOWLEDGMENTS. I have borrowed freely from the published work of many researchers. Although I would like to cite each author separately for the contribution he or she has made, space limitations make this impossible. In selecting references, I attempted to include those that represent a wide variety of topics on aging in lower animals, are reasonably current, and have extensive literature lists for the interested reader. My ideas on the mechanisms of senescence in *Octopus* grew out of work by Martin Wells (Cambridge University, England) on the hormonal control of reproduction in *Octopus;* Alain Richard's (Institute de Biologie Maritime et Regionale de Wimereux, France) work on environmental control of growth and reproduction in cuttlefish; and my own research on growth, reproduction, and life-span of octopuses. Information presented on sea hares is largely from current research headed by Michael Hadfield and Marilyn Switzer-Dunlap in which Dale Sarver and I have participated, at the Pacific Biomedical Research Center, University of Hawaii. My studies on growth and life-span of the octopus were supported by the University of Hawaii NOAA Sea Grant Aquaculture Program, Grant Nos. 2-35243 and 04-3-158-29, and the State of Hawaii. Support for research on aging of octopuses was provided by a University of Hawaii Biomedical Sciences Support Program Grant to M. G. Hadfield. While writing this chapter, I was partially supported by an NIH Research Contract, #N01-RR-4-2168, also to M. G. Hadfield. I am grateful to R. E. Johannes, and M. Switzer-Dunlap (University of Hawaii) and J. P. Trinkaus (Yale) for critical readings of the manuscript.

BIBLIOGRAPHY

Baker, G. T. III. 1976. Insect flight muscle: maturation and senescence. *Gerontology* 22: 334–361.

Briegel, H., and C. Kaiser, 1973. Life-span of mosquitoes *(Culicidae, Diptera)* under laboratory conditions. *Gerontologia* 19; 240–249.

Brock, M. A. 1975. Circannual rhythms. III. Rhythmicity in the longevity of hydranths of the marine cnidarian, *Campanularia flexuosa. Comparative Biochemistry and Physiology* 51A: 391–398.

Comfort, A. 1964. *Ageing—The biology of senescence.* Holt, Rinehart, and Winston, New York.

Gartner, L. P., and R. C. Gartner. 1976. Nuclear inclusions: a study of aging in *Drosophila. Journal of Gerontology* 31: 396–404.

Hieb, W. F., and M. Rothstein. 1975. Aging in the free-living nematode *Turbatrix aceti.* Techniques for synchronization and aging of large-scale axenic cultures. *Experimental Gerontology* 10: 145–153.

Klass, M., and D. Hirsh. 1976. Non-aging developmental variant of *Caenorhabditus elegans. Nature* 260: 523–525.

Lints, F. A. 1971. Life span in *Drosophila. Gerontologia* 17: 33–51.

Massie, H. R., M. B. Baird, and T. R. Williams, 1975. Lack of increase in DNA cross linking in *Drosophila melanogaster* with age. *Gerontoligia* 21: 73–80.

Meadow, N. D., and C. H. Barrows. 1969. Studies of aging in a bdelloid rotifer. *Journal of Experimental Zoology* 176: 303–314.

Mertz, D. B. 1975. Senescent decline in flour beetle strains selected for early adult fitness. *Physiological Zoology* 48: 1–23.

Miguel, J., P. R. Lundgren, K. G. Bensch, and H. Atlan. 1976. Effects of temperature on the life span, vitality, and fine structure of *Drosophila melanogaster. Mechanisms of Aging and Development* 5: 347–370.

Nayar, J. K. 1972. Effects of constant and fluctuating temperatures on the life span of *Aedes taeniorhynchus* adults. *Journal of Insect Physiology* 18: 1303–1313.

Pittendrigh, C. S., and D. H. Minis. 1972. Circadian systems: longevity as a function of circadian resonance in *Drosophila melanogaster. Proceedings of the National Academy of Sciences USA* 69: 1537–1539.

Rockstein, M., and J. Miguel. 1973. Aging in insects. Pages 371–478 in M. Rockstein, ed.

The physiology of insecta, Vol. I, Second Edition. Academic Press, Inc., New York.

Schweizer, P., and D. Bodenstein. 1975. Aging and its relation to cell growth and differentiation in *Drosophila* imaginal discs: developmental response to growth restricting conditions. *Proceedings National Academy of Sciences USA* 72: 4674–4678.

Sincock, A. M. 1975. Life extension in the rotifer *Mytilina brevispina* Var Redunca by the application of chelating agents. *Journal of Gerontology* 30: 289–293.

Sohal, R. S. 1976. Aging changes in insect flight muscle. *Gerontology* 22: 317–333.

Soliman, M. H., and F. A. Lints. 1975. Longevity, growth rate, and related traits among strains of *Tribolium castaneum*. *Gerontologia* 21: 102–116.

Soliman, M. H., and F. A. Lints. 1976. Bibliography on longevity, aging, and parental age effects in *Drosophila, Gerontology* 22: 380–410.

Van Heukelem, W. F. 1977. Laboratory maintenance, breeding, rearing, and biomedical research potential of the Yucatan octopus *(Octopus maya)*. *Laboratory Animal Science* 27, Part II: 852–859.

Van Heukelem, W. F. 1978. Environmental control of reproduction and life span in octopus: an hypothesis. In S. E. Stancyk, ed. *Reproductive ecology of marine invertebrates.* Vol. 9 in the *Belle W. Baruch Library of Marine Sciences,* University of South Carolina Press, in press.

Wallach, Z., and D. Girshon. 1974. Altered ribosomal particles in senescent nematodes. *Mechanisms of Aging and Development* 3: 225–234.

Wells, M. J., and J. Wells. 1977. Optic glands and the endocrinology of reproduction. Pages 525–540 in M. Nixon and J. B. Messenger, eds. *The biology of cephalopods. Symposia of the Zoological Society of London,* No. 38. Academic Press, Inc., London.

Wodinsky, J. 1977. Hormonal inhibition of feeding and death in *Octopus:* control by optic gland secretion. *Science* 198: 948–951.

Wohlrab, H. 1976. Age-related changes in the flight muscle mitochondria from the blowfly *Sarcophaga bullata. Journal of Gerontology* 31: 257–263.

8

Exercise and Aging

ROY J. SHEPHARD

People in Western society take less exercise as they become older. Dramatic documentation of the progressive diminution in both sports participation and other forms of vigorous leisure activity has been provided by such agencies as the President's Council on Fitness and Statistics Canada (Table 1). To some extent, the reduction of activity is culturally determined. A person is expected to "act his age," to "slow down a bit," and to "enjoy a well-earned rest" once he has passed the age of 65 years. Nevertheless, the progressive decrease of voluntary activity seems a more general biological phenomenon; the bouncing puppy and frisky kitten are in marked contrast with more sedate 10-year-old animals, and studies at the cellular level suggest a parallel decrement of biochemical activity that commences soon after puberty. It has been suggested that habitual activity is regulated by a specific "center" in a part of the brain known as the hypothalamus; if this hypothesis is correct, one might speculate that the control setting of the "activity center" is adjusted in a downward direction as part of the biochemical slowdown that accompanies aging.

This chapter will cover several aspects of exercise and aging, including alterations in acute responses to exercise, training responses, and possible effects of a lifetime of vigorous activity as seen in continuing athletes and in certain "primitive" tribes. Because data on other species are extremely limited, the discussion will be focused on the reactions of human subjects. Even in man, there have been few "longitudinal" studies, where the same individuals have been tested repeatedly as they become older. The alternative, "cross-sectional" approach is to carry out a single series of measurements on people of various ages. Interpretation of such results is complicated by a progressive improvement in the relative status of older volunteers. Poorly en-

ROY J. SHEPHARD, M.D., Ph.D. ● Department of Preventive Medicine and Biostatistics, University of Toronto, Toronto, Canada

Table 1. Hours per Week Spent in Sport and Other Forms of Vigorous Activity*

Age (years)	Participation (% of sample)			
	1–3 hours per week		More than 3 hours per week	
	Sport	Other activities	Sport	Other activities
14	23.3	26.3	30.5	24.2
20–24	19.5	17.7	14.1	11.0
35–44	12.1	10.9	8.2	5.6
65–69	2.9	3.7	3.1	3.4

*Data for Canadian subjects were collected by Statistics Canada in 1971.

dowed members of the community are eliminated by disease and death, and only the more active older people come forward for "fitness testing." Longitudinal studies, also, have inherent biases. Volunteers for a long-term project inevitably are more health conscious than the general population. Typically, they are lean nonsmokers with an above-average initial level of habitual activity. As they become older, volunteers lose some of their enthusiasm for activity, and the associated deterioration of physical condition may give them the appearance of a faster-than-normal rate of aging. Difficulties in data interpretation are an inevitable concomitant of human experimentation in a free society; we cannot force one group of people to sustain a high level of activity and a second, parallel group, to remain inactive as they become older. However, account must be taken of confounding trends when examining the published reports.

ACUTE RESPONSES TO EXERCISE

Acute responses to physical activity are best classified in terms of the type and duration of exercise. It is necessary to distinguish static or "isometric" effort, where the muscle does not shorten, from rhythmic, dynamic, or "isotonic" contractions, where the muscle shortens with a constant tension. Within each of the two categories of activity, brief all-out efforts have a different physiological basis than more sustained work.

Static Effort

Laboratories commonly measure the maximum isometric force developed by a group of muscles, for example, the hand grip or the knee extension force. In such tests, the effort is sustained for 1 or, at most, 2 seconds. The main factor limiting contraction is probably central inhibition rather than the strength of the muscle fibers; the brain has learned not to command a maximum effort, and force readings can thus be augmented substantially by such procedures as hypnosis. Nevertheless, there is a rough relationship between the cross section of the active muscles and the maximum force that is developed. With aging, a progressive loss of lean tissue and an infiltration of the muscle itself by fat and connective tissue lead to a decline of muscle strength. Our data for male Torontonians indicates a relatively constant hand-grip force of 530 to 540 newtons[1] from 16 to 45 years of age, with a drop to 460 newtons at the age of 55, and to 430 newtons at 65 years. The deterioration of isometric strength limits the ability to lift heavy weights; the associated reduction in muscle bulk also handicaps continued participation in contact sports.

The heart has some difficulty in pumping blood through a muscle if it is contracting at more than 15% of its maximum strength. With efforts over 70% of maximum, the blood vessels are completely occluded by the force of muscle contraction. As soon as oxygen delivery by the bloodstream fails to meet the needs of activity, the muscle must rely on alternative ("anaerobic") processes of energy liberation. These include the use of small quantities of oxygen stored in the red pigment of the muscle (myoglobin), the breakdown of high-energy phosphate compounds (adenosine triphosphate, ATP; creatine phosphate, CP), and the conversion of the carbohydrates, glycogen, and glucose to lactic acid. The last reaction is the most important anaerobic resource, but unfortunately the accumulation of lactic acid is fatiguing for the muscle and eventually inhibits the biochemical processes of energy release. The tolerance of a sustained static effort thus depends on the percentage of maximum force that is exerted. Contractions at less than 15% of maximum force can be held for long periods. The endurance of an isometric effort shortens progressively as the load is increased from 15 to 70% of maximum, and efforts of more than 70% are tolerated for no more than 20 to 30 seconds.

As aging leads to a weakening of the muscles, so a given external load demands the usage of a higher percentage of maximum force, and tolerance of the given loading inevitably diminishes. Other handicaps of the older person are smaller myoglobin stores of oxygen, lesser tissue reserves of ATP and CP, and less active oxygen-using enzymes within

[1]Older hand-grip measuring devices show force in kilograms; a 1-kg force is equal to 9.8 newtons.

the muscle fibers. On all counts, the elderly person must resort to the fatiguing conversion of glycogen to lactate at a smaller work load than someone who is younger.

Dynamic Effort

Dynamic effort can be classified into (1) very brief sprints of 10 seconds or less, which depend on the maximum tolerated rate of anaerobic metabolism ("anaerobic power"); (2) longer sprints of 10 to 60 seconds, which depend mainly on the tolerance of lactate accumulation ("anaerobic capacity"); (3) sustained efforts of 1 to 60 minutes, which depend on the steady transport of oxygen ("maximum oxygen intake" or "anaerobic power"), and (4) very prolonged activities, where thermal balance, fluid and mineral reserves, and the extent and mobility of food reserves become important.

Anaerobic Power. A muscle contracts as its protein constituents, actin and myosin, unite to form actomyosin. The reaction requires substantial energy. Anaerobic power thus depends on the maximum rate at which energy stored in the adenosine triphosphate and creatine phosphate molecules can be applied to the coupling of actin and myosin. The reaction is catalyzed by the myosin-bound enzyme ATPase, concentrations of which diminish with age. The anaerobic power available for external work depends also on the resistance offered to fast movement by the viscosity of the tissues, the speed of relaxation of antagonistic muscles, and the stiffness of the joints, all of which deteriorate with age. One simple overall measure of anaerobic power is to time a sprint up a short flight of stairs; the subject is allowed to accelerate for about 3 seconds over a horizontal distance of 30 yards, and the distance that he climbs in the next 2 seconds is measured accurately by the interruption of a series of light beams. A young adult male can ascend 1.5 meters/sec, a rate of working equivalent to a steady oxygen transport of 165 ml/min/kg of body weight; however, by the age of 65, the maximum possible speed has dropped to 0.9 to 1.0 meter/sec.

Anaerobic Capacity. Stores of adenosine triphosphate and creatine phosphate within the muscle fibers are exhausted in less than 10 seconds. Continued anaerobic effort depends on the liberation of energy by the splitting of glucose or glycogen to pyruvate. The pyruvate is converted to lactic acid, and the latter accumulates within the active muscles; there the buildup of acid metabolites inhibits certain key enzymes, setting a finite anaerobic capacity. One measure of acid accumulation is the concentration of blood lactate attained in maximum effort. In the young adult, a sharp peak of 13 to 14 mmoles/liter is seen 1 to 2 minutes after exhausting effort; in oxygen units, the lactate system can perform work equivalent to a steady oxygen transport of about 68 ml/kg· min, with a capacity of about 45 ml/kg. Terminal blood lactate concentrations are appreciably lower in the 65-year-old adult. Some authors have set the limit at 7 to 8 mmoles/liter. Often, both investigators and subjects are unwilling to pursue a true maximum anaerobic effort, so this may be a conservative estimate. In subjects recruited for a retirement physical activity program, our laboratory has consistently observed readings of 10 to 11 mmoles/liter. Nevertheless, there is some decrease with aging. While the total blood volume of the body remains relatively unchanged in the elderly, the volume of active muscle diminishes. Given a fixed tolerance of acid accumulation within muscle (a lactate concentration of around 25 mmoles/liter), the small muscles of an old person inevitably give a lower blood lactate at exhaustion. The post-effort peak of blood lactate in the elderly may also be blunted by a slower diffusion of the acid from the muscle into the circulation.

Lactic acid may accumulate in the early phases of a particular effort, owing to a slow adaptation of the circulation to the demands of physical activity. In a young person, such adaptation is at least 90% complete within 1 minute, but in the elderly, 4 or 5 minutes may elapse before the blood vessels are fully supplying the oxygen needs of the active tissues. Indeed, if a high proportion of the required effort is sustained by a single muscle group, equilibrium may never be reached; increasing quantities of lactate will accumulate until the exercise is halted by local pain and muscular weakness. One example of this problem is seen during use of the bicycle ergometer. In a young adult, the maximum oxygen intake attained on a bicycle ergometer is some 7% less than that realized on the treadmill; on the bicycle, subjects are restricted not only by the oxygen-transporting power of the heart and lungs but also by local exhaustion of the thigh muscles. In an old person with weaker muscles, the problem is more serious, and bicycle measurements of maximum oxygen transport can be 30% less than the corresponding treadmill result.

Aerobic Power. If dynamic effort is continued for longer than 1 minute, it becomes increasingly dependent on the ability to transport oxygen from the atmosphere to the working tissues, that is, the maximum oxygen intake or aerobic power of the individual. Values decline by some 30% over the course of working life, as can be seen in the results from two large cross-sectional sudies of Canadian populations (Figure 1).

Data are sometimes expressed in absolute terms (liters of oxygen transported per minute), but more commonly (as in Figure 1) they are quoted per kilogram of body weight. One reason for this is that the oxygen cost of external work depends on the body weight of the individual. The total oxygen cost of most activities can be divided into a

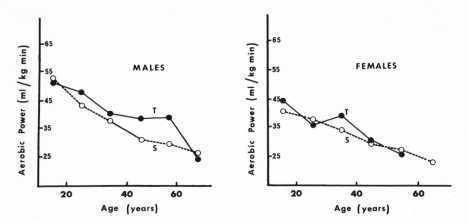

Figure 1. Maximum oxygen intake of Canadian adults in relation to age. Data for large samples studied in Toronto (T) and Saskatoon (S). Saskatoon data collected jointly with D. Bailey, R. L. Mirwald, and G. A. McBride of the University of Saskatchewan.

resting cost, proportional to the 0.75th power of body weight, and a movement cost with a weight exponent varying from zero to 1.0, according to the displacement of the center of gravity of the body. During walking and many heavy industrial activities, oxygen cost is almost directly proportional to body weight. Unfortunately, most Western people show a substantial increase of body weight during the fourth and fifth decades of life (Table 2), and there is a corresponding decrease in the effective working capacity. During the sixth and seventh decades, the overall body weight may stabilize or even diminish, but this is not necessarily a healthy sign. Measurements of skinfold thicknesses show no change, and the weight loss often reflects a wasting of lean tissue rather than a reduction of body fat. One way to demonstrate the loss of muscle is to determine the radioactivity of the subject, using a suitable whole body counter. Radioactivity is due largely to the body content of a naturally occurring isotope, potassium-40. Since this is located mainly in muscle, the total counts recorded reflect the amount of lean tissue in the body. The elderly also lose calcium from their bones. This can be demonstrated if a bony region such as the forearm is bombarded with neutrons. The calcium in the treated area is made temporarily radioactive, and the subsequent emission from the bombarded bones gives an index of their calcium content.

The tolerance of industrial work can be calculated on the basis that each liter of oxygen liberates approximately 21 kJ[2] of energy. A young

---

[2]The kilojoule (kJ) has now replaced the kiloCalorie (kCal) as a unit of diet and energy expenditure; 1 kCal is approximately equal to 4.19 kJ.

Table 2. Changes in Body Weight and Skinfold Thicknesses with Aging*

Age (yr)	Excess weight[+] (kg)	Average skinfold (mm)
16–19	0.8	11.3
20–29	1.7	11.2
30–39	6.4	16.1
40–49	9.3	14.0
50–59	8.8	15.2
60–69	5.1	15.4

*Author's cross-sectional data for adult males living in Toronto in 1966.
[+]Weights are expressed as an excess relative to an actuarial "ideal" weight.

worker with a maximum oxygen intake of 3 liters/min can tolerate a load of 62 kJ/min for a few minutes, while an old worker with an aerobic power of only 2 liters/min is limited to 42 kJ/min. Over an 8-hour shift, the recommended industrial load drops to 40% of aerobic power, implying a limit of some 25 kJ/min for a young employee and 17 kJ/min for an old person. In industrial terms, 21 kJ/min is considered heavy work, but loads below this are rated as moderate work. Thus, the figures for aerobic power suggest that the older worker would become fatigued if he sustained even moderate work over an 8-hour day. In practice, problems are less frequent than the data might indicate. The cumulative development of skills, the selective retirement of those who are unfit, and the use of seniority to find easier jobs are all mechanisms that spare the elderly employee embarrassment from his declining physical capacity.

The decrease of maximum oxygen intake reflects largely a deterioration of circulatory function. In the absence of chest disease, respiration does not limit oxygen transport until an advanced age is reached. Blood leaving the lungs remains almost fully saturated with oxygen even in the elderly, and no great advantage could accrue from any increase of ventilation or pulmonary gas transfer during vigorous exercise. From the physiological point of view, variables limiting aerobic work are the output of the heart, the oxygen-carrying capacity of the arterial blood, and the proportion of arterial oxygen extracted in the tissues.

The cardiac output is the product of heart rate and the output per beat (stroke volume). Whereas the maximum heart rate of a young adult is about 195 beats per minute, the average for a person of 65 years is no more than 170 per minute. A similar drop of maximum heart rate occurs when young subjects migrate to high altitudes, and some investigators have thus suggested that aging leads to oxygen lack in the heart muscle, with a resultant impairment of its contractile function. The main argument against this hypothesis is that while oxygen administration reverses the effects of altitude, it does not correct the age-related slowing. Nevertheless, very low maximum heart rates are sometimes encountered in middle-aged people when the oxygen supply of the heart muscle is compromised by coronary vascular disease. Another factor that limits the heart rate of both young and older subjects is the time needed to fill the cardiac chambers. As the heart ages, its muscular wall becomes infiltrated increasingly by connective tissue and small scars. Contractility deteriorates. It takes longer to expel blood from the heart, leaving relatively less opportunity for filling. A reduction in the elasticity of the heart wall also hampers filling, and there may be less of a stimulus to heart rate from the sympathetic nerves at a given intensity of effort. If an old person is in poor physical condition, he may have a slightly smaller stroke volume than someone who is younger, but in general, the peak output per beat changes relatively little with age. The main difference is seen as maximum effort is approached; perhaps because of problems in filling the heart, the stroke volume of an old person plateaus and even diminishes in very heavy work, whereas in young subjects it usually increases until exhaustion is reached.

The oxygen transported per liter of cardiac output depends on the oxygen-carrying capacity of the blood and thus on hemoglobin level. Provided that nutrition is good, hemoglobin readings vary little over the normal span of working life. However, anemia is not uncommon during retirement. Sometimes poverty or loneliness leads to an inadequate intake of the raw constituents of hemoglobin — iron, B vitamins, and first-class protein; in other instances, atrophy of the gastric mucous lining or bleeding from hemorrhoids is responsible.

The proportion of oxygen extracted from arterial blood during its passage through the tissues usually decreases with age. During maximum efforts, a fit young man extracts up to 160 of the 200 ml of oxygen held by 1 liter of arterial blood. In contrast, the average 65-year-old person often extracts no more than 120 to 130 ml/liter even at exhaustion. Some biochemists have attributed this exclusively to a diminished enzyme activity in the muscle fibers. However, we must remember that the oxygen extraction usually reported is an average value for the entire circulation. Extraction is almost complete in the active muscles but is very slight in the kidneys (where the main function of blood flow is excretion) and in the skin (where the main function is heat elimination). The poor overall oxygen extraction in the elderly may thus reflect a less effective diversion of blood flow from the skin and the viscera to the muscles during physical activity. This is hardly surprising. The elderly need more flow to the skin because they are fatter and have difficulty in dissipating the heat generated by exercise. Declining kidney function

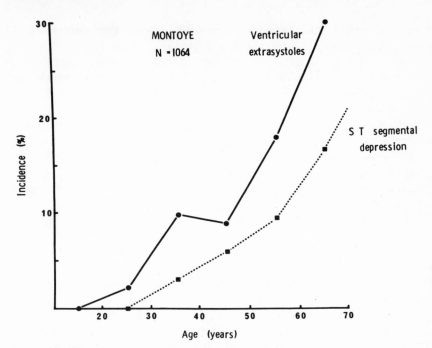

Figure 2. Influence of age upon the incidence of electrocardiographic abnormalities in males. Based on the data of H. Montoye, University of Tennessee, for 1064 men living in Tecumseh, Michigan.

also limits the possibility of restricting renal flow. Finally, a relatively small flow to muscle increases the impact of a given skin and visceral flow on the average oxygen content of mixed venous blood.

With advancing years, abnormalities of cardiac function further limit performance. The subject may stop exercising because of symptoms such as a thump in the chest from a ventricular extra-beat or the vice-like pain of angina, while if a test is being supervised by a physician, he may call a halt because of adverse electrocardiographic findings, particularly a negative voltage between the S and T waves of the ECG tracing ("ST segmental depression") or an abnormality of cardiac rhythm (particularly extra beats originating in the ventricles). The proportion ot subjects showing abnormal exercise electrocardiograms increases steadily after the age of 40, and at 65 years as many as 30% show abnormal records (Figure 2).

One might suppose that an elderly person would perceive effort as being more severe than a younger individual and, in consequence, would limit his physical activity voluntarily. However, in the experience of our laboratory, the perception of effort at a given percentage of maximum oxygen intake is unaffected by age.

Very Prolonged Activity. The performance of very prolonged work is limited by the demands of thermoregulation, the need to preserve fluid and mineral balances, and the availability of food reserves such as fat and glycogen.

We have already noted that an increase of subcutaneous fat makes thermal regulation more difficult for an elderly person. Often, the problem is compounded by poor functioning of the peripheral circulation as a result of conditions such as atherosclerosis and varicose veins. Exposure to heat is most likely in industry but can also arise in prolonged walking and jogging; thus, we have recorded rectal temperatures of 38.3 to 40.2°C in 45-year-old "postcoronary" patients participating in marathon events.

We have seen sweat losses of over 4 liters in these patients, with a corresponding excretion of mineral ions and vitamins. While large fluid losses are well-tolerated in a young person, kidney failure may develop in an older person with impaired renal function.

Glycogen reserves within the muscle fibers tend to be smaller in an old person. This probably reflects both a lesser tissue enzyme activity and a progressive reduction of habitual activity. At the same time, glycogen utilization during effort is increased, since a combination of weaker muscles, poorer peripheral vasculature, and lower tissue enzyme activities increase the proportion of work that must be performed anaerobically. Fat mobilization is probably hampered by a lower lipase activity in adipose tissue, although 65-year-old subjects still secrete substantial quantities of fat-mobilizing growth hormone in response to vigorous exercise.

CHRONIC EFFECTS OF EXERCISE

If a young person undertakes activity of sufficient intensity and duration several times per week, he shows a progressive improvement of exercise tolerance; training is said to have occurred. How far can similar reactions be produced in the elderly?

The hope of inducing modifications of cell number — either a decrease in fat cell count or an increase in the number of muscle fibers — is apparently lost shortly after birth. Nevertheless, there remains scope to improve performance through a reduction in the fat content of individual adipose cells, an increase of contractile protein in both skeletal and cardiac muscle fibers ("hypertrophy"), and an increase of intracellular enzyme activity levels. Both fat loss and muscle hypertrophy can apparently continue until late in life; fat loss depends on striking a negative energy balance, while the key to muscle hypertrophy seems to be the development of a sufficient tension in the active fibers. It is well known that some athletes, particularly contact sportsmen, have taken analogs of the male sex hormone (androgens) in an attempt to increase muscle mass. This is rather pointless in a young man with active testes. However, it is less clear whether synthesis of contractile proteins can be enhanced by the administration of androgens in advanced age.

Relatively few training experiments have been conducted on elderly subjects. However, our research team now has experience on some 600 patients who have sustained a myocardial infarction ("heart attack"). When first seen, the typical patient is in his mid-40s. He has had about 3 weeks in a hospital followed by 2 months rest at home. His aerobic power has thus dropped to 70% of that for a sedentary person of comparable age. However, there is a good response to progressive long-distance training. Starting by covering perhaps 3.2 km in 40 minutes, some members of the group have developed their capacity sufficiently to cover a marathon distance (42 km) in little over 3 hours, with an associated 60% gain of maximum oxygen intake.

With older patients (60 to 70 years of age), some investigators have reported almost no response to training. The most likely explanation is that an insufficient intensity of exercise was used in such experiments. Occasional reports have described very large gains. Here, also, there may be an artifact. It is probable that initial scores were limited by muscle weakness, symptoms, or fears on the part of the supervising physician; as the test population became more fit, they pushed themselves closer to a true maximum effort and were restrained less by their supervisors. Our laboratory has followed a group of some 40 men and women over a year of progressive training. Supervised exercise classes were provided for 1 hour four times per week; the program included a gymnastic "warm-up," followed by alternate fast walking or jogging and slow walking. The exercise prescription called for an initial pulse rate of 120 per minute, with progression to a training pulse of 140 to 150 per minute as the subjects became more fit. Our class automatically categorized itself into four groups, characterized by the intensity and frequency of participation (Figure 3). The fastest gains of maximum oxygen intake were naturally seen in the high-frequency/high-intensity group. However, those who trained regularly at a lower intensity of effort progressed, so that over a longer period of observation, their gains were roughly as large as for the high-intensity group. This seems an important practical observation, since several attempts to train elderly

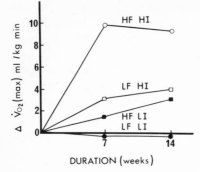

DURATION (weeks)

Figure 3. Influence of various intensities of chronic exercise upon the training response, measured as the gain in maximum aerobic power ($V_{O_2(max)}$, ml/kg·min). Data for 65-year-old men and women. HF HI: high frequency, high intensity; LF HI: low frequency, high intensity; HF LI: high frequency, low intensity; LF LI: low frequency, low intensity. Based on a study conducted jointly with K. Sidney, Department of Physical Education, Laurentian University, Sudbury.

patients have been thwarted by up to a 50% incidence of orthopedic problems (muscle strains, stress fractures, and prolapse of intervertebral discs).

Several authors have indicated that quite moderate levels of activity, such as brisk walking, can reverse normal age-related changes in body composition. Our subjects lost more than 3 mm of subcutaneous fat over 1 year of training; this was about 80% of the burden that they had accumulated since early adulthood. At the same time, we saw a small increase of lean tissue, as indicated by an increased body content of the naturally occurring isotope potassium-40. Changes in bone calcium were also encouraging; the more active subjects showed either an increase or no change, whereas there was a significant loss of bone calcium in those members of the group who took only infrequent low-intensity exercise.

Many cardiologists have hoped that vigorous training might reverse oxygen lack in the heart muscle. There is little argument that, after training, a subject can perform more work before his electrocardiogram shows adverse changes such as ST segmental depression. The main reason for this is that the workload of the heart has been reduced by a decrease in the exercise heart rate and possibly also a drop of systemic blood pressure. However, we have also seen some evidence that the severity of ECG abnormalities at a fixed heart rate can be reduced by very hard training. This could be explained by a thickening (hypertrophy) of the cardiac muscle; at any given blood pressure, the tension within the fibers (and thus their liability to oxygen lack) is inversely proportional to the cross section of the muscle wall. The high incidence of ECG abnormalities in elderly women might have a similar type of explanation; although their coronary blood vessels are in better condition than in men, a relatively thin heart wall increases the tension within individual fibers.

Although there are substantial physiological gains from regular exercise, many health professionals prefer to promote activity for the elderly on the basis that it improves mood — the person who exercises "feels better." Attempts to document subjective reports by formal tests of mood and personality have not been uniformly successful. One study showed a reduction of manifest anxiety but only in patients who were initially anxious because of a recent heart attack. Our 65-year-old exercisers showed small improvements on various measures of anxiety. Some of the group also showed an improvement of body image, but in those who failed to meet their exercise prescription, body image actually worsened.

LIFELONG PHYSICAL ACTIVITY

Many authors have cherished the hope that a lifetime of deliberate and well-disciplined physical activity might slow the inevitable process of aging and increase longevity. In rats, regular swimming confers a small advantage, but only if it is started at an early age; if begun after maturity, the death of the animal is actually hastened by the additional activity.

There have been occasional reports that continuing athletes enjoy a slower deterioration of maximum oxygen intake than the general population. However, the losses cited in such studies (0.5 to 0.7 ml/kg·min/year) are of the order normally encountered in cross-sectional studies of the general population. Where athletes have appeared to fare well, the problem has been an unusually rapid deterioration in the comparison group. Possibly, the controls themselves decreased their physical activity, with an accumulation of body fat. The influence of athletic participation is very difficult to determine, since even the continuing athlete may train less intensively as he becomes older. Further, there is the question of genetic selection; at least two-thirds of the unusual initial capacity of the athlete is probably a peculiarity of inheritance, and if so, there is no fundamental reason to compare his rate of aging with that of the general sedentary population. Nevertheless, the majority of data suggest that the rate of loss of maximum oxygen intake is similar for the athlete and the average person. Because the sportsman starts his career with a much greater working capacity, his ultimate condition is far superior to that of a sedentary contemporary.

The middle-aged athlete also manages to avoid obesity. A Scandinavian study took a cross-sectional look at orienteers.[3] At 25 years of

[3]Orienteering is a vigorous sport that involves cross-country running and use of compass bearings.

age, such sportsmen weighed an average of 70 kg. Among continuing participants, the weight at age 45 was still 70 kg, dropping to 64 kg at 55 years and 63 kg at 65 years. Protection was lost when the sport was dropped; inactive orienteers were 5 kg heavier than their contemporaries at 45, 12 kg heavier at 55, and 7 kg heavier at 65 years.

Some investigators have argued that athletes maintain a large heart volume, irrespective of whether they continue to participate in sport. A Swedish group examined young women who had been champion swimmers a few years previously but now were ordinary housewives. Their aerobic power had dropped to below the population average, but their heart volumes remained large. Nevertheless, cardiac tissue is lost with longer periods of inactivity. Thus, the heart volume of the orienteers was 15.0 ml/kg in active 45-year-old participants compared with 11.1 ml/kg in former sportsmen of the same age. At 65 years, the cardiac volume had decreased to 13.2 ml/kg even among the continuing participants.

There is not a great deal of information regarding the muscular strength of older athletes. A Danish report examined the right-hand-grip force of physical education teachers at the time of graduation, repeating the measurements after 25 years. In the men, the grip force had fallen from 54 to 44 kg, while in the women the corresponding values were 37 and 24 kg. Unfortunately, no information was given on the activity habits of the teachers, although they were presumably more active than the general population; nevertheless, losses of strength were at least as large as in sedentary groups.

Losses of lung capacity seem very small in physical education teachers. In the Danish study, vital capacity decreased by 7.7 ml/yr in the male teachers and 8.9 ml/yr in the females. Swedish investigators found no change of lung capacity when measurements made on physical education teachers were repeated after the lapse of 21 years. In contrast, cross-sectional studies of the general population have shown vital capacity to decrease by 17 to 22 ml/yr. One might thus infer that physical activity had some protective effect upon aging of the lung. However, interpretation is complicated by the association between exercise and other habits beneficial to health. In particular, the proportion of smokers is much smaller among physical education teachers than in the general population.

Early mortality statistics suggested that athletes gain some advantage of life expectancy over their nonathletic counterparts. A classical study of Oxford and Cambridge oarsmen showed an advantage of 2 years relative to the average Englishman from life-insurance statistics, and in a parallel study of Harvard oarsmen, longevity was 2.9 years

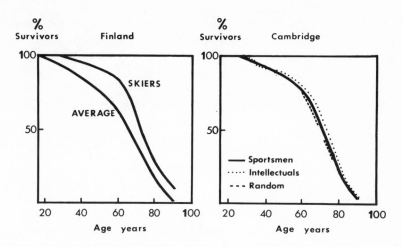

Figure 4. Influence of athletic participation on mortality. Based on data of M. Karvonen, Institute of Occupational Health, Helsinki. Left panel, data comparing Finnish cross-country skiing champions and average men. Right panel, data for Cambridge University graduates, including sportsmen who gained a "blue" for their college, "intellectuals" who obtained an honors degree, and a random sample of students.

greater than for the average American. However, both studies neglected the privileged life enjoyed by the university graduate. In more recent investigations, controls have been drawn from nonathletes attending the same university; mortality curves have, then, shown no advantage to those who participated in sports while at university (Figure 4). This is hardly surprising, for in middle age, the typical university athlete is less active than his nonsporting contemporary; he is also heavier and more likely to smoke and to drink than his supposedly sedentary control. Observation of university athletes, then, cannot answer the question of possible advantages of a lifetime of physical activity. One study often cited in this connection compares the mortality of champion cross-country skiers and sedentary Finns. The champion skiers apparently gain 6 to 7 years of longevity relative to sedentary members of the same communities (Figure 4). Unfortunately, questions of selection still complicate data interpretation; those in poor initial health are unlikely to become cross-country skiing champions, and those developing heart disease almost certainly drop out of the events once their condition becomes clinically manifest.

Electrocardiographic studies of athletes have yielded conflicting information. Some authors have encountered at least as many abnormalities of rhythm and ST segmental depressions as in the general population. Our group has examined Masters class[4] athletes, some in the tenth decade of life, and we were impressed with the low frequency of abnormal ECG records. However, it is likely that those with abnormalities are encouraged to drop out of competition by their physicians; further, few track athletes smoke, and a lifetime of abstinence from cigarettes could in itself account for the healthy electrocardiograms.

AGING OF "PRIMITIVE" GROUPS

The International Biological Program has collected much information on the aging of functional capacity in "primitive" populations. Particular interest has attached to circumpolar populations, where a high level of energy expenditure (14 to 15 MJ[5]/day) is necessary to sustain the traditional hunting life-style. Unfortunately, most of the data is cross sectional in type, and it can be misleading. The population pyramid is very steep, and only exceptional individuals survive the combined assaults of disease, famine, and natural hazards. Elderly people are so scarce that aging must be calculated over the span 25 to 55, or even 25 to 45 years, a procedure that necessarily favors the "primitive" group. Nevertheless, the average rate of loss of aerobic power in four circumpolar communities (0.47 ml/kg·min per year in the men and 0.41 ml/kg·min per year in the women) is of the same order as that encountered in sedentary "white" groups.

Obesity is not usually an expression of aging in underdeveloped countries. However, the relative absence of fat is more a consequence of a limited food intake than a high level of physical activity. In two South American nations where the daily nutrient intake exceeded 10.5 MJ (Uruguay and Chile), body weight was found to increase with age much as in the developed nations. However, in nine other South American countries, where nutrient intake averaged only 8.9 MJ, the men lost 2.0 kg and the women 1.8 kg between the ages of 22 and 62 years.

In male Eskimos, we have found that the aging of muscle force proceeded at least as fast as in "white" communities; between 25 and 55 years of age, hand grip force declined from 54 to 39 kg, and leg-extension force dropped from 88 to 77 kg. On the other hand, figures for the Eskimo women remained relatively constant over the adult span; hand-grip readings decreased from 30 to 27 kg, but leg-exension force re-

---

[4]The Masters competitions provide an opportunity for age-classified track and field events. The Toronto World Masters competition of 1975 attracted more than 2000 elderly participants from many of the "developed" nations.
[5]A megajoule is 1000 kJ.

mained unchanged at 69 kg in 25- and 55-year-old women. Factors preserving the strength of the women included the limited availability of furniture and domestic equipment and the tradition of carrying small children on the back in an "amauti."

Some attempts have been made to determine the anaerobic power of "primitive" groups, using the staircase sprint (page 134). The results have been extremely variable. One problem is that, in many of the populations studied, society is based on cooperation rather than competition. Even if they have seen a staircase before, the idea of running up it faster than their friends does not appeal to members of a "primitive" tribe. For some groups, the rate of aging has seemed faster than in the "white" population, but in other groups, where the young have performed badly, the subsequent deterioration of scores has been slower than expected.

SUMMARY

All facets of the acute response to both static and dynamic exercise deteriorate with aging. Most published reports concern males of the human species, although occasional studies suggest that comparable changes occur in females and in other species.

The loss of muscular strength and endurance accelerates in the sixth and seventh decades of life. Anaerobic power decreases from an oxygen equivalent of 165 ml/kg·min at age 25 to some 100 ml/kg·min at age 65. Over the same age span, anaerobic capacity (as indicated by the blood lactate in exhausting effort) diminishes from 13 to 14 to 10 to 11 mmoles/liter of blood. Aerobic power drops steadily from 3 liters/min at 25 to 2 liters/min at 65 years. Decreases in the potential for industrial work may be aggravated by a parallel accumulation of body fat. The changes of aerobic power reflect both a decrease of maximum heart rate and a smaller extraction of arterial oxygen in the tissues. After retirement, hemoglobin levels may also drop, and in some subjects, effort may be limited by oxygen lack in the heart muscle. In very prolonged work, thermal regulation is made more difficult by obesity, while continued effort is hampered by smaller glycogen reserves and less ready mobilization of depot fat.

Chronic exercise can induce substantial gains of aerobic power and losses of body fat as in a younger person. Regular training also restores lean tissue and slows calcium loss from the bones.

Athletes apparently lose working capacity at about the same rate as the general population, and they have no advantage in terms of longevity. Primitive tribes that sustain high levels of physical activity also have a similar rate of functional loss as the general population. Interpretation of data for both athletes and primitive tribes is difficult to interpret because of many unavoidable factors of sample selection.

ACKNOWLEDGMENTS. The technical data presented are drawn widely from the scientific literature. Detailed references will be found in *Physical activity and aging* (Shephard, 1978), but specific acknowledgment must be made of the contributions of a former graduate student (K. Sidney, Laurentian University, Sudbury), Don Bailey, University of Saskatchewan (coinvestigator in the Saskatoon survey of Figure 1), Henry Montoye, University of Tennessee, Knoxville (principal investigator of the Tecumseh study, Figure 2), Marti Karvonen, Institute of Occupational Health, Helsinki, Figure 4, and Terence Kavanagh, Toronto Rehabilitation Centre (coinvestigator in studies of postcoronary patients and Masters class athletes).

BIBLIOGRAPHY

General

Brunner, D., and E. Jokl. 1970. *Physical activity and aging.* University Park Press, Baltimore.
Shephard, R. J. 1972. *Alive man! The physiology of physical activity.* Charles C Thomas, Springfield, Ill.
Shephard, R. J. 1977. *Endurance fitness,* 2nd ed. University of Toronto Press, Toronto.
Shephard, R. J. 1978. Age and working capacity. In *Human physiological work capacity.* Cambridge University Press, New York.
Shephard, R. J. 1978. *Physical activity and aging.* Croom-Helm, London.

Technical

Adams, G. M., and H. A. DeVries. 1973. Physiological effects of an exercise training regimen upon women aged 52 to 79. *Journal of Gerontology* 28: 50–55.
Asmussen, E., and P. Mathiasen. 1962. Some physiologic functions in physical education students re-investigated after 25 years. *Journal of American Geriatric Society* 10: 379–387.
Åstrand, I., P. O. Åstrand, I. Hallbäck, and Å. Kilböm. 1973. Reduction in maximal oxygen uptake with age. *Journal of Applied Physiology* 35: 649–654.
Barry, A. J., J. W. Daly, E. D. R. Pruett, J. R. Steinmetz, H. F. Page, N. C. Birkhead, and K. Rodahl. 1966. The effects of physical conditioning on older individuals. I. Work capacity, circulatory–respiratory function, and work electrocardiogram. *Journal of Gerontology* 21: 182–191.
Benestad, A. M. 1965. Trainability of old men. *Acta Medica Scandinavica* 178: 321–327.
Brown, J. R., and R. J. Shephard. 1967. Some measurements of fitness in older female employees of a Toronto department store. *Canadian Medical Association Journal* 97: 1208–1213.

Chiang, B. N., H. J. Montoye, and D. A. Cunningham. 1970. Treadmill exercise — study of healthy males in a total community — Tecumseh, Michigan: clinical and electrocardiographic characteristics. *American Journal of Epidemiology* 91: 368–377.

Cumming, G. R., and L. M. Borysyk. 1972. Criteria for maximum oxygen intake in men over 40 in a population survey. *Medicine and Science in Sports* 4: 18–22.

Cumming, G. R., L. M. Borysyk, and C. Dufresne. 1972. The maximal exercise ECG in asymptomatic men. *Canadian Medical Association Journal* 106: 649–653.

Dill, D. B., S. Robinson, and J. C. Ross. 1967. A longitudinal study of 16 champion runners. *Journal of Sports Medicine* 7: 4–32.

Dublin, L. I., A. J. Lotka, and M. Spiegelman. 1949. Chapter 6 in *Length of life, a study of the life table*. Ronald Press, New York.

Durnin, J. V. G. A. 1966. Age, physical activity and energy expenditure. *Proceedings of Nutritional Science* 25: 107–113.

Fisher, M. B., and J. E. Birren. 1947. Age and strength. *Journal of Applied Psychology* 31: 490–497.

Forbes, G. B., and J. C. Reina. 1970. Adult lean body mass declines with age: some longitudinal observations. *Metabolism* 19: 653–663.

Grimby, G., and B. Saltin. 1966. A physiological analysis of physically well-trained middle-aged and old athletes. *Acta Medica Scandinavica* 179: 513–526.

Hartley, L. H., G. Grimby, Å. Kilböm, N. J. Nilsson, I. Åstrand, J. Bjure, B. Ekblom, and B. Saltin. 1969. Physical training in sedentary middle-aged and older men. III. Cardiac output and gas exchange at submaximal and maximal exercise. *Scandinavian Journal of Clinical and Laboratory Investigation* 24: 335–344.

Hollman, W. 1965. Körperliches Training als Prävention von Herz-Kreislauf Krankheiten. Hippokrates Verlag, Stuttgart.

Holmgren, A., and T. Strandell. 1959. Relationship between heart volume, total hemoglobin and physical work capacity in former athletes. *Acta Medica Scandinavica* 163: 146–160.

Karvonen, M. J., H. Klemola, J. Virkajarvi, and A. Kekkonen. 1974. Longevity of endurance skiers. *Medicine and Science in Sports* 6: 49–51.

Kavanagh, T., and R. J. Shephard. 1978. The effects of continued training on the aging process. *Annals of New York Academy of Science* 301: 656–667.

Sidney, K. H., and R. J. Shephard. 1977. Maximum and submaximum exercise tests in men and women in the seventh, eighth and ninth decades of life. *Journal of Applied Physiology* 43: 280–287.

Sidney, K. H., and R. J. Shephard. 1977. Attitudes towards health and physical activity in the elderly. Effects of a physical training programme. *Medicine and Science in Sports* 8: 246–252.

Sidney, K. H., and R. J. Shephard. 1977. Perception of exertion in the elderly. Effects of aging, mode of exercise and physical training. *Perceptual and Motor Skills* 44: 999–1010.

Sidney, K. H., and R. J. Shephard. 1977. Activity patterns of elderly men and women. *Journal of Gerontology* 32: 25–32.

Sidney, K. H., and R. J. Shephard. 1977. Training and e.c.g. abnormalities in the elderly. *British Heart Journal* 39: 1114–1120.

Sidney, K. H., R. J. Shephard, and J. Harrison. 1977. Endurance training and body composition of the elderly. *American Journal of Clinical Nutrition* 30: 326–333.

Sidney, K. H., and R. J. Shephard. 1978. Frequency and intensity of exercise training for elderly subjects. *Medicine and Science in Sports*, in press.

Sidney, K. H., and R. J. Shephard. 1978. Growth hormone and cortisol — age differences, effects of exercise and training. *Canadian Journal of Applied Sports Science* 2:189–194.

9

Diet and Nutrient Needs in Old Age

VERNON R. YOUNG

There are a number of environmental factors that significantly affect our health and well-being; among them, food is of greatest importance. Beginning with conception and continuing into advanced old age, the foods that are consumed play a dominant role in promoting a normal, healthful existence. Furthermore, earlier and current dietary practices affect the health status of the elderly human subject. The effects of the diet during earlier stages of life on life-span and health in the older organism have been discussed elsewhere in this book. In this chapter, we will be concerned with diet and nutrient needs during the later stages of life.

There are numerous physiological and biochemical processes that occur within the body, and they require energy and the various nutrients present in our foods. Thus, it is appropriate to survey first the essential nutrients and the functions they perform before an assessment is made of the amounts of these nutrients that must be supplied by the diet if health is to be maintained in the elderly. This latter aspect of our discussion will include an account of the methods used to determine quantitative nutrient requirements in adult humans and the way in which these estimates are used to arrive at the recommended dietary allowances (RDAs). These RDAs then serve as a guide in the planning and evaluation of diets for the elderly population. Much of what is said in this chapter applies to the nutrition of humans of all ages, but the particular nutritional problems of the older person will receive emphasis wherever appropriate.

THE ESSENTIAL NUTRIENTS

In order to facilitate a review of about 50 essential nutrients that we require throughout life, a summary of the functions and the con-

VERNON R. YOUNG, Ph.D. • Department of Nutrition and Food Science, Massachusetts Institute of Technology, Cambridge, Massachusetts 02139

sequences of deficient or excess intakes of each nutrient is shown in Table 1. It is convenient, for this purpose, to divide the nutrients into six general classes: carbohydrates, fats, proteins, vitamins, minerals, and water.

Carbohydrates, fats, and proteins serve as the major sources of food energy, with alcohol sometimes accounting for sizable proportions of total energy intake. However, the primary need for protein is to supply the essential amino acids and a source of utilizable nitrogen for meeting the needs for tissue protein synthesis. The proteins of cells commonly consist of about 20 amino acids, but only half of these can be made by the body. The remainder, or the essential amino acids, must be present in the diet; otherwise, nutritional deficiency disease will develop. Some food protein sources, such as eggs, milk, meat, and other animal protein foods, contain high concentrations of these essential amino acids, while some vegetable or plant protein sources, such as cereal grains and legumes, may contain low concentrations of one or more of these amino acids. However, a suitable mixture of plant protein sources can supply the essential amino acids in amounts and proportions to adequately meet the nutritional requirements. Thus, expensive meats, for example, are not nutritionally essential. Nevertheless, animal and dairy products serve as palatable sources of good-quality dietary protein, and in addition, they contain sources of other essential vitamins and minerals that make them a desirable component of the diet of older individuals. Additionally, some amino acids serve as precursors for metabolically active compounds, such as the neurotransmitters concerned with the activity of the brain and nervous system and the various hormones that regulate the metabolism of cells and organs.

Fats or lipids are concentrated sources of dietary energy, and they are also carriers of the fat-soluble vitamins discussed below. Dietary fats are also the source of the *essential fatty acids,* which are intimately involved in the maintenance of normal membrane structure and function. These fatty acids cannot be made by the body, but are present as part of the unsaturated fats in vegetable oils. They are parent substances for the formation of "local" hormones (the prostaglandins) that play roles in maintaining the blood platelets, smooth muscle metabolism, and nervous system activity. An adequate intake of these fatty acids is necessary to maintain good blood circulation and to assure adequate heart function.

Of course, excessive dietary intakes of fat can be detrimental to health. Too high an intake of fat, coupled with inadequate intakes of unsaturated fatty acids, may lead to increased coronary heart disease and various types of cancer.

Dietary carbohydrates are quantitatively important sources of our

Table 1. Summary of the Dietary Sources, Functions, and Effects of Deficient or Excess Intakes of the Individual Essential Nutrients*

Nutrient	Dietary sources	Major body functions	Deficiency	Excess
Essential amino acids				
Aromatic Phenylalanine Tyrosine		Precursors of structural protein, enzymes, antibodies, hormones, metabolically active compounds		
Basic Lysine Histidine	From proteins	Certain amino acids have specific functions:	Deficient protein intake leads to development of kwashiorkor and, coupled with low energy intake, to marasmus	Excess protein intake possibly aggravates or potentiates chronic disease states
Branched chain Isoleucine Leucine Valine	Good sources Legume grains Dairy products Meat Fish Adequate sources Rice Corn Wheat Poor sources Cassava Sweet potato	(a) Tyrosine is a precursor of epinephrine and thyroxine (b) Arginine is a precursor of polyamines (c) Methionine is required for methyl group metabolism (d) Tryptophan is a precursor of serotonin		

*Slightly modified from N. S. Scrimshaw and V. R. Young (Scientific American (Sept.) 235: 50–64, 1976).

(continued)

Table 1 (continued)

Nutrient	Dietary sources	Major body functions	Deficiency	Excess
Essential amino acids (continued)				
Sulfur-containing				
Methionine				
Cystine				
Other				
Tryptophan				
Threonine				
Essential fatty acids				
Arachidonic	Vegetable fats (corn, cottonseed, soy oils)	Involved in cell membrane structure and function: Precursors of prostaglandins (regulation of gastric function, release of hormones, smooth-muscle activity)	Poor growth Skin lesions Altered blood circulation	Not known
Linoleic	Wheat germ			
Linolenic	Vegetable shortenings			
Vitamins				
Water-soluble Vitamin B_1 (thiamine)	Pork, organ meats, whole grains, legumes	Coenzyme (thiamine pyrophosphate) in reactions involving the removal of carbon dioxide	Beriberi (peripheral nerve changes, edema, heart failure)	None reported

Nutrient	Food sources	Function	Deficiency	Toxicity
Vitamin B_2 (riboflavin)	Widely distributed in foods	Constituent of two flavin nucleotide coenzymes involved in energy metabolism (FAD and FMN)	Reddened lips, cracks at corner of mouth (cheilosis), lesions of eye	None reported
Niacin	Liver, lean meats, grains, legumes (can be formed from tryptophan)	Constituent of two coenzymes involved in oxidation–reduction reactions (NAD and NADP)	Pellagra (skin and gastrointestinal lesions, nervous, mental disorders)	Flushing, burning and tingling around neck, face, and hands
Vitamin B_6 (pyridoxine)	Meats, vegetables, whole-grain cereals	Coenzyme (pyridoxal phosphate) involved in amino acid metabolism	Irritability, convulsions, muscular twitching, dermatitis near eyes, kidney stones	None reported
Pantothenic acid	Widely distributed in foods	Constitutent of coenzyme A, which plays a central role in energy metabolism	Fatigue, sleep disturbances, impaired coordination, nausea (rare in man)	None reported
Folacin	Legumes, green vegetables, whole-wheat products	Coenzyme (reduced form) involved in transfer of single-carbon units in nucleic acid and amino acid metabolism	Anemia, gastrointestinal disturbances, diarrhea, red tongue	None reported
Vitamin B_{12}	Muscle meats, eggs, dairy products, (not present in plant foods)	Coenzyme involved in transfer of single-carbon units in nucleic acid metabolism	Pernicious anemia, neurological disorders	None reported

(continued)

Table 1 (continued)

Nutrient	Dietary sources	Major body functions	Deficiency	Excess
Vitamins (continued) Water-soluble *(continued)*				
Biotin	Legumes, vegetables, meats	Coenzyme required for fat synthesis, amino acid metabolism, and glycogen (animal-starch) formation	Fatigue, depression, nausea, dermatitis, muscular pains	Not reported
Choline	All foods containing phospholipids (egg yolk, liver, grains, legumes)	Constituent of phospholipids; precursor of putative neurotransmitter acetylcholine	Not reported in man	None reported
Vitamin C (ascorbic acid)	Citrus fruits, tomatoes, green peppers, salad greens	Maintains intercellular matrix of cartilage, bone, and dentine; important in collagen synthesis	Scurvy (degeneration of skin, teeth, blood vessels, epithelial hemorrhages)	Relatively nontoxic; possibility of kidney stones
Fat-soluble				
Vitamin A (retinol)	Provitamin A (beta-carotene) widely distributed in green vegetables; retinol present in milk, butter, cheese, fortified margarine	Constituent of rhodopsin (visual pigment); maintenance of epithelial tissues; role in mucopolysaccharide synthesis	Xerophthalmia (keratinization of ocular tissue), night blindness, permanent blindness	Headache, vomiting, peeling of skin, anorexia, swelling of long bones

	Sources	Functions	Deficiency	Excess/Toxicity
Vitamin D	Cod-liver oil, eggs, dairy products, fortified milk and margarine	Promotes growth and mineralization of bones; increases absorption of calcium	Rickets (bone deformities) in children; osteomalacia in adults	Vomiting, diarrhea, loss of weight, kidney damage
Vitamin E (tocopherol)	Seeds, green leafy vegetables, margarines, shortenings	Functions as an antioxidant to prevent cell membrane damage	Possibly anemia	Relatively nontoxic
Vitamin K (phylloquinone)	Green leafy vegetables; Small amount in cereals, fruits, and meats	Important in blood clotting (involved in formation of active prothrombin)	Conditioned deficiencies associated with severe bleeding, internal hemorrhages	Relatively nontoxic; synthetic forms at high doses may cause jaundice
Minerals				
Calcium	Milk, cheese, dark-green vegetables, dried legumes	Bone and tooth formation; blood clotting; nerve transmission	Stunted growth; rickets, osteoporosis; convulsions	Not reported in man
Phosphorus	Milk, cheese, meat, poultry, grains	Bone and tooth formation; acid–base balance	Weakness, demineralization of bone; loss of calcium	Erosion of jaw (fossy jaw)
Sulfur	Sulfur amino acids (methionine and cystine) in dietary proteins	Constituent of active tissue compounds, cartilage, and tendon	Related to intake and deficiency of sulfur amino acids	Excess sulfur amino acid intake leads to poor growth
Potassium	Meats, milk, many fruits	Acid–base balance; body water balance; nerve function	Muscular weakness; paralysis	Muscular weakness; death

(continued)

Table 1 (continued)

Nutrient	Dietary sources	Major body functions	Deficiency	Excess
Minerals (continued)				
Chlorine	Common salt	Formation of gastric juice; acid–base balance	Muscle cramps; mental apathy; reduced appetite	Vomiting
Sodium	Common salt	Acid–base balance; body water balance; nerve function	Muscle cramps; mental apathy; reduced appetite	High blood pressure
Magnesium	Whole grains, green leafy vegetables	Activates enzymes; involved in protein synthesis	Growth failure; behavioral disturbances; weakness, spasms	Diarrhea
Iron	Eggs, lean meats, legumes, whole grains, green leafy vegetables	Constituent of hemoglobin and enzymes involved in energy metabolism	Iron-deficiency anemia (weakness, reduced resistance to infection)	Siderosis; cirrhosis of liver
Fluorine	Drinking water, tea, seafood	May be important in maintenance of bone structure	Higher frequency of tooth decay	Mottling of teeth; increased bone density; neurological disturbances
Zinc	Widely distributed in foods	Constituent of enzymes involved in digestion, nucleic acid metabolism	Growth failure; small sex glands	Fever, nausea, vomiting, diarrhea
Copper	Meats, drinking water	Constituent of enzymes associated with iron metabolism	Anemia, bone changes (rare in man)	Rare metabolic condition (Wilson's disease)

	Sources	Function	Deficiency	Excess/Toxicity
Silicon, vanadium, tin, nickel	Widely distributed in foods	Function unknown (essential for animals)	Not reported in man	Industrial exposures: silicon — silicosis; vanadium — lung irritation; tin — vomiting; nickel — acute pneumonitis
Selenium	Seafood, meat, grains	Functions in close association with vitamin E	Anemia (rare)	Gastrointestinal disorders, lung irritation
Manganese	Widely distributed in foods	Constituent of enzymes involved in fat synthesis	In animals: poor growth; disturbances of nervous system, reproductive abnormalities	Poisoning in manganese mines: generalized disease of nervous system
Iodine	Marine fish and shellfish, dairy products, many vegetables	Constituent of thyroid hormones	Goiter (enlarged thyroid)	Very high intakes depress thyroid activity
Molybdenum	Legumes, cereals, organ meats	Constituent of some enzymes	Not reported in man	Inhibition of enzymes
Chromium	Fats, vegetable oils, meats	Involved in glucose and energy metabolism	Impaired ability to metabolize glucose	Occupational exposures: skin and kidney damage
Cobalt	Organ and muscle meats, milk	Constituent of vitamin B_{12}	Not reported in man	Industrial exposure: dermatitis and diseases of red blood cells
Water	Solid foods, liquids, drinking water	Transport of nutrients; temperature regulation; participates in metabolic reactions	Thirst, dehydration	Headaches, nausea; edema; high blood pressure

energy intake. The nutritionally important carbohydrates are the simple sugars (glucose, fructose, sucrose, and lactose) and the complex sugars or polysaccharides, particularly starch. In human nutrition, cellulose, a complex sugar, and other plant polysaccharides are not used for meeting energy requirements because they cannot be digested by secretions of the human gastrointestinal tract. They are called unavailable carbohydrates. However, they can have both beneficial and detrimental consequences for health. Currently, there is considerable interest in the role that *dietary fiber* (unavailable carbohydrate plus lignin) may play in the maintenance of normal gastrointestinal function. The decrease in the intake of dietary fiber during the past quarter century has been thought to be a major reason for the increase in digestive diseases in our society. Alternatively, dietary fiber may reduce the availability and absorption of some trace nutrients, and if the latter are already in low amounts in the diet, this may have untoward health effects.

Food contains substances other than the essential nutrients, but these may not be without effects on the general well-being of the body. The occurrence of natural toxins in some foods or their appearance in spoiled foods also emphasizes that there is a greater health significance to food than that represented solely by its constituent essential nutrients. However, this subject is beyond the general scope of the present discussion.

The vitamins are another class of essential nutrients; 13 of them are essential dietary constituents for humans. They are required for the normal maintenance of cells and organs, and their metabolic functions are diverse. For example, thiamine (vitamin B_1), pyridoxine (vitamin B_6), riboflavin (vitamin B_2), and niacin participate in enzyme reactions within cells and organs. Deficient intakes of specific vitamins in experimental animals lead to symptoms, such as graying of the hair (pantothenic acid deficiency), reduced physical and sexual activity (vitamin E deficiency), and shorter mean life-span (vitamin E deficiency). Such findings have often been used by those who are inadequately informed on principles of nutrition to promote higher intakes of vitamins than are provided by an adequate, well-balanced diet, and to claim that this would result in significant improvements in the physical and mental health of older individuals. Unless evidence of a vitamin deficiency was determined by an adequate medical checkup, there would be no indication that the intakes of these or other vitamins should be increased or that they would be beneficial for the elderly person. Indeed, the continued overuse of some vitamins can lead to toxicity.

The mineral elements constitute the fifth class of essential nutrients mentioned earlier. The macroinorganic elements (sodium, potassium, chlorine, calcium, magnesium, phosphorus) are present in the body in significant quantities, their combined mass being about 3 kg in adult men. The microelements, or trace minerals, on the other hand, are required in smaller amounts in the diet, usually less than 30 mg/day, and their total body content is about 30 grams.

The major mineral elements — sodium, potassium, calcium, magnesium, and chloride — fulfill electrochemical functions. Calcium, potassium, magnesium, copper, and zinc participate as catalysts in enzyme systems and some minerals — for example, calcium, phosphorus, and fluorine — play a role in maintaining the structure of the bony tissues, skeleton, and teeth. Other essential mineral elements include iodine, used for formation of thyroid hormones, and iron, required for the formation of hemoglobin, the latter serving as an oxygen carrier to tissues and organs. Dietary iron inadequacy, therefore, leads to low blood levels of hemoglobin, muscular weakness, and a reduced capacity to carry out physical work.

Improved methods for the analysis of the elements present in very small amounts in body fluids and tissues, and the development of refined systems for rearing experimental animals in "clean" environments have led to the identification of a number of newer essential minerals. These are nickel, tin, vanadium, and silicon, but not many years ago they were considered to be only health hazards. They are widely distributed in foods, and a primary dietary deficiency is probably unlikely unless they are present in forms unavailable to the body. Nevertheless, changes in the balance among them and among the other mineral elements may have important consequences for health, and it is thought that imbalances among the mineral elements might be factors responsible for the development of certain chronic diseases, such as diseases of the heart and arteries. We lack sufficient knowledge of the significance of nutrient interactions in human health.

METHODS OF ESTIMATING THE REQUIREMENTS FOR ESSENTIAL NUTRIENTS IN ADULTS AND ELDERLY PEOPLE

Although we know that the essential nutrients are required to promote full health, this condition cannot be measured in the same objective way that the farmer, for example, can assess how much of a feed or nutrient is required to give a certain milk yield by the dairy cow. Therefore, different approaches and criteria are used in an attempt to assess how much of a given nutrient is needed by human subjects to maintain their health.

To describe briefly the methods for estimating human nutrient requirements, the sequence of events that would be expected to occur during the development of nutritional deficiency is schematically depicted in Figure 1. From the stages shown here, it is apparent that one approach is to determine by a survey of diets the nutrient intake of individuals in populations free of nutritional disease. These intakes may then be compared with nutrient intake data obtained from individuals in populations where the nutritional deficiency occurs. This method will give an estimate of the approximate range of nutrient intake that is associated with health maintenance. However, this is not a precise method for estimating minimum requirements, and there are many problems in determining precisely the levels of nutrient intake by free-living subjects. Furthermore, the stages of nutritional deficiency that precede clinical symptoms are often difficult to detect.

If, however, the dietary content of a nutrient is low and if this diet continues, biochemical and pathological changes will develop eventually. These changes are expressed in the form of altered activities of blood enzymes, reduced rates of formation of blood proteins, or by lower levels of excretion of the nutrients or their products of metabolism. Some of these changes can be readily measured, and so this provides a basis for the design of metabolic studies that may be conducted to quantify nutrient requirements in human subjects. In these metabolic studies, volunteers — perhaps young adults or elderly subjects — receive experimental diets in which the level of the nutrient under study is altered during the course of the experiment. The investigator can then determine the minimum intake level of the nutrient that is necessary to meet a given criterion, such as the maintenance of a blood protein within a normal range, or the intake that is just sufficient to prevent the appearance of symptoms of deficiency of a nutrient.

Although this approach is a more precise method than the dietary method mentioned above, metabolic studies with human beings are laborious, time consuming, and expensive. Furthermore, these studies may involve the use of monotonous experimental diets and impose strict requirements, such as complete collection of urine and feces over many days, making them difficult for many volunteers to undertake. For this reason, such studies are frequently of short duration and include only a few volunteers. Thus, the data obtained will be quite limited in terms of application to populations. Nevertheless, this technique, or modifications of it, has been used extensively in the determination of human nutrient requirements.

For some of the essential nutrients, it has not been possible to utilize these or other satisfactory approaches. We must rely, therefore, on careful extrapolation of data obtained in animal experiments or on dietary and epidemiological information to approximate dietary intakes that might be considered adequate for human health.

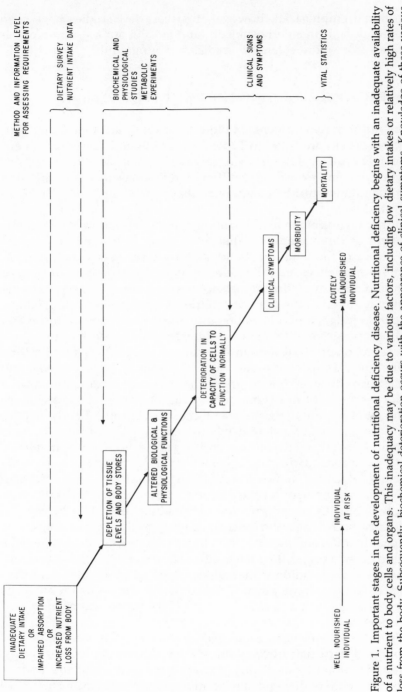

Figure 1. Important stages in the development of nutritional deficiency disease. Nutritional deficiency begins with an inadequate availability of a nutrient to body cells and organs. This inadequacy may be due to various factors, including low dietary intakes or relatively high rates of loss from the body. Subsequently, biochemical deterioration occurs with the appearance of clinical symptoms. Knowledge of these various phases aids the design of different approaches that may be used to assess human nutrient requirements. (Modified from Beaton, G. H., and V. N. Patwardhan. 1976. Pages 445–481 in G. H. Beaton and J. M. Bengoa, eds. *Nutrition in preventive medicine.* World Health Organization, Geneva.)

It should emphasized, however, that very few studies have been carried out directly in elderly subjects, and most of our present information comes from more extensive studies in young adults.

FACTORS AFFECTING NUTRIENT REQUIREMENTS

The nutrient requirements of older individuals are affected by various factors. These are listed in Table 2, according to host, environmental, and agent (dietary) factors. The variable importance of these factors, interacting in complex ways, for different individuals makes a definition of the quantitative nutrient requirements of the elderly an even more difficult task.

Nutrient requirements differ among individuals, and a fundamental component of this variation is that introduced by genetic differences. Not everyone of the same age, build, and sex has the same requirement, and these differences may be due, in part, to variations in genetic background. However, the importance of this factor with respect to the minimum requirements for specific nutrients among elderly populations is difficult to judge, and this will require much more comparative information on nutrient requirements of different population groups.

In terms of practical human nutrition, it is generally thought that the effects of various environmental, physiological, psychological, and pathological influences are of greater importance in determining the variability in nutrient needs among individuals. Thus, the growing infant and child require higher nutrient intakes, per unit of body weight, than the young adult. Nutrient needs are relatively high during the early growth and developmental phase of life, and they fall with the attainment of adulthood. However, other than for energy, where the daily requirement declines due to lowered physical activity, it is uncertain whether the requirement for the essential nutrients changes in the healthy individual with the progression of the adult years. On the basis of quite limited knowledge, the nutrient needs in healthy, aged subjects do not appear to differ significantly from those of young adults. However, a characteristic of aging is the increased incidence of disease and morbidity, and these are conditions that appear to be far more important than age per se in determining practical differences in the needs for nutrients between young adults and elderly people. This aspect will be discussed further below.

Also, there are numerous dietary factors that determine the amounts of a particular nutrient sufficiency to meet the body's needs. For example, all forms of dietary iron are not equally available and, in addition, the type of diet and composition of individual meals influence

Table 2. Agent, Host, and Environmental Factors That Influence Nutrient and Nutritional Status in the Elderly

Agent (dietary) factors
  1. Chemical form of nutrient
  2. Energy intake
  3. Food processing and preparation (may increase or decrease dietary needs)
  4. Effect of other dietary constituents

Host factors
  1. Progressive old age
  2. Sex
  3. Genetic makeup
  4. Pathological states
      (a) Drugs
      (b) Infection
      (c) Physical trauma
      (d) Chronic disease, cancer
  5. Psychological states

Environmental factors
  1. Physical (unsuitable housing, inadequate heating)
  2. Biological (poor sanitary conditions)
  3. Socioeconomic (poverty, dietary habits and food choices, physical activity)

the availability of the iron consumed. Iron in meat products is more readily available than the iron in vegetable foods. Hence, iron intakes that are adequate with a diet consisting of a generous amount of animal foods may not be sufficient if the diet consists largely of cereals.

Many other examples of the important influence of dietary factors could be cited, including the effects of calcium and dietary fiber on the availability of zinc, the increased requirement for pyridoxine (vitamin B_6) in high-protein diets; alcohol interferes with absorption of various vitamins; the retention of calcium and requirements for this mineral are influenced by the level of dietary protein intake. Since we eat foods rather than nutrients, interactions of these kinds are of considerable significance in practical human nutrition, but unfortunately they are difficult to quantify precisely.

Another factor of particular importance in our considerations of diet and nutrient needs during old age is the effect of stressful stimuli, such as those arising from infection or physical trauma, or even those of psychological origin. Thus, early in the infectious episode, there is an increased rate of synthesis of imunoglobulins and of other proteins characteristic of the initial metabolic response to the infectious agent. This is followed by a net catabolic response that results in increased losses of body nitrogen and of some vitamins and minerals and in de-creases in blood levels of these nutrients. Furthermore, absorption of nutrients may be interfered with if the gastrointestinal tract is significantly involved by either acute or chronic infections. The net re-sult of these processes is depletion of body nutrients, followed by a physiological increase in the need for nutrients during the recovery phase to promote recovery and to compensate for the earlier losses. However, while it is appreciated that acute and chronic infections and other stressful stimuli, including anxiety, pain, and physical trauma, generally increase the requirement for many essential nutrients, there are inadequate quantitative data to help determine how much nutrient intakes should be increased to meet the additional nutritional demands created by these conditions that may be frequent in elderly people.

Finally, various drugs may have profound effects on nutrient re-quirements by decreasing nutrient absorption or by altering the utiliza-tion of nutrients. The effects of drugs on nutrient requirements will depend upon the dose and period of administration, and furthermore, the multiple administration of drugs may have synergistic effects, thus increasing further nutrient needs. Reduced appetite is a frequent con-sequence of drug therapy, and this will exaggerate the effects of drug treatment on the individual's nutritional status, particularly if the diet is marginal in adequacy to begin with. Mention of alcohol (ethanol) might be made here because nutritional deficiency is often seen in alcoholics. Although this is due, in part, to an inadequate diet, ethanol interferes with the absorption and/or utilization of various nutrients, thereby effec-tively increasing nutrient needs above those required by healthy indi-viduals.

For many of these factors, their quantitative effects are not known, and they are likely to have a varying impact in different individuals. This adds complexity when attempting to define the needs for individual nutrients in the elderly. However, a knowledge of their quantitative effects is critical for developing rational dietary allowances. In spite of all the problems faced in determining the requirements of human popula-tions, there is a continual need for this information so that RDAs can be developed and used as guides for planning meals and diets and as a basis for evaluating the adequacy of nutrient intakes from dietary data.

RECOMMENDED DIETARY ALLOWANCES

Because the minimum requirement for a nutrient to maintain health varies among apparently similar individuals, the RDAs are designed so that they would be adequate to meet the nutritional needs of practically all *healthy* persons within a particular population. The question of the extent of the variation in nutrient needs among individuals is of particular importance, therefore, in the development of appropriate dietary standards. In healthy populations of adults, this variation is assumed to be normally distributed (Figure 2). Provided that the mean requirement for the nutrient is known, together with an estimate of the distribution in requirements about this mean, a recommended dietary allowance can be developed. Because it is not practical to formulate recommendations sufficient to cover *all* members of a population group, since a few normal individuals in the population will require considerably more of a given nutrient than the other members of the population, the mean requirement, plus 2 standard deviations above the mean, is now considered to

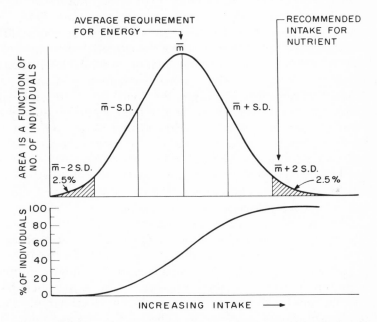

Figure 2. Relationship between requirements and recommended nutrient intakes. Nutrient requirements of individuals within a given population group, e.g., healthy elderly people, vary around the mean requirement (m) for the population according to the distribution curve shown in the top panel. To cover the nutrient needs of practically all the individuals, it is necessary to determine the range of this variation and an objective approach is to set the recommended intakes at a level equal to the mean requirement *plus* 2 standard deviations (S. D.) so that the requirements of 97.5% of the population will be met. This point is emphasized by the curve in the lower panel of this figure, indicating that dietary intakes at recommended levels meet the requirements of nearly all individuals. For reasons stated in the chapter, the allowances for dietary energy are based on the mean requirement for the population.

be a reasonable objective for establishing an RDA. This should then be sufficient to cover the requirements of about 97.5% of the population. Therefore, if the requirements for the specific nutrient have been determined adequately, most individuals will have actual requirements below this level.

For energy requirements, the situation is different because it is usual that the RDA for energy is based on the average requirement for the population. Intakes of energy either well above or well below the individual's true requirement would result eventually in a deterioration of health, and it is assumed that most individuals select diets providing energy intakes that meet or approximate their actual needs. Thus, the average energy requirement is given as a guideline rather than recommending the intake level for those few individuals in a population whose energy requirements are much higher than the mean.

Table 3 summarizes the most recent recommended dietary allowances for adults as proposed by the United States Food and Nutrition Board. It should be noted that not all of the essential nutrients are listed in this table because insufficient information is available on human requirements on which to make a reliable recommendation for all the known essential nutrients. However, the Food and Nutrition Board discusses these other nutrients in the text of their report on dietary allowances, and the reader may refer to this document for information concerning the nutrients not listed in Table 3. Where there are differences shown for the allowances for men and women, they are related largely to differences in body weight and body composition for the two sexes.

It must be strongly emphasized that the recommended allowances are amounts considered *sufficient for the maintenance of health in nearly all adults*. These recommendations are concerned with *health maintenance*, and they are not intended to be sufficient for therapeutic purposes. They are not designed to cover the additional requirements that may occur during and following recovery from infection, or under conditions of malabsorption, trauma, metabolic disease, or other significant stress. The possible benefits that might occur with considerably higher intakes of individual nutrients in a variety of clinical situations are not relevent to RDAs, and proposals for very much higher intakes of nutrients, relative to those intakes proposed in Table 3, as a normal dietary practice cannot be justified for healthy old people on the basis of current information.

Some investigators have suggested higher-than-usual intakes of dietary calcium as a means of preventing or minimizing the marked loss of bone calcium (osteoporosis) that develops with increasing adult age, particularly in elderly women. Although calcium intakes should be maintained at or above the current RDA for calcium, there is insufficient evidence to suggest that markedly higher intakes than those represented

Table 3. *Food and Nutritional National Academy of Sciences–National Research Council Recommended Daily Dietary Allowances, 1974, for Adults of 51 Years and Older**

	Males	Females
Weight (lb)	154	128
Height (in.)	69	65
Energy (kcal)	2400	1800
Protein (g)	56	46
Fat-soluble vitamins		
Vitamin A (IU)	5000	400
Vitamin D (IU)	Not stated	Not stated
Vitamin E activity (IU)	15	12
Water-soluble vitamins		
Vitamin C (mg)	45	45
Folacin (μg)	400	400
Niacin (mg)	16	12
Fiboflavin (mg)	1.5	1.1
Thiamin (mg)	1.2	1.0
Vitamin B_6 (mg)	2.0	2.0
Vitamin B_{12} (μg)	3.0	3.0
Minerals		
Calcium (mg)	800	800
Phosphorus (mg)	800	800
Iodine (μg)	110	80
Iron (mg)	10	10
Magnesium (mg)	350	300
Zinc (mg)	15	15

*These intakes are designed to be sufficient for the maintenance of good nutrition in practically all healthy persons. Diets should be based on a variety of common foods in order to provide other nutrients for which human requirements have been less well defined. Further details regarding these allowances are described in *Recommended dietary allowances*, 8th ed., Food and Nutrition Board, National Research Council, National Academy of Sciences, Washington, D.C. 1974.

by the RDA would result in health benefits. Indeed, the adequacy of dietary calcium during the earlier stages of life may be a highly important factor in determining the degree of bone calcium loss during the later period of adult life.

SPECIFIC CONSIDERATIONS OF DIET AND NUTRIENT REQUIREMENTS IN THE AGED

Much of the foregoing has been concerned with the nutrient needs of adult human beings of all ages. Also, as previously stated, there are no good indications that old age changes nutrient requirements substan-

Table 4. Some Factors That May Lead to Inadequate Nutrition in the Elderly

Apathy
Loneliness, psychological problems
Physical disability: immobility at home, poor vision, arthritis, etc.
Disease: infection, cancer, and other chronic illness
Malabsorptive and gastrointestinal disorders and discomfort
Poverty
Mental deterioration
Inadequate knowledge of dietetic principles: food fads, poor dietary habits
Alcoholism
Drugs
Increased requirements?
Unsuitable housing situation

tially, providing the individual is physically and mentally healthy. However, there are a broad range of conditions that accompany progressive old age, and these may have important effects on the nutritional status of the elderly. A. N. Exton-Smith, University College Hospital, London, has discussed a number of factors that occur particularly frequently in elderly persons, and these are listed in Table 4.

Physical and mental disabilities affect the mode of living, and these may lead to changes in dietary pattern and a deterioration of nutritional status. Underlying medical problems, emotional disturbance, loneliness, and poverty are all factors that may reduce the desire or ability to consume an adequate, well-balanced diet. Thus, risk of nutritional deficiencies will increase under these circumstances, and these are common causes for inadequate nutrition in elderly people. Because the energy requirements are reduced with inactivity, an adequate intake of other essential nutrients may not be met without a change in the dietary pattern toward foods with higher nutrient content. This problem may be compounded by the decrease in taste sensitivity that occurs with old age, and the poor health of the oral tissues that may restrict the selection of foods to those that are bland, soft, and readily masticated.

The net result is a further worsening in the condition of the oral tissues and an increase in the intensity of the vicious cycle leading to nutrient depletion, with a reduced capacity to resist infection and disease and a gradual deterioration in health.

The vulnerable groups of old people must be identified, and means must be developed for improving their nutrient intakes. Only a small percentage of the elderly population may be affected, and of course, many elderly individuals will never experience nutritional deficiencies. However, with the increasing numbers of elderly persons in our society, now estimated to be about 30 million, this small percentage figure translates into more than 1 million lives. This presents a significant medical, nutritional, and public health problem.

CONCLUSION

Although the available data are quite limited, they suggest that the nutrient requirements of the aged in good health are essentially the same as those of younger adults. However, old age is characterized by increased disease and a range of social, cultural, and economic adjustments that increase the risk of nutritional deficiencies in the elderly population. Approaches must be improved and innovative methods developed for exploring many aspects of health in the elderly, including the determination of food and nutrient requirements, and evaluation of nutritional status in the aged. Unless the large gaps in our knowledge of the nutrient requirements of human subjects — of all ages — are reduced significantly, the potentially far greater role that nutrition could play in preventive health measures in the elderly population is unlikely to be achieved.

BIBLIOGRAPHY

General Readings

Burton, B. T. ed. 1965. *The Heinz handbook of nutrition,* 2nd ed. McGraw-Hill Book Company, New York.
Davidson, S., R. Passmore, J. F. Brock, and A. S. Truswell. 1975. *Human nutrition and dietetics.* Churchill Livingston, New York.
Pike, R. L., and M. L. Brown. 1975. *Nutrition: an integrated approach.* John Wiley & Sons, Inc., New York.
Scrimshaw, N. S., and V. R. Young. 1976 The requirements of human nutrition. *Scientific American* (Sept.) 235: 51–64.

Technical Reviews

Hegsted, D. M. 1975. Dietary standards. *Journal of the American Dietetic Association* 66: 13–21.
National Research Council. 1974. *Recommended dietary allowances,* 8th rev. ed. National Academy of Sciences, Washington, D.C.
Rockstein, M., and M. L. Sussman. 1976. *Nutrition, longevity and aging.* Academic Press, Inc., New York.
Watkin, D. M. 1964. Pages 247–263 in H. N. Munro and J. B. Allison, eds. *Mammalian protein metabolism,* Vol. 2. Academic Press, Inc., New York.
Watkin, D. M. 1976. Pages 681–710 in R. S. Goodhart and M. E. Shils, eds. *Modern nutrition in health and disease,* 5th ed. Lea & Febiger, Philadelphia.

Winick, M., ed. 1976. *Nutrition and aging.* John Wiley & Sons, New York.
World Health Organization. 1974. *Handbook of human nutritional requirements* (R. Passmore, B. M. Nicol, M. Narayana-Rao, G. H. Beaton, and E. M. De Maeyer, eds.). WHO, Geneva.

10

Nutritional Regulation of Longevity

MORRIS H. ROSS

"To what do you owe your long life?" is a question that has been asked over the centuries of people fourscore and ten or older. It is not unusual in reading the old uncritical, sometimes metaphysical prescriptions for a long life to find that dietary habits were considered important. The advice, however, was often contradictory and often nonsensical (sometimes intentionally so).

Diet does have much to do with the functional well-being of the individual and much has been accomplished by nutritionists to alleviate the life-shortening effects of dietary inadequacies and imbalances. In economically advanced societies, nutritional upgrading is cited along with improvements in environmental conditions and progress in the medical arts for the dramatic increase in average life expectancy during the past 100 years. Can life expectancy, which is a measure of the length of life of the average individual, be improved even more by dietary means? Can it be done in a way that will increase the life-span of all individuals, including those that are already considered to be long-lived? Will this be accompanied by a reduction in the frequency and severity of disease? Or will the added years mean that the quality of life will suffer — that there will be even more cases of the debilitating diseases commonly seen among aging members of a population?

Actuarial and demographic data could be useful toward this end in so far as they might indicate which dietary factors appear to be associated with a long life-span. But other than the consequences of overfeeding and resulting obesity, or of deficiencies or excesses of specific dietary components in terms of disease risk, the data now available are too general to provide the type of guidelines needed. Because aging is a progressive and cumulative process that ends in death, it would seem logical that any investigation concerned with the modification of life-

MORRIS H. ROSS, V.M.D. ● The Institute for Cancer Research, Philadelphia, Pennsylvania 19111

span should begin with the young. Only with a complete life history will it be possible to learn the extent to which nutritional status at different periods of life affects the duration of life. Problems in using human beings as subjects for long-term dietary intervention studies are self-evident, and the investigator must turn to the laboratory animal. Among choices of animal models, the most relevant would be the one with dietary habits and age-related diseases that are to some extent common to man. For practical purposes, the duration of life should be shorter than his own. Most gerontologically oriented nutritionists prefer the white rat, an omniverous animal with a life expectancy between 600 and 700 days and with life-spans that under conventional dietary practices rarely exceed 1000 days.

Several investigators, working with rats, attempted to study the effects of supplementing a basal diet with a variety of natural food substances: beef, green beans, butter, wheat germ, apples, citrus fruit peelings, bran, cellulose, skim and whole milk, coffee, and sugar solution were some of the substances tested. On the basis of our present knowledge, the rationale for incorporating these materials into the diet is questionable. In any event, too few animals were used to obtain meaningful information and frequently disease episodes terminated the studies prematurely.

EFFECTS OF THE QUANTITY AND QUALITY OF THE DIET

Early experimentalists showed that the life-span of a variety of invertebrate animals could be increased by either a lowering of environmental temperature or a reduction in food intake. Clive McKay at Cornell University was the first to successfully demonstrate that the length of life of a warm-blooded animal could be extended by dietary manipulation beyond what had been considered to be the maximum limit of the species. In the intervening 40 years since his study was published, no other method for prolonging the span of life has been found to be as effective. McKay worked on an old premise, extending back as far as Aristotle, that lengthening the time required to reach maturity would increase the length of life. The most practical method of retarding growth was to limit the amount of food the rats were allowed to eat. The basal diet contained all the essential food substances known at that time so that maintenance requirements were satisfied but not enough of it was given for growth. After some 300, 500, 700, and 900 days, these stunted rats were permitted to consume as much of the diet as they wished. Despite disease episodes and difficulties in maintaining adequate environmental conditions, a number of the rats in the experimental group were still alive long after all the rats permitted to grow normally had died. This finding was entirely contrary to the prevailing notion that in order to maximize the probability of long life, the dietary regimen should be the one that was best for growth and development. Even today, the concept of the bigger the better is widely held. Perhaps an even more important observation made by McKay and his colleagues was that tumors did not appear during the period of semistarvation. The frequency of lung infections and chronic ailments in general were also lower.

The life-extending effects of underfeeding, together with the results of studies carried out on primitive life forms, renewed speculation about whether the maximum life-span of a species was fixed. Were earlier arbitrarily set limits too low? Was this method for extending life applicable or even advisable for man? However, more pragmatic questions were also being asked. A. J. Carlson of the University of Chicago undertook a study to determine whether retardation of growth was the essential factor. His less stressful regimen, alternating periods of "feast and famine," also led to an increase in life expectancy without severe growth stunting. Accompanying the increase in average life-span was a reduction in the development of mammary tumors. A series of other long-term studies followed involving vitamins, minerals, and trace elements. In no case, other than the beneficial effects directly attributable to the correction of deficiences in the basal diet, did the increase in life expectancy or longevity of any individual compare with effects obtained by simply limiting the intake of an adequate diet.

Diets characteristic of a geographical area were also examined. For example, L. L. Sperling, a colleague of McKay, fed a mixture of foods typical of the diet of people living in the northeastern United States, with and without additional amounts of vitamin A and most of the known B vitamins. The food mixture, fed in ad libitum amounts, consisted of eggs, milk, margarine, butter, beef, pork, white bread, potatoes, tomatoes, beans, carrots, apples, salt, and sugar. However, the average lengths of life of both groups of rats were similar. This was also the case when the food mixture was modified so that it contained liver, whole wheat bread, and more milk.

Animal studies of lifetime duration require a comprehensive analysis of multiple intrinsic and extrinsic factors. Each, separately or in combination, may, at any stage of life, influence the events that lead to the termination of life. It would be unrealistic to believe that we can define the countless number of variables or even to assert a direct cause–effect relationship for a single variable. A start can be made, however, to study in a systematic manner the effects of nutrition on the condition and length of life.

The use of mixtures of complex natural foods, which vary in quality depending on their sources, season of the year, and treatment, present problems in formulating diets suitable for lifetime study and in assessing the results. Purified semisynthetic diets, comprised of ingredients that are uniform in quality, offer the means of controlling this variability. Even with purified diets, caution must be exercised in attributing an effect wholly or in part to the absolute level of intake of a component, to alterations in the relative proportions among all the ingredients that result when a change is made in the content of one or more of the ingredients, to differences in caloric value of the diet, or to any appetite-modifying effect that may accompany a change in dietary composition.

In our laboratory, we have dealt with questions relating to the quality of the diet under ad libitum and restricted conditions of feeding, the effects of dietary intervention at different stages of life, and the consequences of the dietary practices of animals permitted to select their own diets. In all of this work, male rats were used; each was housed individually, and the levels of the vitamins, minerals, and trace elements in the purified diets were always adequate and kept constant.

In order to establish base-line values for the strain of rats we were to use, we first reexamined the effects of underfeeding, in varying degrees, a complete diet throughout postweaning life. As anticipated, the life expectancy of these animals was considerably greater than that of animals fed ad libitum. With advancing age, the difference in the rate of attrition between the control and experimental groups became progressively more pronounced. When the amount alloted to the animals in an experimental group was less than 50% of normal intake, a few animals died during the first several months. The more severe the restriction, within limits, the greater the initial loss but the higher the expectation of life of the survivors. By restricting intake to 30% of normal, the expectation of life at 21 days of age was more than 1100 days. It was approximately 600 days for rats permitted to consume the same diet in ad libitum amounts. A number of the restricted rats survived more than 1800 days, life spans far beyond any so far recorded. In terms of the human being, if such an extrapolation was to be made, this would be equivalent to 180 years. In contrast, the last of the rats fed ad libitum died some 700 days earlier. Significant reductions in the prevalence of several age-related diseases were also found. In general, these occurred at chronologically later ages. For example, among well-fed rats, 40% developed, in varying degrees of severity, a highly debilitating disease of the kidney, but in restricted rats, this condition was rare. Among those few cases that were found, the disease occurred at advanced ages and was mild. The decreases in risk of several other age-related affections, including tumors attributable to lifelong restriction in food intake, are shown in Table 1.

Rats are subject to a broad spectrum of spontaneous cancers. Some strains of rats are more likely to develop specific neoplasms than other strains. The relatively small size of the animal populations used by McKay and other investigators limited their analysis of the effects of undernutrition to the types of tumors that were most commonly found; fewer tumors of the anterior pituitary gland and of the lung developed in their restricted rats than among the ad libitum rats. In subsequent studies, particularly work done by A. Tannenbaum and his colleagues while at Michael Reese Hospital in Chicago some years ago, it was shown that lymphoid leukemia and tumors of the mammary gland, liver, and lung occurred less frequently in inbred mice, highly prone to these tumors, when they were subjected to a regimen of chronic restriction in food or caloric intake. The more severe the underfeeding within limits, the greater the decrease in incidence. In the case of mammary gland and liver tumors in mice, a restriction in food intake of only 35 to 40% reduced the incidence of these tumors by nearly 100%. Other workers have shown that for these two tumors, excessive food intake evokes a response opposite to that of restriction. While tumor-promoting effects of induced overeating in the rat have not been investigated, we found that among rats consuming the same diet, the more obese the animals, the more susceptible they were to nearly every tumor type than were less heavy rats. On the basis of these data, it would seem logical to

Table 1. *Long-Term Effects of Restriction in Food Intake Early in Life on Risk of Age-Related Diseases of Selected Organs of the Rat*

Site of disease*	Dietary regimen [Relative morbidity ratio (X100)]†		
	AL‡	R§	R-AL¶
Tumors, all sites	100	10	63
Kidney	100	0	62
Heart	100	2	76
Lung	100	11	61
Prostate	100	5	47

*Disease present — kidney: glomerulonephrosis; heart: myocardial fibrosis; lung: peribronchial lymphocytosis; prostate: prostatitis.
†Number of cases, all ages/number of expected cases, all ages; values < 100 indicate extent to which the disease risk relative to standard population (AL rats) was reduced.
‡Rats fed ad libitum throughout postweaning life.
§Rats restricted in food intake throughout postweaning life.
¶Rats restricted in food intake from 21 to 70 days and then fed ad libitum for remainder of life.

make the generalization that chronic underfeeding is an effective and consistent means of reducing the frequency of spontaneous tumors.

Since large numbers of rats were used in our experiments, a larger body of tumor-type data was available for analysis. We found that there was a sufficient number of exceptions to the above proposal to question its validity. While the frequency of tumors of the pituitary gland, lung, and of the pancreatic islet cells was reduced, the frequency of tumors of the thyroid gland, urinary bladder, and those of soft tissue origin were not significantly altered. There was, however, a delay in time of occurrence of these tumors. More important, there was a significant increase in the frequency of adrenal and parathyroid gland tumors and of reticulum cell sarcomas of the lymphoid organs. Malignant epithelial tumors were also more prevalent among the restricted rats than in those fed ad libitum.

The composition of a diet can modify the life-extending effects of underfeeding. As the proportion of the protein component in a series of diets of equal caloric value increased from 10 to 51% (at the expense of the carbohydrate component), life expectancy increased. This result was obtained whether the rats were fed in restricted, but identical, amounts or in ad libitum amounts. At all of these levels of dietary protein, the restricted rats were longer-lived than the ad-libitum-fed rats. The life-prolonging effect of restriction was accentuated when the diet was high in protein and low in carbohydrate content and minimized when the diet was low in protein and high in carbohydrate. Rats fed a high-protein diet in restricted amounts had an average life-span that was 75% longer than rats consuming a low-protein diet in ad libitum amounts (934 vs. 540 days).

The quantity and composition of the diet together had a marked effect on the risk of a number of age-related diseases. Several diseases occurred in significant numbers in some dietary groups, whereas in other groups, they occurred rarely or not at all. The sum effect was that the spectrum of disease susceptibility differed from dietary group to dietary group even though the caloric values of the diets and/or the levels of caloric intake were identical. For example, the effect of restriction on the incidence of prostatitis was negligible when the diet was low in protein content but not when it was high. The opposite held for some tumors in which the effects were more pronounced at low dietary protein levels. Protein undernutrition and protein overnutrition did not always elicit diverse responses. For some tumors, the relationship between the proportion of protein in the diet and tumor frequency were biphasic. This type of pattern was found for tumors of the anterior pituitary gland, thymus, and β cells of the pancreas. These neoplasms occurred in greater number among rats fed, ad libitum, a diet most conducive to rapid growth than rats fed diets that were either low or excessively high in protein.

This biphasic relationship is in contrast to two other dietary protein tumor-risk responses. For tumors of the urinary bladder, the risk progressively increased as the level of dietary protein increased; for tumors of the thymus gland, the risk decreased. Without respect to tissue origin or site, the probability of incurring a malignant tumor increased almost linearly as the level of dietary protein decreased. Again we can infer that no two dietary groups exhibit the same disease susceptibilities. This in turn suggests that for some age-related diseases, there is for each a degree of nutritional specificity that promotes its development.

Before concluding that a high-protein diet consumed in severely restricted amounts throughout life is of advantage to the organism, one must consider several adverse effects. In addition to stunting of growth and greater vulnerability to malignant tumors, animals maintained on restricted intake suffered higher infant mortality, impairment in functional and behavioral development, and an increase in susceptibility to bacterial and parasitic diseases. Separately or together, these are sufficient to discourage the use of severe undernutrition as a means of extending life-span.

EFFECTS OF DIETARY INTERVENTION AT DIFFERENT STAGES OF LIFE

Some of the deleterious effects of restriction in food intake may be the result of disturbances in maturational processes. If so, it may be possible to avoid them by delaying the time at which a dietary regimen is begun. Significant correlations between length of life or risk of some diseases and mature body weight, irrespective of the animal's dietary background, encourage this line of thought; on the other hand, correlations between the rate of growth early in life and the duration of life are of a significantly higher order. This also holds for associations between life-span and the rate of change in size and number of the cells of certain organs and in their biochemical activities. Since the dietary regimens were always started at weaning age and maintained throughout life, the associations for the early period and for the later period could be real, or either one could be coincidental. In order to resolve this question, it was necessary for us to separate any long-lasting influences of dietary experiences of the young animal from the immediate effects of the dietary regimen on the mature and aging rat. Two approaches were taken, both involving a change in regimen. In one series of experiments, the importance of nutritional status in early life was emphasized and changes in the regimen were therefore made when the animals had achieved approximately 50% of their adult weight. In the second series, changes in regimen were introduced after the animals were fully mature in order to bypass the postweaning growth period.

When rats were fed ad libitum throughout life, except for a 7-week period immediately after weaning, there was an overall reduction of approximately 20% in mortality risk. When the restricted ad libitum feeding schedule was reversed, the average span of life was increased, but not to the same extent as was obtained when the regimen of restriction was imposed throughout postweaning life. In both experiments, the nutritional experiences during the 49-day period immediately following weaning were sufficient to modify the life-prolonging or -shortening effects of the dietary regimens that followed.

The influence of early nutritional status on the frequency of age-related diseases was particularly striking. A brief period of underfeeding significantly reduced the incidence of the more common diseases of age (Table 1), including most but not all tumor types. The extent of reduction in tumor risk varied from 14 to 90%. For the highly diet-sensitive anterior pituitary gland tumors, short-term restriction was nearly as effective in reducing risk as lifetime restriction. This suggests that the processes through which nutrition influences the events of later life are operative during early stages of life. In support of this is the observation by W. S. Hartroft and C. H. Best of Canada that a 1-week period of choline deficiency, immediately after weaning, led in later life to hypertension even though the animals were fed an adequate diet throughout the remainder of their lives. The fact that the beneficial effects of restriction in food intake on mammary tumor frequency diminishes progressively with increasing age at initiation of the regimen also strengthens this conclusion. Perhaps the old adage, "We are what we eat," should be changed to read, "We are what we ate."

When the identical regimen of restriction was started at 300 days of age, instead of being prolonged, the lives of all the animals were sharply curtailed. At still greater ages, the life-shortening effects were even more drastic (Table 2). Although life expectancy of mature rats could still be extended, the composition of the diet and level of restriction proved to be highly critical and age-dependent. Of the number of diets tested, in all cases but one, restriction in intake accentuated the life-shortening effects of the new regimen. The single diet which improved life expectancy did so only when the level of restriction in intake was moderate. Under these conditions, the life expectancy of the mature rats exceeded by 35% that of rats fed ad libitum throughout life. When the dietary change was delayed until midlife, the degree of restriction had to be still further reduced if life-extending effects were to be obtained.

Table 2. *Influence of Age on Life-Span-Modifying Effects of Restriction in Food Intake*

Dietary regimen	Age when regimen of restriction begun (days)	Relative mortality ratio (X100)*
Ad libitum†, ‡	—	100
Restricted ‡ §	21	29
Ad libitum — restricted§	70	35
Ad libitum — restricted§	300	243
Ad libitum — restricted§	365	320

*Mortality ratio values, all ages. Values < 100 indicate extent to which mortality risk relative to standard population (ad libitum fed group) was reduced and values > 100 the extent to which it was increased.
†Average daily food intake, 19 g.
‡Regimen maintained through postweaning life.
§Daily food allotment during period of restriction in food intake, 6 g.

E. Stuchlikova in Czechoslovakia has recently extended these observations to include mice and hamsters.

The importance of the age of the animal when a dietary regimen is begun has also been shown under ad libitum feeding conditions. For example, high-fat diets are usually associated with a lowering of life expectancy; the extent varies with strain, sex, and age of the animal. M. and R. Silberberg of the University of Washington learned that male mice fed a high-fat diet throughout postweaning life were shorter-lived than were mice fed the same diet from early adult life onward. But, when the same regimen was started late in life, the life-shortening effect was considerably greater than it was for those whose regimen was changed when they were young adults.

While the evidence is scanty, the beneficial effects on life-span attributable to restriction begun late in life would seem to result from an altered rate of decrease in functional activities or from a reduction in the number of new instances of age-related diseases and in the rate of development of existing pathological processes. A brief anecdotal report by E. A. Vallejo of Spain is highly suggestive that dietary intervention had a significant effect on the state of health and survival of aged humans. His institutionalized subjects, all over 65 years of age, were fed institutional food either daily or on alternate days. On the other days they received milk and fresh fruit only. The caloric intake on these days was on the average 60% less and the level of protein intake was lower by 28%. The allotment of fat was identical. During the 3-year period of the study, the length of time spent in the infirmary by the fully fed subjects was twice that of the restricted subjects, and they suffered twice as many deaths.

In a somewhat similar vein, the involuntary limitations in the amount and type of food consumed during World War II in several European countries was associated with a reduction in the number of deaths due to diabetes and in the rate of development of arteriosclerosis.

The results of these longevity studies demand that we no longer ignore the fact that nutritional requirements change with age. An unvarying regimen may be eminently satisfactory during one stage of life but not during others. The extent of contrition of a diet to longevity might well be determined in large part on whether the dietary conditions suitable for extending life were in force during the early, middle, or late phases of life.

The problem of establishing the progression of dietary modifications that should be instituted at different stages of life so as to be most conducive both to optimal development and to longevity is complicated by the lasting effects of early dietary experiences, together with the specificity in dietary conditions later in life. How, then, can we design a program of feeding that incorporates the sequence of changes in age-related requirements which maximizes the probability of a long life? Since the amount and composition of the diet consumed by man and animals living under natural conditions differ from individual to individual, this aspect should also be taken into account in the experimental design.

DIETARY PREFERENCE AND LONGEVITY

The self-selection method of feeding proved to be an excellent way of answering these questions. By substituting the convenient but artificially fixed dietary system of feeding with a mode of feeding that allows each animal to make its own choice, several advantages are gained. The free-choice method simulates natural conditions more closely; the diet chosen has a physiological basis, and the consequences of different dietary practices at different stages of life can be assessed on an individual basis.

The rationale for choosing the self-selection method came from the observations of investigators in the behavioral sciences who were dealing with the phenomena and mechanisms of appetite and hunger. The young of a number of species, including the rat, unless trained to do otherwise, apparently have the ability to select essential food substances in such quantities as to sustain life and promote growth and reproduction. There are several reports indicating that the newly weaned human infant also exhibits the ability to select diets commensurate with normal development and maintenance of good health, but this anecdotal literature is difficult to assess. In view of the many factors that can bias the selection of food in the adult human, it is not surprising that the question of whether instinct guides selection of food is still debated.

The dietary preference and changes in preference that accompany aberrations in the regulation of metabolism early in life are found almost consistently to be of the type that tend to maintain normal physiological equilibrium. For example, diabetic rats eating a complete diet will die within a short period, but they survive if they are given free choice. These animals avoid consuming carbohydrate, thereby ameliorating or eliminating the symptoms of diabetes, whether this condition is of genetic origin or is induced. Adrenalectomized rats ingest large quantities of NaCl; parathyroidectomized rats modify their (Ca:P) intake, and hypophysectomized and thyroidectomized animals similarly adjust their dietary practices in compensatory ways. Changes in appetite have also been seen in rats experiencing a variety of environmental and physical stresses. During pregnancy and lactation, changes in dietary practices also occur. On the other hand, animals prone to obesity compound this problem by their increasing intake of dietary fat.

Because of the logistics of feeding large numbers of individually housed rats, we elected to make available to each animal full diets rather than individual dietary components. The diets differed only in the proportion of protein to carbohydrate so as to represent diets low, intermediate, or high in protein content. All other constituents were present in adequate amounts. Thus, each rat received three isocaloric diets in separate containers. By measuring the amount of food consumed from each container, it was possible to compute the relative proportion of protein to carbohydrate of the composite diet, the absolute intake of protein, carbohydrate, and of calories. Body-weight information was collected for us (1) to determine whether the growth characteristics were indeed related to longevity, (2) to express the level of intake of the different dietary constituents on a body-weight basis, and (3) to derive gross food efficiency and other factors. The self-selection regimen was begun at 21 days of age and maintained for life.

The life span varied from approximately 300 to over 1000 days. Notwithstanding that dietary practices differed from rat to rat, ranging from diets low to excessively high in protein content, their life expectancy on a group basis was similar to that of rats fed a fixed diet high in protein. Their average life-span was, therefore, significantly longer than rats fed diets low or intermediate in protein content.

In our earlier studies, age-specific mortality rates, survival curves, relative mortality risk ratios, and life-expectancy values provided a measure of the effects of an imposed dietary regimen. These actuarial methods are employed to detect similarities or differences between populations. In order to apply them to the data obtained from the self-selection rats, animals had to be classed into subgroups according to their dietary behavior. While the class ranges were set arbitrarily, several general trends were revealed. Rats that consumed relatively large amounts of food were generally shorter-lived than rats with smaller intakes. This finding conforms with and extends conclusions reached with regimens that depend on the stresses associated with imposed under- and overfeeding. A more provocative observation was the inverse linear relationship between food or caloric intake and length of life that was detected at a very early age, the seventh week of life. An average difference of 10% in the amount of food consumed when the animals were 8 weeks old was associated with an 8% difference in the average length of life.

Inasmuch as the dietary practices differed from one animal to another (in fact, no two animals were alike), other methods of analyses were needed which would permit the use of ungrouped data; methods were required that would permit quantitative assessment of the relationship between any one variable or of the combined effects of two or more variables and length of life. When correlation analysis was applied, it was found that the level of intake of food was inversely correlated with the duration of life. The extent of influence varied with age. Between the 14th and 28th week of life, based on linear regression analysis, a difference of 1 g in food intake was associated with a 26-day difference in life-span. By midlife, however, the effects attributable to the amount of food consumed decreased to such an extent as to be negligible. At still later ages, no significant correlation could be detected between the level of food intake and life-span.

The protein/carbohydrate ratio of the diet, as well as the absolute intake of protein and of carbohydrate correlated significantly with length of life. There were for each, however, temporal specificities involving the earliest age at which an association was found, the age at which the correlation coefficient value was maximum, and the age beyond which a correlation was no longer found. For the absolute and relative intake of protein, there was, in addition, a reversal in the direction of the correlations (i.e., initially the correlations were direct; at later ages they were inverse). This indicates that a diet conducive to long life at one age may produce the opposite at another age. The direct relationship between protein/carbohydrate ratio of the diet and length of life was detected as early as the sixth week of life; rats that preferred a high-protein diet at this age were, therefore, more likely to be long-lived. At greater ages, beyond 200 days, the lower the protein content of the composite diet chosen, the longer-lived are the animals. D. S. Miller and P. R. Payne in England imposed this type of regimen on rats and found that length of life was indeed improved over the control animals whose regimens were uniform throughout the course of the study.

The relationship between life-span and the quantity of protein and of carbohydrate consumed by animals permitted to select their own diet were separable, each apparently exerting an independent influence. It appears that, prior to 50 days of age, the protein content of the diet was the major determining factor. Food intake took on an important life-span influencing role between 50 and 300 days. After 300 days and before midlife was reached, the quantity and composition jointly influenced life-span. Thus, the effects of the level of food intake throughout much of an animal's lifetime is diminished or enhanced depending upon the composition of the diet that an animal chooses to eat at a critical period early in life and during early mature stages of life.

Several parameters involving body weight also correlated significantly with life-span. This was anticipated, since the progression of changes in body weight throughout life is in large part determined by the amount and type of food consumed. Unexpectedly, the growth response of an individual to the diet it preferred correlated more closely with length of life than did the dietary factors separately or in combination. For each relationship, there was a unique temporal specificity pattern: the age of the animal when a correlation was first detected, when the coefficient values were maximum and when it was lost. In every case, the weight or change in weight of the animal after maturity had been reached did not correlate with life-span.

Of the various ways of expressing changes in body weight, the highest correlation coefficients were obtained for the time required by an individual to double its weight or to go from one preassigned weight, irrespective of the age when that weight is reached, to another. These measures differed from the absolute and relative growth rate expressions in that they were age-independent derived values. Significant relationships with life-span were also obtained when the dietary intakes were computed relative to body weight. The age for maximum correlations did not necessarily coincide with those of dietary or weight variables used in the calculations.

MODEL FOR PREDICTING THE LIFE-SPAN OF THE INDIVIDUAL

With almost 20 interdependent indexes linked with length of life, it was necessary to establish which were the important ones and in what sequence they become important. A close correlation between an independent and dependent variable does not necessarily indicate causality.

A correlation may be only coincidental or indirect because that independent variable is closely associated with another variable that is directly related. The construction of mathematical models, necessitating the use of a computer, was resorted to as a means of learning which, when, and to what extent the different variables acting in combination explain the variance in life-span. An accurate model comprised of a number of factors, each of which contributes significant information, can be presumed to show a dependence relationship. In so doing, it could reveal the conditions that maximized the probability of a long life. The model can serve another function — as a predictor of life-span. Needless to say, countless environmental forces can affect length of life and it should not be expected that an estimate of life-span mathematically derived from an equation consisting of a limited number of variables would precisely match the actual life-span. The extent to which the estimated life-span differs from the actual life-span represents errors. The smaller the error, the more accurate the model.

The dietary practices and the growth responses of the animals were both included in the combination of factors that best explained the longevity of the animals. The multiple regression equation consisted of seven independent variables, each contributing significant information. Of these, the four variables that characterized the growth of the animals contributed more information as predictors of future events than did the dietary factors. The important weight-related variables were: (1) the weight of the animal at two age periods prior to 150 days of age; (2) the age at which a weight of 500 g was attained, and (3) the interval required by the animal to double its weight (250 to 500 g). The two dietary variables in this system describe the relative content of protein of the diet and the absolute intake of protein. In both cases, information for the 43- to 49-day period was critical, since substitution with data from different age classes led to a less informative equation. The seventh variable represents a measure of the efficiency with which the diet consumed by an animal between 49 and 98 days was used for growth. In every case, the data necessary to explain best the variance in longevity were readily obtainable early in life. All the dietary data needed were obtained before the animals were 100 days old and, except for a few animals, all the weight-related data before 150 days. Apparently, when the self-selection mode of feeding is maintained throughout life and the essential food elements are always available, the dietary habits and body-weight characteristics after maturity has been reached are of little importance. This interpretation should not be construed to be in conflict with the effects of dietary intervention at later ages. Rather, the sharply delineated temporal specificities reinforce the view that the early days of life represent a highly critical stage. If this conclusion is valid, clinicians concerned with the origins of human gerontological medical problems would be well advised to take a pediatric approach. S. Hejda, in a survey of the dietary history of a sizable number of very old people in Czechoslovakia, found that all of them came from large, poor families and, by today's standards, had a very modest or even insufficient diet during their childhood.

The seven factors acting in combination accounted for approximately 50% of the variance in length of life. The average absolute error between the predicted and actual number of days lived was 12%. The larger part of the error was due to an overestimation in the length of life of the shortest-lived rats and to an underestimation in the length of life of the longest-lived rats. Even so, a rat's longevity could be predicted to be low, intermediate, or high.

A new series of analyses was carried out to determine whether the predictive capability of the model could be upgraded by eliminating several of the shortest-lived animals. This proved to be so; the average absolute error in estimating life-span decreased to 10%. This observation suggested that, as the size of the population decreased because of attrition, a stepwise redefining of the best combination of variables for the survivors would increase the accuracy of the predictions. The most precise prediction was found for those animals which survived longer than 700 days; the average absolute error was less than 4%. When each of the conditional regression equations was applied in a sequential manner at the appropriate age period, the weighted overall absolute error in the estimated life-span from the actual was only 8%.

In the course of this sequential series of analyses, the level of contribution of the dietary variables to the model progressively diminished, so that ultimately only the weight-related variables supplied significant information. Again, only weight data prior to 150 days were important. The increasing emphasis on body weight and the decreasing and eventual elimination of the dietary variables indicate that the type of information needed to explain why one rat has a short life-span is not necessarily the same as that needed to explain why another rat has a long life-span. Thus, animals that early in life choose and efficiently utilize a low-protein diet and complete the greater part of their growth before 100 days of age will probably be short-lived. By contrast, young healthy rats that grow slowly, particularly between 50 and 150 days, will likely be long-lived.

The roles of fat, other food substances, and fiber under free-choice conditions were not investigated by us, and we therefore have no knowledge of their relative importance or whether these potential life-span-modifying dietary factors, if included as variables, would further improve the ability to predict life-span. Nor can we state whether different sources of carbohydrate or protein would, through their effect on appetite and growth, alter the relative importance of the variables used in the present study.

Perhaps the most basic question that can now be posed is whether the variables act as determinants or whether they represent signs of underlying genetically determined processes whose effects are recognized only late in life. Data supporting either interpretation come from short-term self-selection studies on the changes in preference accompanying modification in metabolism and from the studies on the dietary behavior of different strains of animals.

Along practical lines, we can now ask whether a regimen designed to conform with the practices of the long-lived rats will be effective in increasing the longevity of short-lived rats. At the risk of being simplistic, the effect of such intervention may depend to a large extent on the interactions of the stage of life when the imposed regimen is begun and of such genetic variables as growth potential, specificity in dietary requirements, and efficiency of food utilization. These factors, to some degree, may be responsible for the variation in life-span of animals maintained under identical environmental conditions fed a single fixed diet in the conventional manner.

The pathological processes that contribute to the death of an animal were not considered in the preceding analyses. There were, however, obvious differences among the animals in the type, age of occurrence, rate of progression, and site of the occurrence of life-terminating diseases. In fact, many of the animals had several diseases at time of death. It must be that the mathematical model that links the dietary and growth behavior of the individual early in life with its date of death incorporates the effects of these diseases. While the present analyses and conclusions centered around longevity, it seems reasonable to propose that the mathematical model is a composite of a number of models, each identifying the individually specific early-life conditions which govern the susceptibility and progression of the several age-related diseases, which in turn determine the duration of life. If this interpretation is valid, then dietary intervention tailored to the individual animal, as opposed to the population, offers a means of regulating the risk of specific affections.

It is unlikely that the dietary and growth behavior patterns associated with long life are confined solely to laboratory rodents. Although promising, the available information is far too incomplete to suggest that the data from animal models serve as guidelines for man. The impact and interactions of diverse environmental experiences and activities on appetite and on the course of aging have not been investigated. However, the means are now available by which to assess their influence and importance at different stages of life.

ACKNOWLEDGMENTS. I wish to acknowledge the contributions of my colleagues, Gerrit Bras of the Faculty of Medicine, University of Utrecht, The Netherlands, and Edward Lustbader of this laboratory. I also thank M. Cahalan and L. Sweeny for their help in collecting and organizing the data and M. Schleifer for preparing and executing the computer programs. I am grateful to Amstar Corp., Best Foods, a division of CPC International, Hoffman–LaRoche, Inc., Lederle Laboratories, Merck Institute for Therapeutic Research, and Mead Johnson Co. for donating some of the ingredients used in the diets. This work was supported in part by U.S. Public Health Service Grants CA-16442, CA-06927, and RR-05539 from the National Institutes of Health, Grant DT-52 from the American Cancer Society, a grant from Stichting Koningin Wilhelmina Fonds, and an appropriation from the Commonwealth of Pennsylvania.

BIBLIOGRAPHY

Dublin, L. I., A. J. Lotka, and M. Spiegelman. 1949. *Length of life,* rev. ed. The Ronald Press Company, New York.

Furnival, G. M., and R. W. Wilson. 1974. Regression by leaps and bounds. *Technometrics* 16: 499–511.

Gruman, G. J. 1966. A history of ideas about the prolongation of life. The evolution of prolongevity hypothesis to 1800. *Transactions of the American Philosophical Society,* Philadelphia, Vol. 56, Part 9.

Lat, J. 1967. Self-selection of dietary components. Pages 367–386 in C. F. Code, ed. *Handbook of physiology* Vol. 1, Sec. 6, Alimentary Canal. American Physiological Society, Bethesda, Md.

McCay, C. M. 1952. Chemical aspect of aging and the effect of diet upon aging. Pages 139–202 in A. I. Lansing, ed. *Cowdry's problems of aging,* 3rd ed. The Williams & Wilkins Company, Baltimore.

Overmann, S. R. 1976. Dietary self-selection by animals. *Psychological Bulletin* 2: 218–235.

Richter, C. P., L. E. Holt, Jr., and B. Barelare, Jr. 1976. Nutritional requirements for normal growth and reproduction in rats studied by the self-selection method. Pages 166–178 in E. M. Blass, ed. *The psychobiology of Curt Richter.* York Press, Baltimore.

Ross, M. H., E. Lustbader, and G. Bras. 1976. Dietary practices and growth responses as predictors of longevity. *Nature* 262: 548–553.

Silberberg, M., and R. Silberberg. 1955. Diet and life span. *Physiological Review* 35: 347–362.

Tannenbaum, A. 1953. Nutrition in relation to cancer. Pages 451–501 in S. P. Greenstein and A. Fadden, ed. *Advances in cancer research,* Vol. 1. Academic Press, Inc., New York.

Walker, A. R. P. 1968. Can expectation of life in Western populations be increased by change in diet and manner of life? *South African Medical Journal* 42: 944–950; 43: 768–775.

White, F. P. 1961. The relationship between underfeeding and tumor formation, transplantation and growth in rats and mice. *Cancer Research* 21: 281–290.

11

Consequences of Alcohol and Other Drug Use in the Aged

MICHAEL J. ECKARDT

Older people use more drugs than any other age group. In 1967, for example, approximately 19 million people over the age of 65 in the United States had an estimated 198 million prescriptions filled. An additional 26.9 million prescriptions were obtained from hospitals or mail-order pharmacies. In 1971, it was estimated that about 50% of older Americans drank alcohol, and it has been well documented that the elderly self-administer excessive quantities of nonprescription drugs. These figures emphasize the fact that as we grow older we use large quantities of many different types of chemical compounds. In the following pages, a number of facets that are associated with this enhanced drug use in the aged will be discussed. Specifically, I shall examine possible reasons underlying the greater reliance on, the consequences of, and potential misuse of drugs in the aged. Additional emphasis will be placed on the behavioral and physical effects of alcohol.

A drug can be defined as any chemical compound that is administered so as to achieve either a therapeutic or a self-desired effect. It is also important to point out that each drug has multiple consequences, some of which are desirable and some of which may be harmful. How often have we purchased a nonprescription drug and actually read all that small print on the box? However, if one reads that material carefully, it becomes quickly apparent that even such commonly used drugs as aspirin have multiple effects. Generally, drug effects are categorized as being either therapeutic and beneficial or adverse and harmful. Returning to our aspirin example, the relief of a headache or reduction in a fever would constitute a beneficial effect, and any accompanying

---

MICHAEL J. ECKARDT, Ph.D. • Clinical and Biobehavioral Research Branch, Division of Intramural Research, National Institute on Alcohol Abuse and Alcoholism, Rockville, Maryland 20857

stomach upset or irritation would be an adverse effect. There is a great deal of individual variability in physical and mental responses to a specific drug. Thus, for any particular individual, the pragmatic choice between two equally effective drugs would depend on which had the least side effects for that person. Obviously, individuals are seldom aware of the totality of consequences resulting from the taking of a drug, but rather are concerned only with getting better. However, it will become apparent later that potentially adverse side effects become more likely in the physiologically compromised elderly.

Why do older people use more drugs? It has been estimated that the per capita expenditure for prescription drugs for people over 65 years of age is four times more than for people under 65. Although the reasons for this increased reliance on drugs are undoubtedly complex, it may be conceptually useful to divide them into physiological and psychological categories. Some of the physiological antecedents of increased chemical use derive from the more frequent medical intervention associated with an increased incidence of physical illness. As we grow older, our body's capacities to remove toxic substances, effectively obtain energy as well as manage energy reserves, and maintain adequate immune defense systems decrease. In addition, the functional capabilities of many organ systems become reduced from years of constant work. For these reasons, as well as a number of less obvious ones, our bodies become more susceptible to illness and other debilitating conditions, which often have life-threatening consequences. Given the above, it is not particularly surprising that approximately 80% of people over 65 years of age have one or more chronic diseases or conditions, in contrast to 40% of those under 65. Physicians have relied heavily on chronic drug medication to alleviate or control many of the symptoms associated with physical illness, particularly in such debilitating diseases as arthritis and rheumatism, heart disease, and blood-pressure abnormalities.

There is another group of related, but conceptually different, ailments which are associated with the psychological dilemmas and stresses which accompany forced retirement, loss of loved ones, economic uncertainty, isolation, decreased feelings of self-worth, poor physical health, and so on. The manifestations of these pressures are often reflected in the aged as depression, difficulty in sleeping, anxiety, and other related emotional responses. In attempting to deal with these very real pressures, the aged, as well as their families and physicians, have often come to rely on psychoactive medication as a means of coping or escaping. An added concern is that continued heavy use of some of these chemical compounds can result in psychological or physical addiction.

Another strategy employed by the elderly in an attempt to alleviate the symptoms which accompany impaired physical health and increased mental stress is to self-medicate with nonprescription drugs. The television advertisements directed toward the older sufferer of arthritis and rheumatism and those with "occasional irregularity" attest to the popularity of such medication. The utilization of the media by drug companies to produce an increased awareness of body ailments that are presumably ameliorated with nonprescription drugs emphasizes the vulnerability of the elderly. Even psychoactive medication is available in the form of ethyl alcohol and some of the antihistamine compounds present in "sleep-encouraging" drugs.

Thus, the elderly come to rely on physicians and medication to relieve the physical and psychological debilitations that often accompany old age. In addition, the process of seeing a physician, taking medications on a regular basis, and being overly concerned with one's bodily functions tends to become a realistic focal point for these disenfranchised and vulnerable people.

CONSEQUENCES OF PRESCRIPTION DRUG USE

The most common consequence resulting from the use of a physician-prescribed drug is a therapeutic or anticipated therapeutic one. If this were not generally the case, then physicians would not continue to prescribe medications, and we would not continue to buy and use them. In fact, our reliance on pharmacological medicine has contributed to a steady increase in the number of available drugs. The number of drugs has also increased as knowledge in medicine and pharmacology has expanded, and as the goals of enhancing the beneficial consequences, while reducing harmful side effects, became more feasible.

Drugs are expensive, particularly for people living on fixed incomes. It has been reported that 20% of the total costs for health services are spent on medication. This percentage evokes great concern when coupled with the observation that people over 65 spend four times more for prescription drugs than those under 65. This is an important point, considering older people's limited financial resources and inordinate health needs.

Another consequence of increased drug use in the aged is the greater potential for adverse side effects. Two major factors are involved in this enhanced likelihood of drug-related side effects. The first of these is related to changes in our physiological systems with age, including both normal and illness-associated alterations in capacity and function. Age-dependent reductions occur both in the reserve capacity of different organ systems and in the adaptive responsiveness to body-chemistry alteration. The extent of these reductions varies in each individual and thereby serves to increase the variability in individual responsiveness to a specific drug. In addition, the added complications associated with an active disease process produce greater variability in the responses of the elderly to medication. The problems of drug selection and dosages are compounded even further when it is noted that the standardization of dosage and the ratio of therapeutic to harmful effects for new drugs are determined on the responses of young and middle-aged adults, not the aged. Thus, it is not surprising that the elderly have a greater proportion of adverse or paradoxical drug effects than any other age group.

A simplistic example of the above involves an individual who developed high blood pressure in middle age. Response to prescription medication at a standard dose was good, and the high blood pressure remained controlled until suddenly, after 25 years of taking the drug, the patient complained of always being sleepy and experiencing nausea after taking the medication. These are two of the most common side effects reported by patients when the dosage is too high. Therefore, the physician reduced the dosage, which served to alleviate the sedation and nausea while still controlling the high blood pressure. This example emphasizes two very important principles. First, adverse side effects in the aged are often manifest exaggerations of known pharmacological actions of the drug in question. Second, normal physiological changes which accompany aging, such as a less efficient liver function, resulting in decreased drug metabolism, often can produce an accumulation of the drug, resulting in much larger dosage effects than when the person was younger. Thus, responsible physicians will routinely prescribe lower doses for older people than the recommended standard dose.

The second major factor which has been correlated with an increased incidence of adverse drug effects in the aged is the widespread use of multiple-drug therapy. Physical illness in older people is more likely to involve a larger number of organ systems than in younger people. Thus, the use of more than one drug is often indicated. For example, hospitalized elderly patients take approximately 10 drugs per day, whereas nursing home residents average around 5 drugs per day. Also contributing to the greater incidence of adverse side effects with polypharmacy are physician errors, such as overprescribing and incorrect medications, and errors in administering the medication, including wrong drug, dosage, and/or time period between doses. Taking 16 to 20 drugs per day can result in up to 40% adverse effects, in contrast to 7% with 6 to 10 drugs.

An example of a polydrug-associated side effect occurred in a 64-year-old woman who had been receiving medication for chronic high blood pressure for some 15 years without incident. The woman suddenly developed severe blood clots in her left leg and was immediately placed on an oral anticoagulant. The physician knew that the blood-pressure medication potentiated the anticoagulant effect and so reduced the dosage accordingly. Unfortunately, the woman also suffered from rheumatoid arthritis, for which she treated herself with large doses of aspirin. Aspirin also potentiates the effect of the anticoagulant, and the woman returned the next day and was found to be hemorrhaging internally. Clearly, nonprescription drugs can also contribute to serious adverse side effects.

In concluding this section, it is important to reemphasize that even though there are more risks and financial hardships involved in treating the elderly with drugs, our reliance on pharmacological medicine has steadily increased. It is unfortunate that little effort has been spent on the development of alternative techniques or methods to medication.

ALCOHOL

Between 1964 and 1971, several relatively large-scale surveys were conducted to describe the drinking behavior of the American people. Much of what we know about the incidence of the use of ethyl alcohol and its abuse in the elderly is derived from these surveys. In general, it appears that drinking declines with age. Seventy-eight percent of people between the ages of 30 and 39 consume alcohol, in contrast to only 53% in people 60 years old and older. Within this population of alcohol users is a subgroup of people who consume relatively large quantities of alcoholic beverages. Examination of these statistics reveal that the incidence of heavy drinkers also decreases in advancing years from approximately 20% in ages 50 to 64 to 10% in the group over 64.

The same general trend holds for both males and females, with the exception that only one-fourth to one-fifth as many women as men are heavy drinkers. In the 64+ category, the percentage of women is only around 2%. Thus, the overall trend is one of declining popularity of alcohol with age, with heavy drinking also displaying a decrease. However, alcohol is the most commonly used nonprescription drug, with the possible exception of aspirin.

A variety of reasons have been proposed to explain the reduction in alcohol use and misuse in the aged. One explanation is based on the observation that long-term alcohol abuse is correlated with decreased longevity, which would thereby serve to eliminate from consideration an entire group of alcohol users. Another explanation proposes that people give up drinking with increasing age for reasons related to physical and mental health.

Even though the overall percentages decrease with age, problem drinking and alcoholism is still found in over *1 million* elderly people in the United States today. Some geographical areas, such as the retirement communities in Miami, Arizona, and southern California, have relatively high concentrations of elderly persons, and older problem drinkers may be overrepresented in our hospitals. There has been some discussion concerning the extent of problem drinking in the elderly in the future, with a consensus that it will get worse. This projected rise is based on the notion that, as more and more people consume alcohol, the likelihood of problem drinking increases. There is no question that the total consumption of alcohol is rising rapidly, and the recent alarming increase in the use of alcohol among our youth is a matter of concern for the years ahead. In addition, it has been suggested that some of today's elderly were raised in the abstinence tradition, which may be a reason for a decreased incidence of drinking. Even if the percentage of aged problem drinkers is somehow maintained at its present rate, the absolute numbers will be larger because the elderly population is expected to continue to grow in size in future years.

What are the consequences of alcohol consumption in old age? The answer to this question is complex, in part because it obviously depends on how much a person drinks. An occasional drink is not usually considered detrimental to older people in good health, and in fact, in some instances has been associated with beneficial effects. A nursing-home situation is one environmental setting in which the occasional social imbibing of 1 to 2 drinks has been associated with such beneficial consequences as increased morale and better sleeping habits. It has also been reported that in poor urban areas, inhabited primarily by minority groups, the most physically and socially active aged were also found to be those who drank alcoholic beverages. In addition, most of us are familiar with the social lubricant effects of alcohol within comfortable group settings, and it appears that alcohol functions the same way for the elderly.

However, as soon as an older person begins to consume even moderate amounts of alcohol on a somewhat regular basis, the consequences become more harmful than beneficial. For instance, moderate alcohol consumption has been repeatedly shown to affect adversely or complicate many disease processes. Also, the therapeutic properties of other drugs have been shown to be altered by even moderate alcohol consumption. The therapeutic effects of many different classes of drugs are altered by alcohol: pain relievers, anticoagulants, antibiotics, anticonvulsants, antihistamines, antihypertensives, diuretics, hypoglycemics, major tranquilizers, antidepressants, sedative-hypnotics, and even stimulants. An example of one of the most widely known adverse interactions occurs when both alcohol and a normally tolerated dose of a barbiturate or tranquilizer are taken within several hours of one another. The combination of these two types of drugs results in a potentiation of each of their independent effects and has, on occasions, resulted in death. Thus, it should be readily apparent that alcohol can be a dangerous drug when taken alone or with other chemical compounds. The inescapable fact is that ethyl alcohol is a drug and should be considered as one at all times. Lay people are not the only ones who forget this. It has been documented that physicians do not always inform their older patients — even known heavy drinkers — that alcohol will interfere with the beneficial effects of other drugs they are taking.

The drinking patterns observed in elderly alcohol-abusers are different than in other age groups. The aged problem drinker is more likely to drink alone, drink at home, and be a binge drinker. These behavioral characteristics have been interpreted as maladaptive attempts to cope with and/or escape from the role of growing old in America.

Older people who are either problem drinkers or alcoholics can be divided into two groups, based on the age at which they start to consume excessive amounts of alcohol. This point is of interest because it has been suggested that people who begin to drink alcohol excessively only in their later years may be more responsive to treatment. This particular group of older drinkers appears to possess fewer deep-seated psychological problems than younger alcohol-abusers. Furthermore, the observation that their drinking appears to be related to commonly shared factors associated with aging increases the probability of successful treatment if we can help the aged to cope with their socioeconomic plight, feelings of worthlessness, and disenfranchisement. The second and smaller group of aged problem drinkers is composed of individuals who have consumed alcohol in excess for a number of years. Treating them is less likely to be successful because of their severely reduced socioeconomic capability, poor physical and mental health, and resistance to developing new life-styles or to break old habits.

Appropriate treatment for older alcoholics has been full of problems. It is very difficult to spot these people because of their relatively few "brushes" with the law and the reluctance on the part of relatives and friends to seek treatment for them. This sounds deplorable, but consider your reluctance in taking your grandmother to the local alcohol treatment center that helps those "drunks." Once an elderly problem drinker is detected, the next obstacle is finding an appropriate treatment facility. The social agencies designed to assist the elderly are poorly equipped to deal with alcohol problems, and many alcohol treatment centers are reluctant to work with the elderly. However, there are some indications that the treatment of alcohol problems in the elderly is improving.

Alcohol has also been implicated as a compound which can deleteriously influence the normal physiological changes which accompany aging. It was noted by Cyril Courville, formerly associated with the Los Angeles County Hospital, in 1955 that the brains of relatively young alcoholics were very similar in postmortem appearance to those of older nonalcoholics. On the basis of these observations, he suggested that the chronic use of alcohol may "speed up" the normal aging process which takes place in the brain. Also in the 1950s, psychologists began to develop tests for measuring cognitive functions that rely on normal functioning of the nervous system. Shortly after the development of these tests, it was reported that certain cognitive functions (e.g., visual–spatial abstraction) decreased in a predictable manner with increasing age. More recently, it has been observed that the aged perform at the same level on some of these tests as do middle-aged chronic alcoholics. At first glance, these findings add credence to Courville's suggestion that the chronic abuse of ethanol can produce premature aging. However, consideration of the facts that abstinence often results in relative improvement of cognitive function, and the physical concomitants of aging are an irreversible process, leads one to conclude that premature aging is not the most accurate description. Rather, the real relationship is the similarity between the behavioral concomitants of heavy alcohol use and of advancing age.

Another variable of interest is the age of onset of excessive drinking. Differences in performance on intellectual and psychological tests between young and old alcoholics who were equated for duration and amount of drinking have been reported. Older alcoholics performed better on tests which required accumulated information and experience, whereas young alcoholics were superior on tests of immediate adaptive ability. It has also been noted that older alcoholics were inferior to age-matched controls on visual–spatial abstracting tests, whereas no such findings were observed in young alcoholics. On the basis of the above, it has been suggested that an older brain is more susceptible to alcohol than a younger one. This makes sense intuitively because: (1) the same dose of alcohol adjusted for surface-area differences results in higher peak blood alcohol levels in the aged, (2) the older brain has fewer nerve cells, and (3) the efficiency of neural cell function in older people is probably less than in young people. It is not surprising, then, that the aged human brain is less able to tolerate a chemical stress, such as excessive alcohol consumption, than a younger one.

Ralph Ryback of the National Institute on Alcohol Abuse and Alcoholism has suggested that alcohol's effects on memory can be described as a continuum from cocktail-party deficits to the almost total memory loss for recent events accompanying the alcohol-induced Wernicke–Korsakoff syndrome. This syndrome is primarily found in chronic alcoholics, and afflicted individuals are virtually unable to store new memories. In addition, Korsakoff patients exhibit disturbances in orientation and excessive confabulation. Ryback's suggestion of an alcohol-related continuum of memory deficits is a provocative one, and there are several interesting examples which involve age. It has been shown that sober older alcoholics have memory impairments as severe as Korsakoff patients when tested the day after admission, but improve to the levels of younger alcoholics within a month of abstinence. It has also been reported that ethanol intoxication of nonalcoholic subjects produced greater memory deficits in aged subjects as well as prolonging their reaction time. Not only were the immediate behavioral effects of alcohol greater for the elderly men, but their recovery of normal function took longer than for younger men.

Similarly, alcohol's effect on cognitive functioning may also be represented as a continuum. It has been well documented that impaired cognitive performance accompanies chronic alcoholism. Recently, Elizabeth Parker and Ernest Noble of the National Insitute on Alcohol Abuse and Alcoholism noted that the more alcohol reportedly consumed per occasion by social drinkers, the poorer their performance on cognitive tests. Their data are consistent with the suggestion that the older social drinker's thinking processes may be more susceptible to the effects of alcohol than those of younger drinkers.

MISUSE OF OTHER DRUGS

Although the incidence of illicit drug abuse seems to decline with age, it does not disappear, as can be exemplified by the increasing number of older people in methadone maintenance programs. In fact, some people have suggested that illicit drug abuse will represent a serious problem for the elderly in the years ahead. It is hard to imagine older drug addicts, because they are hidden from the public eye and are generally considered to have high mortality rates. However, several reports have documented that many individuals between 45 and 75 still use opiates and, furthermore, have changed their life-styles so as to be less easily recognized. In addition, there are economically motivated changes in the amount and types of drugs used.

There is considerable misuse by the elderly of legal psychoactive (mood-modifying) drugs such as tranquilizers, sedatives, and other depressants. In fact, some people have gone so far as to label this class of drugs as the most often misused or abused. People that use these drugs to excess run the risk of developing psychological and, in some cases, physical dependency. Furthermore, their potential danger was highlighted recently by the report that 80% of drug overdoses in the elderly involved either sedatives or tranquilizers. Given their addictive potential, as well as sometimes harmful side effects, one might wonder why these drugs are so often prescribed for the elderly. The reason is a simple and somewhat persuasive one; they are an easy, but perhaps too transient, solution to many of the pressing problems which accompany aging: insomnia, anxiety, irritability, and depression. Unfortunately, the therapeutic expectations of physicians, patients, and patients' families are often unrealistic.

It has been observed that over one-third of all patients receive either diazepam (Valium) or chlordiazepoxide (Librium) for anxiety while in the hospital. In fact, these two drugs are the most commonly prescribed medications in the United States. It has been demonstrated that intravenous administration of Valium, or the related compound Lorazepam, although done rarely, can produce amnesia for events that occur after the drug is injected. One of the major concerns of older people is their purportedly reduced cognitive and memory capacities. Most of them are able to recall events from their remote past, but recent events are difficult to remember. Thus, the widespread use of these minor tranquilizers is of great interest, especially in light of the recent report by Louis Gottschalk of the University of California at Irvine that disorganization in cognitive and intellectual processes, as reflected in speech, accompanies orally administered therapeutic doses. It goes without saying that the utilization of minor tranquilizers to reduce anxiety, hostility, and to alleviate human distress is of value. But the possibility of compromising an older individual's mental competence is a consequence of real concern.

Another group of drugs that has both beneficial and adverse effects for the elderly are the antipsychotic drugs, with chlorpromazine (Thorazine) being the most commonly prescribed. These drugs have been effectively used to diminish hyperactivity, decrease the expression of delusions, and depress psychotic symptoms without producing undue sedation. Unfortunately, the elderly, especially women, may be at a greater risk of developing severe and specific involuntary motor movements, called dyskinesia, with long-term therapy. These symptoms are persistent and may be irreversible. Parkinsonism-like characteristics have also been noted with these drugs, but such symptoms usually disappear when the dosage is decreased. The minor tranquilizer and the antipsychotic examples above should serve to reemphasize the fact that multiple consequences are associated with any drug.

Errors of medication in the elderly occur much more frequently than one might imagine. In one study of a sample of older people, it was noted that approximately 60% of the sample made errors in self-medication, including errors of omission, self-prescribing, incorrect dosage, and incorrect timing between doses. Such errors have also been documented in institutional settings, with some investigators reporting up to 20 to 40 percent of the drugs administered in error. The prescribing physician contributes to errors of self-medication. Although physicians have been accused of overprescribing, as well as prescribing the wrong drug, perhaps their most serious error has been not explaining to the elderly in sufficient detail how much and how often to take prescribed medication. In addition, the aged should be given enough information so that they will be able to recognize any early signs of an adverse drug reaction. Ideally, this information should also be supplied to the patient in written form in *large* print. This is essential because the elderly have visual deficits as well as memory lapses.

The last topic to be discussed under drug misuse concerns the attempt by the aged to use chemical compounds on a regular basis that are claimed to prolong life-span. This attempt to find "the fountain of youth" in the form of a pill is becoming increasingly widespread. Some of the most commonly used drugs of this sort are antioxidants, hydrocortisone, and procaine (Gerovital H3). It is important to state emphatically that no chemical compound has yet been shown to increase human life-span within the rigors of a scientifically sound study. On the other hand, there are a number of drugs that produce mood elevation, reduce depression, and ease the physical pain and mental anguish associated with physical and mental illness. Such medications are justified since they result in a more optimistic outlook on life, but this does not mean they increase life-span. The distinction to be made is between a drug used for therapeutic purposes versus one used to recapture one's "lost youth." The real crime is not necessarily in the taking of so-called life-extending compounds, but rather is in the false hopes and financial expense in their use.

CONCLUSIONS

Although a number of different issues have been discussed, all too briefly, in the preceding pages, it seems appropriate to emphasize several of the more important issues by indicating areas in which future effort might be profitably invested.

1. Given the increased reliance by the elderly on prescription medications, it would seem desirable to require all new investigational drugs to be tested clinically for therapeutic and adverse side effects on older people or at least on aged subhuman animals. We require this procedure for the young and the pregnant; why not for the elderly?

2. Greater emphasis should be placed on the training of primary care physicians so that they will be able to recognize the physical and mental problems which accompany aging, and furthermore, to be aware of the benefits and dangers of polypharmacy for these people.

3. Pharmaceutical companies can improve drug packaging so that it will be easier for the elderly to understand how much of a drug they should take and how often. Written information should be in large print.

4. There is very meager epidemiological information available on drug use in the elderly. The extent of drug misuse and abuse is, therefore, one largely of conjecture from a few isolated investigations. Since the elderly use inordinate amounts of drugs, it would appear imperative that accurate and detailed epidemiologic information be collected and analyzed.

5. Social help agencies utilized by the aged must be staffed with people capable of recognizing and assisting in instances of drug abuse and alcoholism.

6. The elderly must be made aware that alcohol is a drug and that it, as well as other drugs, interacts with medication and can produce life-threatening situations.

7. Greater emphasis should be placed on the development of alternative methods to medication in treating the mental concomitants of growing old in the United States. Some physical ailments may also be amenable to treatments other than medication.

8. The newly formed National Institute on Aging should be encouraged to continue to develop, as well as expand, its efforts in determining the consequences of alcohol and other drug use in the aged. In addition, increased interactions among the National Institutes on Aging, Alcohol Abuse and Alcoholism, and Drug Abuse would be desirable.

The use of prescription medication has helped to enable us all to live longer and more complete lives. However, in a nation where approximately 30% of its elderly live below poverty levels, it is discouraging to note that these are the people who must spend the most for medication, as well as bear the brunt of adverse side reactions. In my family, neither my 4-year-old-son, my wife, nor I take any regular medication. On the other hand, my mother-in-law, who is 67, takes six different prescription drugs on a somewhat regular basis. Is the situation the same in your family?

ACKNOWLEDGMENTS. Elizabeth Parker and Ralph Ryback of the National Institute on Alcohol Abuse and Alcoholism, and my wife, Johneen, made valuable suggestions for the content of this chapter.

BIBLIOGRAPHY

General

Bozzetti, L. P., and J. P. MacMurray. 1977. Drug misuse among the elderly: a hidden menace. *Psychiatric Annals* 7: 95–107.

Davis, R. H., ed. 1973. *Drugs and the elderly.* University of Southern California Press, Los Angeles.

Freeman, J. T., ed. 1965. *Clinical features of the older patient.* Charles C Thomas, Publisher, Springfield, Ill.

1968 HEW Task Force on Prescription Drugs. 1971. *The drug users and the drug prescribers.* Government Printing Office, Washington, D.C.

Technical

Cahalan, D. 1970. *Problem drinkers.* Jossey-Bass, Inc., Publishers, San Francisco.

Cermak, L. S., and R. S. Ryback. 1976. Recovery of verbal short-term memory in alcoholics. *Journal of Studies on Alcohol* 37: 46–52.

Cherkin, A., and M. Sviland, eds. 1975. Procaine and related gero-pharmacologic agents — the current state of the art. *Proceedings of the First Workshop of the Veterans Administration Geriatric Research, Education, and Clinical Centers.*

Coleman, J. H. and W. E. Evans. Drug interactions with alcohol. *Alcohol Health and Research World* Winter: 16–19.

Courville, C. B. 1955. *Effects of alcohol on the nervous system of man.* San Lucas, Los Angeles.

Craik, F. I. M. 1977. Similarities between the effects of aging and alcoholic intoxication on memory performance, construed within a "levels of processing" framework. In I. M. Birnbaum and E. S. Parker, eds. *Alcohol and human memory.* Lawrence Erlbaum Associates, Hillsdale, N.J.

Duckworth, G. L., and A. Rosenblatt. 1976. Helping the elderly alcoholic. *Social Casework* 57: 296–301.

Gottschalk, L. A. 1977. Effects of certain benzodiazepine derivatives on disorganization of thought as manifested in speech. *Current Therapeutic Research* 21: 192–206.

Jones, B., and O. A. Parsons. 1971. Impaired abstracting ability in chronic alcoholics. *Archives of General Psychiatry* 24: 71–75.

Mishara, B. L., R. Kastenbaum, F. Baker, and R. D. Patterson. 1975. Alcohol effects in old age: an experimental investigation. *Social Science and Medicine* 9: 535–547.

National Clearinghouse for Alcohol Information. 1975. Older problem drinkers. *Alcohol Health and Research World* Spring: 12–17.

Parker, E. S., and E. P. Noble. 1977. Alcohol consumption and cognitive functioning in social drinkers. *Journal of Studies on Alcohol* 38: 1224–1232.

Parker, E. S., and E. P. Noble. 1977. Alcohol and the aging process. Paper presented to the NATO Conference on Experimental and Behavioral Approaches to Alcoholism, Bergen, Norway.

Pascarelli, E. F., and W. Fischer. 1974. Drug dependence in the elderly. *International Journal of Aging and Human Development* 5: 347–356.

Robertson, E. A., D. Arenberg, and R. E. Vestal. 1975. Age differences in memory performance following ethanol infusion. Paper presented to the 10th International Congress of Gerontology. Jerusalem, Israel.

Ryback, R. S. 1971. The continuum and specificity of the effects of alcohol on memory: a review. *Quarterly Journal of Studies on Alcohol* 32: 955–1016.

Schukit, M. A., and P. L. Miller. 1976. Alcoholism in elderly men: a survey of a general medical ward. *Annals of the New York Academy of Sciences* 273: 558–571.

12

Physical Changes of the Aging Brain

ROBERT D. TERRY

Of course, one must either grow old or die young. The implicit choice would seem more obvious than it really is or might be in our society. Realistically viewed today, the late decades of human life are hazardous in their course and have but a single outcome. One can perhaps accept with some equanimity most aspects of aging, such as decreased agility or muscle strength, or lessened acuity of hearing. It is not possible, however, for most of us to regard impassively our own prospects in regard to significant loss of mental ability.

That proportion of our population which reaches the age of 70 or 80 in reasonably good physical health is increasing. The numbers manifesting the changes of cerebral aging are also increasing, and their deficits present one of the great, but still largely unrecognized, health problems in all the advanced countries of the world.

There can be little doubt that at least some structural and functional changes involving the nervous system are brought about by aging in essentially all human beings. Investigators have found that there is a measurable decline in some mental abilities after an age as early as 25 years. The great majority of the population over 65, probably about 85%, continue, however, to function adequately in the community. About 11%, according to an English survey of people over 65, were found to display mild to moderate senile dementia, and about 4.5% of this age group are very seriously involved with this disorder. It is the 85% that we refer to as the "functional elderly."

While it is not the purpose of this chapter to deal in any considerable detail with the clinical observations in either group, a brief summary is surely in order. Very common to normal aging are various mild losses

ROBERT D. TERRY, M.D. • Department of Pathology, Albert Einstein College of Medicine, Bronx, New York 10461

in sensory acuity. Vision, for example, requires more illumination than in the young. Hearing is diminished, especially for tones in the higher-frequency range. Olfaction and taste are lessened. The elderly quite frequently seem to lack self-confidence, so that they require a stronger probability of success prior to accepting particular tasks. Slowing of reaction time is very common. While verbal intelligence tests are often well done by elderly subjects, the results of performance tests usually fall. Mild forgetfulness is a very frequent complaint of the elderly, and is often first noted as a difficulty with names.

Most such measurements have been made on age-matched cohort groups or cross sections of the population, and therefore these studies are subject to errors introduced by altered educational systems, changed nutritional habits, and modifications of society during the extent of human life. Longitudinal studies of a single group of subjects frequently provide data indicating less change than those displayed by such cross-sectional analyses. These longitudinal studies, however, are subject to other errors, the most prominent of which is the survivorship pattern of the group under study. Relevant to the subject of this chapter, it has been shown that a significant decline in mental function is associated with a significantly shortened life-span. Therefore, the survivors in a longitudinal study group will have deteriorated less in the later measurements, and the analysis might erroneously minimize age changes.

Senile dementia of the common Alzheimer type is, to some extent, an exaggeration of the relatively mild mental deficits noted above. It is characterized by increasing forgetfulness involving especially short-term memory, failing judgment, decreased abstract thinking, inability to perform simple calculations, and lessened language facility. Disorientation in time and place are common. Ultimately, the patient is wholly unable to deal with the environment and loses interest in it. The personal habits of a lifetime dissipate, hygiene deteriorates, and extensive nursing care is required. The genetic pattern of senile dementia is a complex one, but there can be little doubt that close relatives of proven cases are significantly more likely than random, age-matched people to develop the same disorder. This may reflect a dominant gene with incomplete penetrance, or a polygenic situation. Life expectancy of the demented patient is, mercifully, shortened by one-half to one-third of normal so that the disorder must be viewed as a malignant one, although death usually intervenes from secondary events such as pneumonia.

The real purpose of this chapter is to demonstrate some of the organic changes of the brain which are associated with, and may cause, these functional alterations in normal and abnormal aging. The question as to whether these changes or lesions in the brain tissue represent a disease or whether they are an exaggerated part of the normal aging process is only a semantic one. The point is that when the lesions are present in quantity, they may well cause abnormal brain function; and that if the process responsible for so altering the brain tissue can be retarded or reversed, we will have fewer people manifesting the lessened brain function of old age. Implicit here is the concept that brain tissue is in control of both mind and muscle, and that changed brain tissue — be it by chemical, structural, or immunologic forces — has an altered functional capacity.

ARTERIOSCLEROSIS AND BLOOD FLOW

It goes almost without saying that arteriosclerosis is common in the aged, and that it all too often affects the vessels of the brain. This, in no sense, is normal aging, and it is not, furthermore, the most frequent cause of senile dementia, despite the widely held notion that it is. Narrowing of the vascular canal by atherosclerosis and its subsequent closure by a superimposed blood clot is the usual cause of sudden destruction of brain tissue with consequent symptoms of stroke. Signs similar to the Alzheimer form of senile dementia, in addition to focal motor, sensory, and speech deficits, may result when enough tissue in relevant areas is destroyed because of inadequate blood supply. It is said that about 50 ml of brain tissue must be lost before signs of dementia become apparent. That 50 ml represents about 5% of the weight of the cerebral hemispheres, and an even smaller proportion of the whole brain.

Blood flow to the brain can currently be measured in a variety of ways. It has been amply shown that the amount of blood going to the brain is reduced in the aged, and is further lessened in the presence of senile dementia of the Alzheimer type. Whether this lowered blood supply is the cause of brain shrinkage has often been debated. The normal brain, however, has an effective regulator of this supply; the flow is reduced when metabolic needs are low, as in dementia. Most students of this physiologic phenomenon are now convinced that the cerebral tissue changes are primary in aging, and that the cerebral blood supply is reduced as a consequence of cerebral atrophy.

BRAIN SIZE

The human brain undergoes slowly accelerating shrinkage as it ages, ultimately losing 10 to 12% of its maximum weight in the course of a normal long life. The gross effect of this shrinkage is to narrow the convolutions and widen the sulci between them (Figure 1), thus enlarging the fluid-filled space between the brain and its coverings. A diagnostic test, called pneumoencephalography, involves the injection of air into this space, temporarily replacing the cerebrospinal fluid over the cerebral hemispheres so that x-rays of the head can reveal the pattern of convolutions and size of sulci for comparison with normal. The more modern computerized axial tomography shows this cerebral shrinkage without air injection and is, thus, preferable.

When the age-shrunken brain is examined at autopsy, the cortical mantle or gray matter of the cerebrum is seen to be a little thinner than normal, and the white matter, both in the convolutions and in the deeper centrum, is smaller. Both gray and white matter are thus involved in the loss but at somewhat different rates. Some English investigators have recently shown, by the use of a computerized image analyzer, that the gray matter diminishes faster than the underlying white from age 20 to age 50, changing from a gray/white ratio of 1.28 at 20 to a minimum of 1.13 at 50; thereafter, more white matter is lost than gray, so that the ratio rises to 1.55 by age 100. The deep gray masses, thalamus and basal ganglia, seem to contract less, but they have not really been measured. As a result, presumably, of the loss of tissue in gray and white matter, the cerebral cavities, especially the lateral ventri-

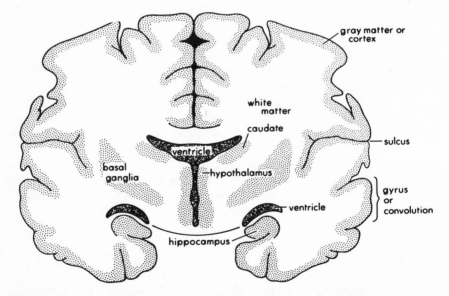

Figure 1. Simplified diagram of a cross section of the normal human brain cut in the coronal plane (parallel to the face), a little less than halfway back. In aging, the gyri are shrunken, and both gray and white matter are smaller than normal. Reduced to about 60%.

cles, enlarge to a modest degree and become blunted at their lateral angles. There is no obstruction to the flow of cerebrospinal fluid, and it maintains normal pressure. The cerebellum also shrinks a little in the aging process, again losing both gray and white matter, as does the spinal cord.

EXTRACELLULAR SPACE

Prior to about 1960, light microscopists had long believed that a very large proportion of the brain's volume was space between cells. Their fixatives and stains clearly showed an empty zone around each neuron and each capillary. The neuropil, that is, the area between the identifiable cell bodies, myelinated axons, and vessels, appeared like a sponge, with countless minute vacuoles traversed and divided by the filamentous processes of neurons and supporting cells or glia. Of course, no one thought these spaces were empty but rather that they were filled with fluid, perhaps the cerebrospinal fluid or with a more stable material, a mucopolysaccharide ground substance. Chemical compounds such as inulin or thiocyanate remain largely outside the cell membrane in the extracellular zone and thus can, by dilution techniques, be used quite simply to measure the volume of this compartment.

With electron microscopy in the late 1950s and early 1960s came much improved chemical fixatives, fixation by perfusion, and closer examination of thinner sections. It quickly became known that, with these methods, the extracellular space appeared to be much smaller than previously accepted, since the cells and their processes were much less shrunken and seemed to fill the interstices almost completely, their surfaces being only 10 to 20 nm (nanometers) apart. An unusual technique for preparing tissue for electron microscopy is by freeze substitution. Here the fresh tissue is very quickly frozen, and while solidified, the tissue water is replaced first with fixative (osmic acid) and then with dehydrating agents, after which the tissue is impregnated with plastic, polymerized, and sectioned. The advantage is thought to be that the technique prevents shifts of tissue water between compartments such as may take place during fixation at temperatures above freezing.

Estimates of the size of the cerebral extracellular space, by different methods, vary widely, from a low of 5 or 6% by ordinary electron microscopy, through 8 to 18% by chemical methods, to about 20% by freeze-substitution electron microscopy, to 30 to 70% by ordinary light microscopy. Today, there seems to be some degree of consensus around 12 to 15%.

Studies of extracellular space as a function of aging are very few, and they must necessarily be performed on experimental animals rather than on human beings. A pair of investigators have used freeze substitution to prepare rat brain for electron microscopy and have found that the cerebral cortex of the 3-month-old rat has 20.8% extracellular space, and that this declines to 9.6% in the 26-month-old animal. They have also shown that extracellular fluorescent tracers diffuse at a slower rate in the old living animal than in the young.

As with much experimental data, drawing inferences about function is easier than it is certain. The data may be somewhat in error in that freeze substitution is not universally accepted as providing a wholly reliable image of the living architecture. Molecular markers have not yet been utilized to confirm the shrinkage of the cerebral extracellular space in aging. Usually, a loss of cellular components as is expected in the aged brain causes an increase in extracellular space rather than a decrease. If, however, this reported decrement is confirmed, there would be important physiologic implications, since it is generally believed that significant movement of metabolic precursors and products takes place, probably by diffusion, through this pathway. Were it to be constricted in age, the resulting reduction in transport capacity might be harmful to cerebral tissue where metabolic needs are high.

NEURONAL LOSS

As indicated above, the human brain by old age has lost about 10 to 12% of its youthful weight. The major cause of this decrement is presumed to be a progressive loss of neurons. Not all parts of the central nervous system behave equally in this regard, and there are data to the effect that some species of aged animals suffer no such change at all. It has been shown that several functional groups of neurons (such a group is called a nucleus) in the human brainstem, including the vestibular nucleus (a sensory area for head posture), maintain a constant number of cells throughout life. The same durability of cells has been demonstrated in the inferior olive, which is another large group of cells in the hindbrain with a complex motor function in association with the cerebellum. The locus ceruleus is a small but important group of pigmented nerve cells in the floor of the upper fourth ventricle, and it, on the other hand, does indeed lose neurons from about 18,000 in youth to about 12,000 in old age. This cell group is very widely connected throughout the brain and may have a bearing on the stages of sleep as well as other functions.

The human neocortex, that is, the gray cortical mantle over the cerebral hemispheres, undergoes quite a remarkable loss in the years beyond 50. This process has been measured extensively on presumably normal brain specimens from autopsies, from which particular regions were embedded in paraffin, sliced, and stained. Neurons were recognized as seen in a light microscope equipped with an eyepiece scored with a grid. Despite the tedium of the task, the investigator counted cells in several specimens from each decade of life and reported that only about 50% of neurons remained in some areas by the ninth decade. He indicated that the greatest loss was among the small neurons of the second and fourth layers — the external and internal granule laminae of the typical six-layer cortex. Furthermore, he pointed out that the greatest loss was in the frontal and superior temporal regions. These are also the areas of greatest gross atrophy.

More recently, these data have been largely confirmed in other laboratories by the use of computerized, semiautomated television scanning devices. The scanner looks through a microscope at a slide similar to those used for hand counts. The automated device, however, can measure diameter or area of each cell in the field of view, and then move to the next very quickly. Since the apparatus does not differentiate between neurons and other brain cells, this classification is performed only according to the size of each cell. Actually, this is about as satisfactory as skilled visual recognition, since it is virtually impossible to distinguish the smallest neurons from glial cells on the basis of anilin stains and light microscopy. The field of view in the scanner can also be edited with a light pencil directed at the viewing screen, so that contiguous cells and artefacts can be properly taken into account. The apparatus thus speeds up and details the counting procedure so that one can deal with larger samples and more specimens.

The most striking information thus far gained from the image-analysis apparatus has not yet been published in detail, but has been identified in two laboratories. It seems probable that the specimens from patients with nonarteriosclerotic senile dementia of the Alzheimer type still have as many cells per unit area as do normal, age-matched controls. This is contrary to previous teaching.

Given the data on neuronal loss in normal aging, we can now return to the changing ratio of gray and white matter as a function of age. One might explain these findings by proposing that the predominant loss is among small neurons prior to age 50. These small cells do not project their axons into the white matter, so that during these early decades, the cortex shrinks more than the white matter, and the ratio goes down. Later, predominantly larger cells die off, and these cells have myelinated axons projecting through the white matter. When the cell body dies, so must its axon and myelin, and since these constitute a larger bulk than the cell body, the white matter shrinks more rapidly than the gray, and

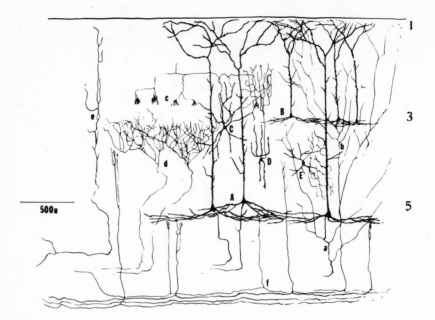

Figure 2. Cerebral cortex, as seen in an idealized Golgi preparation, showing how nerve cells are arranged and interconnected. The surface of the brain is at the horizontal line above 1, and the white matter lies below the lower horizontal line. Only a very few neurons are shown in each layer. The major horizontal bundles of dendrites are seen in layers 3 and 5, and are made up of dendrites from pyramidal cells in these layers. The principal apical dendrites are directed vertically toward the surface, while the axons are primarily directed toward the white matter. [From M. E. Scheibel and A. B. Scheibel, Structural changes in the aging brain, pages 11–37 in *Clinical, morphologic, and neurochemical aspects in the aging central nervous system* (H. Brody, D. Harman, and J. Mark Ordy, eds.), Vol. 1 of *Aging*, Raven Press, New York, 1975.]

the gray/white ratio rises. The older data on cortical cell numbers do not concern this hypothetical changing pattern of cell loss, but other analyses now under way with the scanning device might further confirm the proposed differential susceptibility of cortical neurons.

DENDRITES, SYNAPSES, AND TRANSMITTERS

The method Golgi described for metallic impregnation of neuronal cell bodies and the dendritic branching pattern or arborization (Figure 2) yields beautifully detailed images. It is, however, an extraordinarily capricious technique, successful now and failing tomorrow, even at best impregnating this cell but not that one. Perhaps because of its erratic behavior, the technique was largely neglected during several recent decades. During the past few years, however, the method has again shown its value and, perhaps, nowhere more than in the studies of aging.

Several investigative groups have recently shown that in the aging cerebral cortex the neuronal dendritic arbor shrinks (Figure 3), especially in its basilar region. The normal arbor is made up of numerous delicate, branching fibers which ramify from the cell body into the nearby tissue to receive synapses from other cells (Figures 2 and 3). As well as the loss of dendrites, there is also a further decline in the number of dendritic spines per unit length of remaining dendrite. The spines are very small projections from the dendritic shaft, and they receive the incoming signals from closely apposed axonal terminals through the axodendritic synapse. The particularly affected horizontal dendrites of layers 3 and 5 are thought to be involved primarily with receiving intracortical loops making circuits with nearby neurons. It has also been proposed that the bundles of horizontal dendrites are storage sites for central neural programs. Degeneration of this apparatus must have major effects on brain function.

Recent chemical analyses of brain tissue from autopsies on normal aged and senile patients have shown highly significant abnormalities in

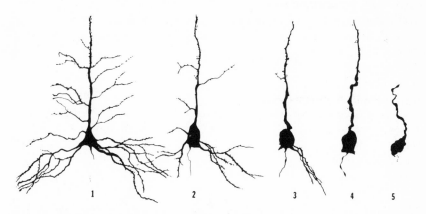

Figure 3. These are five neurons from the cerebral cortex impregnated by the Golgi technique and showing the serial changes said to be associated with aging. In 1, all the dendrites and even their spines are apparent. In 2, there has been a significant loss of all dendrites, especially of the horizontal ones at the base. In 5, the degeneration of the nerve cells is almost complete. [From M. E. Scheibel and A. B. Scheibel, Structural changes in the aging brain, pages 11–37 in Clinical, morphologic, and neurochemical aspects in the aging central nervous system (H. Brody, D. Harman, and J. Mark Ordy, eds.), Vol. 1 of Aging, Raven Press, New York, 1975.]

three classes of enzymes related to neurotransmitters. These transmitters are the agents which activate various sorts of synapses, and are therefore essential to function in the nervous system. Some investigators have found an important decrease in those enzymes involved with the cholinergic transmitters in the specimens from senile patients. Other chemists are more impressed with apparent age changes in enzymes related to catecholamine and gamma amino butyric acid transmitters. All such alterations may well be related to the loss of synapses, and thus the question remains as to what causes these synapses to degenerate.

Changes in both dendritic arbor and in concentration of spines have been noted in some animal species as well as in aged human beings. The rate at which the deterioration occurs, the frequency of involved neurons, and the cerebral areas particularly affected are unknown but well worth studying. The cause of the degeneration is also unknown, and this is the most important question, since its answer might lead to appropriate preventive or remedial therapy. It is possible that aging alters the neuronal cell body so that it synthesizes less of the metabolites necessary to maintain its complex dendritic arbor. Perhaps there is deterioration of the cell's transportation machinery, which is responsible for moving newly synthesized materials from cell body to the distant terminals. Alternatively, there might be an agent, such as an antibody or a toxin in the blood or cerebrospinal fluid, which might injure the cells or their arborizations.

LIPOFUSCIN

Lipofuscin is a yellow, insoluble, coarsely granular material found increasingly in the cytoplasm of most neurons as they age. Not all neurons are equally susceptible, and in fact, those of certain areas do not aggregate lipofuscin at all. The membrane-limited, bosselated granules are made up largely of complex oxidized lipids, high cross-linked and polymerized with protein and unsaturated peptides. They also contain iron and zinc in somewhat more than trace amounts, and significant quantities of hydrolytic enzymes, such as acid phosphatase. It is thought that lipofuscin is largely undegradable waste deriving from partly broken down membranes and other cell components. Because the lipofuscin is ultimately quite prominent in some neurons and occupies so much of the cytoplasmic volume, it can be assumed that it displaces significant quantities of cellular organelles. Indeed, a decreased amount of endoplasmic reticulum, which is the protein-synthesizing apparatus, has been measured in neurons parallel with their increased content of lipofuscin. Although this has led many to assume that lipofuscin is a major element in causing cellular malfunction and death, this has never been directly demonstrated. The inferior olive, a purse-shaped aggregate of neurons in the brain stem, was cited above as one of the groups in which cell number is unchanged by age, and yet these neurons contain more lipofuscin than any others in the brain, and they begin to collect it in early childhood. This, of course, is evidence that here, at least, lipofuscin in large amounts does not cause cells to die. Still possible, however, is a decrease of the dendritic arbor or some loss of function not yet examined. It is also possible that neurons elsewhere in the brain are more susceptible than the olive to the presence of excess lipofuscin. An alternative view is that the precursors of lipofuscin are toxic, and that successful cells can isolate them in the harmless form of lipofuscin. Cells that cannot do this die.

NEUROFIBRILLARY TANGLES AND NEURITIC PLAQUES

Shortly after the beginning of this century, it was found that the senile brain had two particular microscopic abnormalities, the neurofibrillary tangle and the senile or neuritic plaque, which most investigators relatively quickly accepted as being both diagnostic and at least partially causal of the mental deficits. Then, in the second phase, between about 1920 and 1950, it was demonstrated that these lesions were also present in specimens from normal old patients. This lack of specificity justifiably induced skepticism about the significance of these lesions, and those doubts were reinforced by the then current rise in nonorganic psychiatry, which held that personality, intelligence, and most other psychologic phenomena arose from a complex of environmental and personal history, rather than having much, if anything, to do with the brain as a structural and chemical organ. During recent years, it has become increasingly evident that the plaques and tangles, with the other structural and chemical lesions of aging brain tissue, could well account for the functional changes which are so readily found to some degree in the normal elderly and to a much greater degree in the senile. This recognition has again coincided with a major change in the outlook of psychiatry. Acceptance of the organic basis of psychologic functions was brought about in large part by the development of the many drugs that change mood and personality.

At any rate, plaques and tangles are the hallmarks of Alzheimer's disease, and this is the type of senile dementia found in more than 50% of cases. The neurofibrillary tangle is an abnormal mass of fibrillar material, visible, when stained, to the light microscopists as occupying much

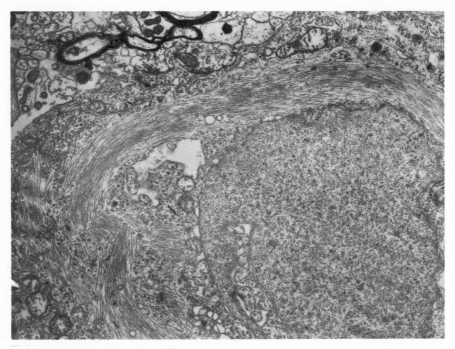

Figure 4. Electron micrograph of a neurofibrillary tangle. The nucleus is at lower right, while the cytoplasm is largely occupied by a mass of abnormal paired helical filaments. Magnification about 10,000 [From R. D. Terry, and H. Wiśniewski, The ultrastructure of the neurofibrillary tangle and the senile plaque, pages 145–168 in *Alzheimer's disease and related conditions* (A Ciba Foundation Symposium) (G. E. W. Wolstenholme and M. O'Connor, eds.), J. & A. Churchill Ltd., London, 1970.]

of the cytoplasmic space in the neuronal cell body. These tangles are found in an increasing proportion of normal people after 60, and in almost all after 80. Only human beings have so far been demonstrated to have these lesions. In the functional aged, there are very few tangles in the cerebral cortex over the convexities of the brain, while they are often moderately numerous in the hippocampal cortex, which is the infolded undersurface of the temporal cortex and is concerned with memory processes (Figure 1). The senile plaques also lie in the cortex and hippocampus, and are also found in small numbers in the normal elderly. Rather than occupying the cell body, as does the tangle, the plaque is situated between and among the neurons and measures 60 to 150 micrometers in diameter.

One recently recognized significance of the plaques and tangles lies in the fact that they are present in small numbers in the functional elderly, and in large numbers in the senile. They seem to concentrate in direct proportion to the mental disability. Furthermore, the abnormal constituents of both lesions represent important components of the cerebral apparatus, such that their malfunction in large numbers could well be important.

The neurofibrillary tangle was demonstrated by electron microscopy (Figure 4) to be made up of abnormal neurofibers present in extraordinary numbers in the cell body (Figure 5). Normal fibers in the adult neuron are of two types: the 24-nm-wide hollow neurotubule made up of a protein called tubulin; and the 10-nm-wide neurofilament, of which the major constituent is the protein named filarin, with a molecular weight of 50,000 daltons. Both structures have short side arms, which may represent a high-molecular-weight component. The fibers making up the neurofibrillary tangle appear to be double helices: a pair of 10-nm filaments twisting around each other every 800 nm. The individual members of these "paired helical filaments" or PHF resemble the normal neurofilament but lack its side arms. The molecular weight and peptide composition of the PHF are also similar to those of the

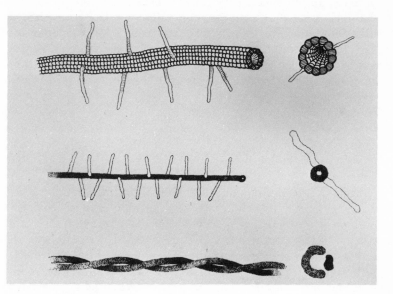

Figure 5. The three major neurofibers under consideration. At the top is the normal 24-nm neurotubule made up of tubulin molecules. Note the short side arms along the length of the microtubule. The middle fiber is the normal neurofilament, 10 nm in diameter, with a very small lumen and side arms similar to those of the neurotubule. The lowest fiber is the paired helical filament characteristic of the neurofibrillary tangle. These are all drawn to approximate scale at a magnification of about 330,000 times for the longitudinal sections, and twice that for the cross sections.

normal filament, but further critical examinations have yet to be completed. If the protein making up the abnormal PHF is indeed identical to a normal protein, investigators must search for the mechanism causing a misassembly. If, on the other hand, it is a new protein, search will be directed toward viral infection or genetic derepression, since new genetic information or activity must have been introduced to instruct the cells to make a new protein.

The neurofibrillary tangle is not unique to the aging brain. Its paired helical filaments are to be found in neurons affected by postencephalitic Parkinson's disease, and this is the long-term result of a particular viral infection. They are also present in adult patients with Down's syndrome, or mongolism, and this is a chromosomal disorder. Again, these lesions are to be found in the brain of the punch-drunk fighter, presumably as the result of repeated trauma. Thus, the chemical comparisons between normal and abnormal fibers take on added significance, since they will help select from among this apparent plethora of causal mechanisms.

Another sort of chemical finding of possibly great significance, and related to the tangle, has to do with the abnormally increased concentration of aluminum in the brain tissue of patients with Alzheimer's disease or senile dementia. This is of particular interest since it is known that one can induce the formation of filamentous aggregates in rabbit neurons by the injection of aluminum into the animal's spinal fluid. The possibly toxic effect of aluminum on the human brain is, nevertheless, unknown.

The senile or neuritic plaque (Figure 6) has a central core of an abnormal, extracellular, fibrillar protein called amyloid, which sometimes resembles gamma globulin. Surrounding this core are numerous abnormal synaptic elements and preterminal axons which are distended with clusters of paired helical filaments identical to those of the neurofibrillary tangle, degenerating mitochondria, and lamellated residual bodies. The mitochondria are normally the site of oxidative metabolism. They are structurally abnormal here, but their chemical function in this situation has not been accessible to analysis. The residual bodies have considerable enzyme activity of the hydrolytic type, and, although they are a type of lysosome, they seem to derive from the degenerating mitochondria. The many abnormal synapses are almost surely not functioning normally, but their structure indicates that their disorder is probably still reversible.

Unlike the neurofibrillary tangle, the plaque is not unique to the human being, having been found in aged monkeys and dogs. In these animals, however, its component neurites lack the abnormal paired helical filaments which are prominent in the human brain. Experimental

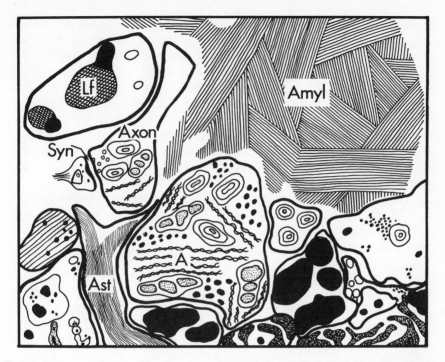

Figure 6. Corner of a neuritic plaque, revealing the central amyloid core (Amyl) surrounded by numerous neurites, most of which are axonal (Axon or A) and contain paired helical filaments, mitochondria, and lamellated residual bodies. We also find lipofuscin (Lf) in some of the cells, and intertwined astrocytic fibers (Ast). A synapse (Syn) is abnormal in that the presynaptic axonal terminal is distended and has abnormal contents. Magnification about 40,000.

plaques identical to those of the dog and monkey have recently been induced in certain strains of laboratory mice by treating them with particular varieties of the scrapie agent, which is a type of slow virus. The implications of this experiment are almost surely very important, but its precise relevance to the aging human brain is still uncertain. Obviously, the relationship between slow virus of this group and senile dementia must be extensively studied.

CONCLUSIONS

The most important structural and related chemical changes seen in the brain as it ages have been described in this chapter. Arteriosclerosis giving rise to inadequate circulation and death of part of the brain should be seen as a separate problem relative to this aging process. The other changes — plaques, tangles, lipofuscin, dendritic shrinkage, neuron loss, decreased extracellular space, and general atrophy of the brain — all go together, and would seem capable of causing the brain to function abnormally. When these cortical lesions are quantitatively advanced, they probably cause senile dementia of the Alzheimer type. What causes the lesions is the focus of widespread laboratory effort. Also note that similar lesions occurring in the hypothalamic area deep in the brain (Figure 1) might have such significant effects as decreasing resistance to stress and inducing a variety of endocrine abnormalities which are prominent in the elderly human being.

The major point is that the altered social environment which surrounds the aged person is acting on an organically altered nervous system to produce the psychologic phenomena and mental deficits which are of such great concern.

ACKNOWLEDGMENTS. I have discussed the work of many investigators from many countries, but without names, and hope that they will forgive me. My own studies have been supported in part by grants from the National Institutes of Health (NS-02255, 03356, and 08180).

13

Aging and Immune Function

WILLIAM H. ADLER, KENNETH H. JONES,
and MARY ANNE BROCK

Immunology has become increasingly exciting as an area for investigations of the cell biology of aging. The reasons for this involve the nature of immunology (1) as a scientific discipline dealing with cell differentiation and function, and (2) as it relates to the adaptive ability of the host in dealing with environmental challenges. In looking at the fundamental mechanisms of normal immune function and that of older individuals, it was discovered that many facets of the immune response apparently decrease with the age of the organism. These findings have led to the intriguing hypothesis that there may be an association between the function of the immune system, aging phenomena, and age-related disease patterns.

THE FUNCTION OF THE NORMAL IMMUNE SYSTEM

The immune system provides the body with a defense against foreign substances. The cells responsible for this defense are called lymphocytes, one of several kinds of white cells in the blood. When bacteria, viruses, fungi, or foreign tissues, such as skin and kidney grafts, enter the body, lymphocytes recognize and react against the foreign invaders in an attempt to neutralize or destroy them.

Lymphocytes circulate through the body in both the blood and lymph vessels, and large numbers of lymphocytes populate lymphoid organs such as the spleen and lymph nodes. These organs are composed of reticular cells, which form a structural, sponge-like meshwork, and

WILLIAM H. ADLER, M.D., KENNETH H. JONES, Ph.D., and MARK ANNE BROCK, Ph.D. ● Gerontology Research Center, Clinical Physiology Branch, Baltimore City Hospitals, Baltimore, Maryland 21224 of the National Institute on Aging, National Institutes of Health, Bethesda, Maryland 20014

lymphocytes, which are found within the cavities of the "sponge." The lymphoid organs filter both blood and lymph and trap foreign particles and tumor cells, thus enabling the lymphocytes to react against the harmful agents. Lymphocytes also accumulate in groups immediately beneath the linings of the respiratory and gastrointestinal tracts and provide a defense against bacteria, fungi, and viruses that enter the body through inhalation or ingestion. The pharyngeal tonsils and the intestinal Peyer's patches and appendix are examples of such lympho-cyte accumulations.

Although lymphocytes are similar in appearance, investigations of the last 15 years have revealed that functionally lymphocytes are heterogeneous subpopulations of cells specialized to interact with each other and to mediate specific immune responses.

During embryonic development, certain cells become "hematopoi-etic stem cells" — cells which divide to produce the body's blood cells. After birth, hematopoietic stem cells are usually found only in the bone marrow. Before birth, however, some of the stem cells produce daughter cells which differentiate into lymphoid precursor cells (Figure 1). The precursor cells colonize the thymus and peripheral lymphoid tissues (spleen, lymph nodes, tonsils, etc.). Further differentiation of precursor cells depends on the microenvironment in which such development takes place, and at least two functionally different main classes of immunocompetent lymphocytes, that is, lymphocytes capable of re-sponding to foreign substances (antigens), mature. Under thymic influence, precursor cells differentiate into a class of immunocompetent lymphocytes called T cells. Morphologically, these cells in the mouse carry a specific antigenic membrane marker called theta (θ). The T cells are distributed in the thymus-dependent areas of the lymphoid organs. Under the influence of a bursa of Fabricius in birds and possibly the bone marrow or a bursa equivalent in mammals, precursor cells can also differentiate into lymphocytes called B cells. Morphologically, these cells carry membrane-bound antibody immunoglobulin or immunoglobulin fragments. B cells populate the germinal centers of lymphoid follicles.

In functional terms, T lymphocytes are responsible for the manifes-tation of cell-mediated immunity, a form of immunity which involves direct contact by lymphocytes for the destruction of foreign invaders. Such immunity can be transferred only by lymphocytes and not by serum antibodies. It has also been hypothesized that a further role for cellular immunity is a defense against neoplastic cells. This concept springs from the understanding that the primary mechanism in the re-jection of tissue grafts is cell-mediated and does not require antibody. Since grafted tissues carry antigenic determinants on their cell mem-

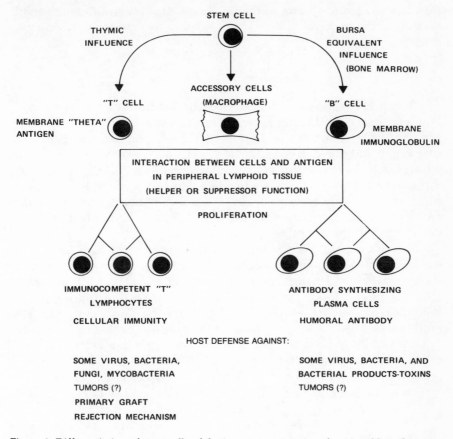

Figure 1. Differentiation of stem cells of the immune system into functional lymphocytes.

branes and since many neoplastic cells have been shown to have tumor-associated membrane antigens, it seemed logical to assume that a cellular recognition of a membrane antigen with subsequent killing of the foreign cell could be a host defense mechanism which evolved to protect the host from the growth of neoplastic tissue. Although this assumption is far from proven, the research in this area has contributed new information on the host–tumor interaction which may be of benefit in constructing rational approaches to cancer prevention and therapy.

The B cells, the second functional class of immunocompetent cells, interact with T cells and accessory cells in the presence of antigen to multiply and subsequently to differentiate into plasma cells which syn-

thesize and secrete specific antibody. T cells may also inhibit the generation of antibody-forming cells, thus helping to regulate antibody production.

Assessment of T-Cell Function

T cells are responsible for cell-mediated immunity, and these functions can be assayed using a variety of procedures.

Delayed Hypersensitivity. Immune animals or human beings in some
cases may have a delayed skin test reaction, redness and swelling,
which reaches a maximum 48 to 72 hours after intradermal injection of a
test antigen. This reaction is considered to be a test of T-lymphocyte
function based on observations in which T lymphocytes were transferred from immune to nonimmune animals and the skin test reactivity
determined in the adoptive recipients.

Transplantation. Transplantation experiments, in which foreign tissues are transferred between donors, were important in demonstrating
the effectiveness of lymphocytes in the rejection of the graft. For example, if a piece of skin from one individual is grafted onto a second
individual who is genetically different from the skin donor, the skin
graft will be rejected. The rejection of the graft is effected by T lymphocytes which have been sensitized to the graft's foreign antigens. The T
cells migrate into the graft to attack and kill the foreign cells. As could be
predicted, skin grafts between identical twins are not rejected. These
experimental findings have clinical significance because of the increased
use of surgical grafting procedures such as kidney grafts.

Cell Culture. A major advance in immunological research was provided with the use of tissue culture techniques. These have allowed
investigators to move their attention from the intact organism to the
tissues and cells of the immune system. Recent tests of T-cell function in
culture have resulted in the findings that lymphoid cells can be stimulated to divide by chemicals called mitogens, such as phytohemagglutinin (PHA) and Concanavalin-A (Con-A), or by foreign cells. Neonatal
thymectomy or genetic absence of a thymus resulted in animals deficient
in T cells and was correlated with an inability of lymphoid cells from
these thymusless donors to respond to PHA, Con-A, or foreign cells.
Further evidence that these thymic-dependent cells were a particular
type was the demonstration that if θ-positive cells were destroyed with
appropriate antisera and complement, the *in vitro* T-cell responses could
be abolished.

Lymphocyte Cytotoxicity. It has been found that an immune T cell,
either taken from a donor immune to tissue antigens or produced by *in
vitro* immunization in tissue culture, will kill target cells bearing the

antigen to which it has been immunized. This T-cell function is under intense investigation because of its possible relation to cancer immunity and the killing of tumor cells.

Lymphokines. Another finding is that T cells, when in contact with antigen or mitogen, will elaborate certain factors or "lymphokines," which can be tested for their biological activity in a variety of *in vitro* assays.

Assessment of B-Cell Function

B cells are responsible for humoral immunity, immunity which can be transferred by immune serum, and assay for their cell functions includes: (1) measurement of serum antibody activity, (2) determination of the numbers of antibody-forming cells in lymphoid tissue, and (3) measurement of the response to B-cell mitogens such as endotoxin lipopolysaccharide (LPS). The use of completely *in vitro* cell culture systems which support the division and differentiation of B cells into plasma cells has allowed the assay of individual antibody-forming cells and has led to an understanding of the kinetics of cellular antibody synthesis. More sophisticated cell separation techniques and morphological criteria will eventually lead to the resolution of the questions of T-cell–B-cell interactions and the mechanisms of control processes.

IMMUNE FUNCTION IN THE AGING ORGANISM

Age-associated immune defects exist, and, although their significance is incompletely understood, certain evidence is available which makes possible an attempt to interpret the defects seen. It has been well established through anatomical studies that lymphoid tissues undergo involutional changes with advancing age. The thymus normally reaches its maximum size in early childhood and begins to involute rapidly during puberty. Such atrophy is characterized grossly by a reduction in organ weight, and histologically by a decrease in the number of lymphoid cells present, chiefly in the cortex, and their replacement by connective tissue. Other lymphoid tissues generally reach their maximum relative size immediately after puberty and undergo more gradual atrophic changes with advancing age.

With a changing morphology of lymphatic tissue, one would assume that there would also be functional changes. If the number of T cells from an older individual is decreased, then decreased T-cell function would be expected. The implications of this finding would be entirely different from a finding that T-cell numbers in young and old individuals were the same but that T-cell function was decreased with age. Preliminary evidence seems to support the latter possibility, although it is possible that better morphologic criteria are necessary in order to quantitate functional T cells.

A further consideration of factors which may influence an individual's immunological functions involves the possible production of a thymic hormone by the epithelial cells of the thymus gland. There is some evidence which indicates that the level of this hormone decreases with age during the course of thymic involution. The mechanism of action of this thymic factor and its various functional abilities needs further investigation.

Clinical and animal studies have revealed a decrease in immunological functions with advancing age. Since an individual's ability to initiate an immune response involves the complex interaction of several subpopulations of lymphoid cells whose functions are, in turn, influenced by general health, nutrition, hormonal balance, and so forth, investigators have turned to *in vitro* assays to study each of the facets of the immune response. The following observations exemplify the general age-related changes in lymphoid cell proliferation, cell-mediated immunity, and humoral immunity.

Humoral Immunity

The first systematic measurements of humoral immunity in relation to advancing age used an *in vivo* cell transfer model, and an age-associated decline in primary antibody production was found. More recently, the technique of culturing plaque-forming cells has allowed assay of antibody synthesis by individual cells. As seen in Figure 2, lymphoid cells are cultured with sheep erythrocytes, resulting in stimulation of the appropriate immunocompetent B cells, followed by proliferation of these cells and the differentiation of their daughter cells into antibody-producing plasma cells. The antibody-producing cells are then recovered from the initial cultures, mixed with fresh sheep red blood cells and complement, and plated onto microscope slides. The slides are then incubated, during which time the plasma cells produce sufficient antibody to the red cells to cause their hemolysis. Such hemolysis appears as clear areas, plaques, in a red background when the slide is examined. Antibody production is quantitated by counting the number of plaques found.

In the B-cell system, the findings can be generalized. The primary immune response — the response of the immune system to an antigen that it is encountering for the first time — has an age-associated decline. This is especially true when one considers the response to those anti-

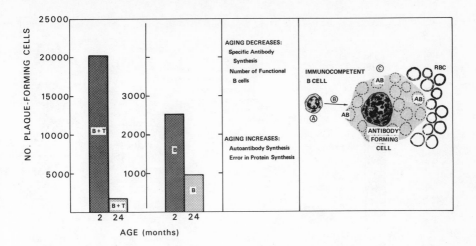

Figure 2. Fundamentally, the immunocompetent cell responds to antigenic stimulation by transformation into a blast cell, proliferation, and subsequent differentiation into effector cells. In the right column, the effector cell (a plasma cell) secretes hemolytic antibody which destroys the surrounding red blood cells, resulting in a "plaque" (shaded area). The left column shows the difference in the ability of spleen cells from 2- and 24-month old C57BL/6 mice to produce antibody against antigens requiring both thymic-derived and bone marrow-derived lymphocytes (B+T) or antigens requiring just bone marrow-derived lymphocytes (B).

gens requiring a T-cell interaction with a B cell. However, if the secondary immune response — the response found after a second or "booster" injection of antigen — is measured throughout life, usually no such age-associated decline is found. The secondary response is usually stronger and lasts longer than the primary immune response, and the efficacy of secondary immune reactions in preventing disease throughout life is demonstrated by the fact that childhood diseases are rarely seen in adults.

Cellular Immunity

Ample evidence exists that T-cell function declines with age. For example, the response of mouse lymphocytes in culture to the mitogens, PHA and Con-A, and antigens shows a marked decrease with age, which, in many instances, occurs relatively early in life, sometimes before the onset of age-related disease. The mitogen assay is one of the *in vitro* tests used widely to measure the ability of lymphocytes to respond to stimulation. Although mitogen stimulation of lymphocyte proliferation is nonspecific, the basic mechanism by which the immunocompe-

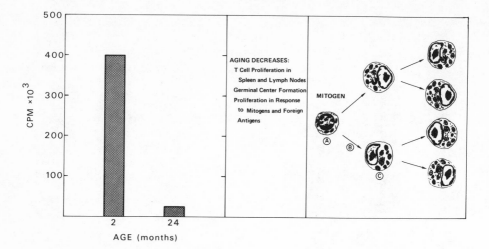

Figure 3. The right column shows the proliferation of C57BL/6 mouse lymphocytes stimulated by mitogens. The left column shows the difference between the incorporation of radioactive thymidine (counts per minute) by proliferating lymphoid cells from young and old mice.

tent lymphocyte is driven *in vivo* into cellular proliferation is thought to be similar. As seen in Figure 3, lymphocytes are placed in culture, and one of several mitogens is added. Generally, PHA and Con-A are known to stimulate T cells, and in certain rodents LPS stimulates B cells. After 48 hours of culture, the stimulated lymphocytes have transformed into cells which will undergo mitosis and begin to synthesize DNA. It is at this time that radioactive, tritiated thymidine is added to the cultures. Cells making DNA incorporate the radioactive thymidine into their genetic material. After 72 hours of culture, the lymphocytes are harvested to measure the amount of radioactivity taken up by the cells. An age-related decrease in cellular proliferation is found when mitogen assays are used to compare the response of lymphocytes from aged individuals to that given by lymphocytes from young adults.

There are several shortcomings of the mitogen assay, however. The results must be evaluated critically because of the wide variation in the response of different individuals' lymphocytes to mitogens. Furthermore, there can be considerable variation between the responses of two different samples of lymphoid cells taken from the same individual. Additionally, cellular proliferation *in vitro* can be influenced by technical problems inherent in the culture system itself. Finally, we are still not certain that the nonspecific mitogen stimulation of cellular proliferation *in vitro* is the same as the specific antigen stimulation of T cells in the thymus-dependent areas and of B cells in the lymphoid germinal centers of the living organism.

Also important in the examination of immunological function is the role of the T cells in their regulatory interactions with B cells which decrease with age and may explain the decrease in primary antibody response. The suppressive effects of T-cell populations on B cells also decline. These data give rise to the hypothesis that a decline in a control mechanism regulating the antibody-forming system results in the appearance of cells which synthesize autoantibody with specificity against "self" antigens.

Another example of the age-related decline in T-cell function is the loss of cytotoxic T-cell activity with age. This is assayed by mixing, in a culture tube, responder lymphocytes from one individual with a stimulating cell population from a different individual. The reactor lymphoid cells "recognize" the foreign histoincompatibility antigens of the stimulator cells and are triggered to undergo cellular proliferation. After 4 or 5 days, the daughter cells of the original responder cells have differentiated into "killer" cells. Radioactively labeled target cells bearing the same histoincompatibility antigens as the original stimulator cells are then added to the culture. The killer cells, which are now specifically sensitized against the stimulator cell–target cell antigens, attack and destroy the target cells, releasing the radioactive tracer. It is then a simple matter to measure the amount of radioactivity which has been liberated into the culture medium and to calculate the number of killer cells generated in the culture. This *in vitro* test is valuable because it most clearly measures killer T cells destroying foreign antigen-bearing cells. The experimental data in Figure 4 show the age-associated decline in the ability of murine splenic lymphocytes to kill target cells to which they have been sensitized. Spleen cells from 30-month-old mice performed poorly compared to middle-aged and young adults.

The decline in cytotoxic T-cell function with age is characteristic of other species, including humans, and can also be measured by more complex *in vivo* tests, such as foreign-tissue rejection, delayed hypersensitivity and graft–vs.–host reactions. These tests are similar to the cell-mediated cytotoxicity assay in that the responding T cells recognize foreign tissue antigens and move in to attack the foreign cells. Although there is considerable variation between individuals (probably genetically controlled) in the intensity to which the individual responds to an antigenic stimulus, nevertheless all the assays generally show decreased T-cell function in the aged.

It is difficult to examine the T-cell system in its entirety, especially the role of T cells in the lymphoid tissue of humans. Most methodologies leading to the concepts of T cells and their function were developed in

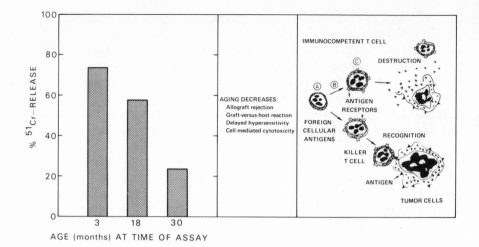

AGE (months) AT TIME OF ASSAY

Figure 4. Proliferation and differentiation of T lymphocyte, "killer" cells are represented in the right column. The immune killer cell (C57BL/6 mouse lymphocytes) must come in contact with the target (P815 tumor cells) so that destruction of the target and subsequent release of its radioactive tracer (*) takes place. The killer cell is then free to attack more target cells. The left column shows the differences in the amounts of radioactive chromium released from tumor target cells when they are incubated with killer cells from young, middle-aged, and old mice. The greater the amount of radioactivity released, the larger the number of killer cells present. The center columns of all three figures list the age-related findings. A, B, and C are explained in the text.

mice, in which assay for T cell activity included lymph nodes, spleen, and thymic lymphoreticular tissue and transfer experiments using inbred strains. In humans, however, most assays are carried out on peripheral blood lymphocytes. There is evidence in animals that peripheral blood lymphocyte performance in assay systems *in vitro* has no correlation with the performance of spleen cells from the same animal in the same type of assay. Since blood is not a lymphoid organ, it may be that the studies carried out on humans using blood lymphocytes have limited relevance in the indication of immune status and immune function. The large variability in the results from these studies may be an indication of this problem and of the evanescent nature of the circulating lymphoid cell population. As human lymphoreticular tissue is inconvenient to obtain, the studies so far have been very limited.

SIGNIFICANCE OF THE DECLINE IN IMMUNE FUNCTION

Aging must be viewed as a normal process with an undefined time of onset, and this makes the cause–effect relationships between aging and immune function difficult to define. However, several theories have been proposed linking immune function to the causes of aging. It is difficult to defend such a position, especially since most decreases in immune activity are seen after the onset of many other aging manifestations. If one must wait for the appearance of an age-related disease to show a decrease in immune function, then the actual effects of the disease may be changing immune function rather than the aging of the immune system itself. In mice, however, a marked decline in some T-cell functions can be seen at a relatively young age, preceding most aging phenomena. It is probable that the causes which lead to aging, whatever they are, also cause a decline in immune function.

A brief analysis of the possible causes of the age-related decrease in immune function can be illustrated by reexamination of Figures 2, 3, and 4. In each of these figures, there are the letters A, B, and C. The letter A has been positioned to coincide with the precursor or progenitor cell. Letter B is associated with the process of division and differentiation of the precursor cell, and letter C is representative of the immunocompetent effector cell. The function of immunocompetent cells is usually what is assayed, and the results of those assays on cells from older animals usually show a loss of function. The theories to explain this deficiency can simply be ascribed to (a) a decrease in number of precursor cells; (b) an inability of precursor cells to divide and differentiate, which can result in (c) fewer and/or defective immunocompetent effector cells.

However, regardless of causation, the association of decreased immune function and old age may lead to an understanding of the causes of some age-related diseases. Figure 5 shows the increase with age in the incidence and mortality of infectious diseases and the incidence of cancer. Again, there must be caution in drawing conclusions as to direct causation of these diseases and decreased immune function. For example, an elderly patient who has a fractured hip and is confined to bed may experience a decrease in lung ventilation and, therefore, may be at a high risk of developing pneumonia, with the possibility of subsequent serious consequences. This may occur with a normal immune function as well as a decreased one, and the infection may have no relation to the state of immunity.

The increases in malignancies with age include diseases of markedly various natures; in some, or possibly all, of these cases, it may be that the immune system would have a protective effect, both in preventing the occurrence of malignant cells or in stopping the growth of tumors by killing them. On the other hand, antitumor antibody may in some way stimulate tumor growth. Therefore, it could be that T cells suppress tumor growth while B cell products encourage tumor growth. If that were true, then during aging one would encounter changing host immune function which could influence interactions with carcinogenic agents and the outcome of these interactions. During the period of nor-

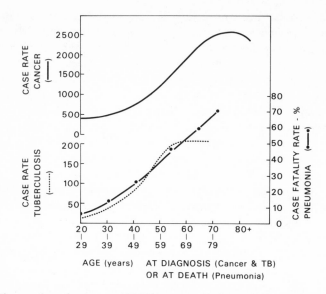

Figure 5. The number of cases per 100,000 persons of all cancers (— upper scale), and of tuberculosis (---- lower scale), and the percentage of fatalities attributed to pneumonia (●——● lower scale) are plotted in relation to the age at diagnosis of cancer or tuberculosis or the age at death from pneumonia. The elderly have 4 to 5 times the case rate for cancer or tuberculosis and 6 to 7 times the rate of fatality from pneumonia than do young adults.

mal immune function, cancer would be uncommon; during the period of decreasing T-cell function and normal B-cell activity (intermediate age groups), cancers would be more common and more virulent; and in the older age groups, when there is decreased B- and T-cell functions, cancers would be most common but less virulent. This hypothesis has inherent problems because there is recent evidence of very low tumor incidence in congenitally immunodeficient mice. Furthermore, it is difficult to correlate the level of immune competence with a specific age because of the large variability in immune function between individuals.

In some tests of age-associated immune dysfunction, autoimmune antibodies, antibodies with specificity against self-tissue, were detected. The significance of this age-associated synthesis of autoantibody is in dispute. Some think it is the cause of several age-related illnesses, such as vascular disease, whereas others believe that autoantibody is not associated with any disease. Some propose that autoantibody is a normal response of the organism to rid itself of damaged tissue by inducing metabolic breakdown of the tissue. There are other hypotheses that autoantibody is antibody directed against viral-induced cell surface antigens on virus-infected cells. It also could be the result of a breakdown in control mechanisms which then allows an "escape" and proliferation of a particular clone of cells which make autoantibody. There is other evidence of age-associated synthesis of abnormal antibody protein which could reflect changes in the proliferative capacity of individual B cells. Whether this results from a deficiency in a T cell control effect on B cells is not certain.

CONCLUSIONS AND OUTLOOK FOR THE FUTURE

This has been a brief review of some of the considerations of normal immune function and the indications of an age-related immunodeficiency. Adding to the difficulties posed by the incomplete knowledge of these factors are practical questions: Specifically, what does a finding of 50% of normal function really mean? What level of immune function can be associated with a deficiency? Is there such a thing as an imbalance between T and B cells, and what would be the effects? How much masking influence have antibiotics had on the expression of an immunodeficiency state in aging? Can data on different species and cells, collected by different methods and measured by different assays, ever lead to a unifying hypothesis? In an aged population of human beings or mice, is the wide variation in the individual levels of immune function significant? What are the most representative tests to determine the level of immune function?

Obviously, there are many problems, but in the study of aging, the immune system may be an excellent vehicle to further our knowledge of a difficult, extensive process. There are few biological systems that offer the advantages of providing a model for research into aging phenomena and also provide a potential beneficial, therapeutic approach to influencing the outcome of age-associated diseases.

ACKNOWLEDGMENT. The authors are grateful to Paul Ciesla for his help in preparing the illustrations.

BIBLIOGRAPHY

General

Cline, M. J. 1975. *The white cell*. Harvard University Press, Cambridge, Mass.
Kohn, R. R. 1971. *Principles of mammalian aging*. Prentice-Hall, Inc., Englewood Cliffs. N.J.
Makinodan, T., and W. H. Adler. 1975. Effects of aging on the differentiation and proliferation potentials of cells of the immune system. *Federation Proceedings* 34: 153–158.

Makinodan, T., E. H. Perkins, and M. G. Chen. 1971. Immunologic activity of the aged. Pages 171–198 in B. L. Strehler, ed. *Advances in gerontological research*. Academic Press, Inc., New York.

Nordin, A. A., and T. Makinodan. 1974. Humoral immunity in aging. *Federation Proceedings* 33: 2033–2035.

Stutman, O. 1974. Cell-mediated immunity and ageing. *Federation Proceedings* 33: 2028–2032.

Weiss, L. 1972. *The cells and tissues of the immune system. Structure, functions, interactions.* Prentice-Hall, Inc., Englewood Cliffs, N.J.

Technical

Adler, W. H., K. H. Jones, and H. Nariuchi. 1977. Aging and immune function. Pages 77–100 in R. A. Thompson, ed. *Recent advances in clinical immunology*. Churchill Livingstone, New York.

14

Genes and Aging

ELIZABETH S. RUSSELL

Genes may be regarded in at least two quite distinct ways in relation to processes of aging. They are units of genetic transmission from parent to offspring and, also, intracellular agents whose products interact with environmental forces to control development, differentiation, metabolism, and degeneration. Much of this chapter will deal with the results of genetic transmission, that is, with genetic similarities and differences between individuals, present throughout their life-spans. I will attempt to demonstrate that the patterns of the aging of individuals carrying an identical array of genes will show greater similarity than will the patterns of the aging of individuals carrying different assortments of genes. To do this, I will have to deal mostly with genetically controlled animals, and I will try to bring home to you why and how these animals with specific genotypes have become important in research on the biology of aging.

We will also be concerned with the nature of the effects of these transmitted genes, considering how their functions may contribute to the processes of aging, which differ between animals carrying different genes.

Before leaving this introductory section, I must point out that deleterious mutations in the somatic cells (i.e., cells of the body rather than reproductive cells) which may occur during the life-span of a single individual could also contribute to his aging. If the gene controlling the structure of an enzyme critical for some aspect of metabolism undergoes somatic mutation in one body cell, that cell and its critically needed potential descendants may disappear. Alternatively, the mutated cell may persist and produce a line of defective cells which do not react appropriately to signals from other parts of the body, or which secrete products harmful to the body. Consequences of somatic mutation would depend on where and when they occur, but there is a good chance that

---

ELIZABETH S. RUSSELL, Ph.D. • The Jackson Laboratory, Bar Harbor, Maine 04609

the net effect of accumulated somatic mutations would be deleterious, and certainly could contribute to aging deterioration. At present, there is a great deal of interest in the possibility of somatic genetic changes as a prime determiner of aging, and several molecular biologists are looking for evidence of somatic mutations and increased presence of abnormal gene products with advancing age. As yet, no one has demonstrated such accumulation unambiguously; so the answer to the question remains in doubt.

In general, somatic, as well as germ-line mutations, occur at random. Thus, if somatic mutations eventually are found to be consistent contributors to aging, one would expect them to result in only chance differences between individuals in the type or site of the aging changes for which these accumulated mutations are responsible. If important genetic effects on the pattern of aging changes of individuals of different genotypes are recognized, these must still be attributed to effects of genes transmitted from the parents, and to be present throughout the life-span of the aging individuals.

GENETIC VARIABILITY IN HUMAN AGING?

Aging in humans is both inevitable and complex, having many and diverse manifestations. Very few people over 50 can count on threading a needle on the first try. Most of us turn gray, but some much earlier than others. Some, but not all of us, are troubled with arthritis. The incidence of constitutional diseases, such as cancer and heart disease, increases markedly with advancing age, but many people survive to a ripe old age without these troubles. A small, but very troubling, proportion of elderly people develop senile dementia. Many of us know families with sprightly, alert 80- to 90-year-old great-uncles, great-aunts, and grandparents. Unfortunately, we also know families whose members tend to die in their sixties. What makes the difference between these families? We may suspect that genetic differences play an important role in human aging, but we cannot be sure. Cultural and economic differences, which could affect longevity, also tend to run in families. Long human life-spans make any observations of successive generations difficult. Also, by our own customs and choice in spouses, we maintain considerable genetic variability within human families. Except for pairs of identical twins, each human being has a unique genetic endowment. Genetic heterogeneity, combined with long life-spans, obviously greatly limits the possibilities to test for genetic effects on patterns of human aging.

Certain special kinds of premature human aging, or progeria, such as Werner's and Cockayne's syndromes, are generally accepted as having genetic causes. Individuals with these diseases show some characteristics of premature aging. They fail to grow properly, become bald and wrinkled in their early teens, and almost invariably die before the age of 20, with extensive circulatory difficulties associated with lipid deposits in the blood vessels. Even in genetically heterogeneous humans, it is possible to establish that the appearance of such rare and sharply defined syndromes can be attributed to effects of one or a very few genes.

A number of researchers are working with cultured cells from progeric humans, trying to determine the nature of their metabolic errors. We hope that their research will lead to understanding of, and successful therapy for, the particular manifestations of premature aging, which appear in these rare progeric human beings.

At the same time, we doubt if these studies can tell us whether or not genetic factors contribute to the great variety of manifestations of "normal" human aging. Why do we feel it is important to see if genes contribute to aging? In my own mind, the reason is *not* to find ways, such as genetic selection, for increasing man's maximum life-span. Such endeavors would be morally and ecologically undesirable on twentieth-century planet Earth, as well as very difficult to carry out.

A much more appropriate aim for research on biological aging is improvement of the health, comfort, and well-being of older people, by reducing old age debilities to the lowest possible levels. If potential aging problems differ genetically among individuals, different prophylaxis and therapy may be required for different people to avoid long-drawn-out aging deterioration. Recognizing the existence of genetic diversity in aging would not immediately identify different specific susceptibilities, but would definitely affect the orientation of future research.

INTERSPECIES DIFFERENCES IN AGING

Although the main thrust of this chapter is intraspecies, rather than interspecies, differences in aging, I would like to digress to point out that genes are responsible for all differences between species. Natural selection has determined the genome (total array of genes) of a species, presumably to fit with an available ecological niche.

Different mammalian species obviously age at different rates, and these interspecific differences obviously have genetic bases. Similar arrays of old-age debilities (neoplasia, arthritis, circulatory problems) occur at greatly different chronological ages in different mammalian species.

Interspecies differences in aging are very intriguing. They suggest, but do not prove, the existence of genes whose specific role is to set the general *rate* of aging. As yet, we have no proven ideas as to how such hypothetical genes might act, although some investigators, particularly molecular biologists, have begun to look for clues. Their conclusions must be to some extent tentative, because they cannot make direct genetic crosses between species.

Very stimulating discussions of possible modes of action of genes, differing between species, which determine *rate* of aging changes, have been presented recently by Richard G. Cutler of the Gerontology Research Center and by Ronald W. Hart of Ohio State University and Richard B. Setlow of Brookhaven National Laboratory. Cutler suggests that very few genetic loci may be involved. Hart and Setlow presented evidence that in seven mammalian species the rate of "unscheduled DNA synthesis was approximately proportional to the logarithm of life-span." Differences in unscheduled DNA synthesis could easily be dependent upon allelic differences at a small number of genetic loci.

George Sacher of the Argonne National Laboratory suggests that natural selection, which determines the characteristic life-spans of species in the laboratory, effectively works on "life assurance mechanisms" (factors which promote survival to adulthood), which are critically important early in life but have secondary consequences for total potential life-span. One complication is, of course, the fact that natural selection can only be effective during the reproductive span of a species, since it must act by affecting the frequency with which a gene, present among animals in one generation of a population, will also be carried into the next generation. In social animals, a possible exception to this generalization may be group "caring" activities of postreproductive individuals, such as elephant "aunts."

Presumably, the genes affecting longevity in a large-bodied, dominant species, which takes a long time to reach sexual maturity, could be quite different from those affecting aging in a small, rapidly growing, early-breeding species near the bottom of the food chain. Certainly, elephants live a long time, mice a short time. Differences in body size and rate of maturation are not, however, the entire story. Although maximum life-spans observed in laboratory-maintained populations of both house mice and rats are not more than 4 years, Sacher reports that laboratory-maintained deer mice (*Peromyscus leucopus*), which are approximately the same size as house mice and mature at the same age as do house mice, regularly live at least 6 years in the laboratory, with many surviving as long as 8 years. Direct natural selection for long potential life-span cannot be very strong, since very few individuals of either house mice or deer mice survive as long as 2 years in the wild.

It may be that genes affecting general rate of aging differ more frequently between than within species. The genes responsible for intraspecific differences in processes of aging may often have specific effects, largely in particular kinds of cells and tissues, rather than general effects common to all cells.

RESEARCH ADVANTAGES OF INBREEDING AND GENETIC FIXATION

Stumbling blocks in the study of human aging have been man's genetic heterogeneity and the difficulty of separating genetic from environmental influences on life-span. Both of these problems can be avoided by studying a variety of different genetically defined strains within one species. Intraspecies genetic differences in aging were first demonstrated in studies of the life-spans of mice from different inbred strains raised together in a common environment. The process by which inbred strains are produced (20 successive generations of single-pair brother–sister matings) and maintained (continuation of identified sib matings) assures that all the members of one inbred strain are as much alike genetically as it is possible for mammals to be; they can become unlike only through the very rare appearance of new mutations in the germ lines. The members of a different inbred strain also all share a unique genome, containing a different array of homozygous gene pairs. If the total life histories of a large number of mice from one inbred strain in a particular animal colony are studied, the range of life-spans observed shows what can happen in individuals of that particular genotype under the environmental conditions of that particular colony. If, at the same time, the life histories of a large number of mice from a different inbred strain are also observed, they will provide information on the potentialities of animals of that second defined genotype under the same environmental conditions.

Significant differences in mean and range of life-spans between two inbred populations studied together must be attributed to genetic factors. In a study of mice from 10 different inbred strains which we carried out long ago (1948 to 1952) under less than ideal environmental conditions, the mean life-span of females in the earliest-dying strain (AKR/J) was only 287 days, while that of females in the longest-lived strain

(C57BL/6J) was 576 days. Thus, genetic differences were important contributors to aging. This demonstration of genetic influences did not, of course, preclude the possibility of environmental influences. Recent improvements in mouse husbandry have eliminated many intercurrent infections, and have improved nutrition, allowing almost all the mice in an inbred strain to survive to their normal old age. They now live in mouse suburbia, instead of in their earlier near-ghetto circumstances. Although AKR/J mice still die very young (with lymphatic leukemia), two recent studies, published in 1973 and 1975, have shown mean life-spans of 794 and 889 days for C57BL/6 females, gains of 200 to 300 days over the 1952 observations. Thus, environmental, as well as genetic, factors can influence the life-spans of mice.

The availability of many individuals from each of several different genetically defined inbred strains has been important in demonstrating both genetic and environmental effects. The use of inbred mice also provides other advantages for research on aging. Investigators in other research areas, such as cancer, developmental biology, physiology, endocrinology, pathology, immunology, and hematology, have used inbred mice very extensively. As a result of this research, the scientific literature is full of information on many characteristics of mice from a great variety of inbred strains. The investigator not only can work with many copies of the same experimental object; he can also know before beginning his experiment a great deal about the potential of that object. He can find in the literature evidence that levels of certain circulating hormones are considerably higher in some inbred strains than in others; that the normal numbers of both red and white blood cells are higher in some than in other strains, and that mice of one strain will become obese on a diet on which other mice remain sylphlike. Immune responses differ markedly among strains, and also change with age. These and many other "differences in normality" could have effects on patterns of aging.

Even differences in learning behavior may provide insight on manifestations of human aging. Richard L. Sprott of The Jackson Laboratory showed differences between mice from two inbred strains in passive avoidance learning. In C57BL/6 mice, *performance* declined with advancing age, but they still maintained complete *ability* to learn, given enough time. Declines in sensory capacity or speed of response should not be mislabeled as declines in mental capacity.

The investigator studying total life histories can profit from knowing that one kind of mouse is very likely to develop mammary cancer, another lung tumors, still others kidney disease, lymphatic leukemia, or Hodgkin's disease. This foreknowledge helps greatly in experimental design and choice of animals for specific projects. It should also make the investigator cautious about making general claims about aging of mice without studying animals from a reasonable variety of different inbred strains.

Observations should also be extended to genetically controlled animals in species other than mice, again in order to avoid false generalizations. Laboratory rats are an obvious choice, particularly because they have been widely used in studies of physiology, growth, nutrition, and learning behavior. Fewer rat than mouse inbred strains have been developed, and data on genetically controlled differences in rat life-spans are only now being collected. Characteristic growth differences between rat "stocks," with differing — but not completely genetically fixed— genomes, are well known. Two examples are Sprague–Dawley and Wistar rat stocks. Sprague–Dawley rats, for example, continue to grow throughout their lives, while Wistar rats stop growing by the age of 1 year. George S. Roth of the Gerontology Research Center in Baltimore found that the number of adipocytes (fat-storing cells) decreased with advancing age in rats from both stocks, but the individual fat cells increased in volume only in Sprague–Dawley rats. Here is a clue to the basis for one manifestation of age in the rat.

Additional evidence for genetic diversity in the processes of aging comes from long-lived, vigorous mice produced in crosses between parents from two different inbred strains. All the offspring of the first generation from such a cross are alike genetically but differ from either parent because they are heterozygous for each of the many gene pairs which differ between the two homozygous parental genotypes. Not all of the most favorable alleles can be fixed by chance in any one strain during inbreeding. It is clear that some of the gene pairs fixed in any one inbred strain are slightly deleterious in their potential effects on life-span, since interstrain F_1 hybrid mice almost always live longer, and eventually develop a different array of pathelogical lesions, than do mice of either parental strain.

This point may be made clearer by referring to an excellent recent study by George S. Smith, Roy L. Walford, and M. Ray Mickey, University of California, Los Angeles, which involved mice from seven inbred strains, including three generally regarded as long-lived (LP, 129, and C57BL/10) and eight different interstrain F_1 hybrid genotypes. All the mice were maintained at the same time in the same good laboratory environment. The four genotypes of mice with the longest lives were of certain interstrain F_1 hybrid types. Particularly instructive are two "sets" of data from this study. The first set involved long-lived strain 129 mice, shorter-lived DBA/2 mice, and their interstrain (129 × DBA/2)F_1 hybrid offspring. The mean life-spans in the 129 strain were 117 weeks for males, 104 weeks for females, and in the DBA/2 strain, 82 weeks for males, 81 weeks for females. In the $(129 \times DBA/2)F_1$ hybrid, the mean life-span for males was 135 weeks, for females, 119 weeks, both values considerably higher than those observed in either parental strain. The maximum *individual* male life-spans were 154 weeks in 129, 125 weeks in DBA/2, and 186 weeks in the $(129 \times DBA/2)F_1$ hybrid. The maximum individual female life-spans were 148 weeks in the 129 strain, 118 weeks in DBA/2, and 169 weeks in the $(129 \times DBA/2)F_1$ hybrid. In the second set of data, mice from two relatively long-lived inbred strains, LP and C57BL/10, were compared with their F_1 hybrid offspring. Mean life-spans for males were LP, 103 weeks, C57BL/10, 118 weeks, and $(LP \times C57BL/10)F_1$ 125 weeks. Mean life-spans for females were LP, 99 weeks, C57BL/10, 99 weeks, and $(LP \times C57BL/10)F_1$, 123 weeks. The longest individual life-spans were an LP male, 142 weeks, female, 127 weeks; a C57BL/10 male, 165 weeks, female, 141 weeks, and an $(LP \times C57BL/10)F_1$ male, 185 weeks, female, 176 weeks. In each set of data, both mean and maximum life-spans were clearly longer in F_1 hybrid mice than in mice of their parental strains. In general, a greater diversity of tumors was observed in F_1 hybrids than in mice of parental strains, but the incidence of a specific type of neoplasm could be either higher or lower in the F_1 hybrid. Thus, an investigator who needs genetic constancy, combined with very long life-span, may well find his best examples in F_1 hybrid mice. If he wants to determine whether genetic factors influence a characteristic which appears only in very old animals, he can search through old mice of a wider array of genotypes by comparing incidences in a variety of different kinds of long-lived F_1 hybrid mice than by depending solely on mice from the longest-lived inbred strains.

These examples from both mice and rats provide considerable evidence that genes transmitted from generation to generation have significant effects on both patterns of aging and duration of life. In some cases, knowledge of differing characteristics appearing earlier in life can provide clues on the "how" of aging changes. Since each gene is believed to have its original effect on one process, the best information on mechanisms for genic effects on aging should come from retrograde tracing of the effects of a single genic substitution back to the time and place of original gene action. In cases of differences in aging between inbred strains and genetically uniform F_1 hybrids, such analyses are very difficult, because any two inbred strains differ at many genetic loci, so that it is usually impossible to single out and study the pathway by which any one particular gene affects processes of aging.

An efficient research tool for tracing the pathway of action of a specific single-gene difference is the congenic line, which carries an identified "foreign" gene segregating against a known uniform genetic background. Using such mice, the investigator may make conventional Mendelian monohybrid crosses and obtain offspring which are genetically essentially identical except at the single known segregating locus. Many such congenic lines have been developed for studies in development, immunology, pathology, physiology, and many other disciplines. The "foreign" gene may be a mutant allele with quite deleterious effects, or it may be a different, presumed normal, allele which has been transferred (by repeated crosses) from its usual home in another inbred strain. The great advantage of comparisons between mice from a particular inbred line and mice from a specific "partner" congenic line, with an identifiable single-gene difference, is that all differences observed consistently between the two groups must be attributable to that specific gene difference.

SINGLE-GENE DIFFERENCES AND THE IMMUNE SYSTEM

Some of the most promising recent analyses of mechanisms controlling particular aspects of aging have made excellent use of comparisons between mice with identified single-gene differences. Particularly pertinent are studies involving inherited differences in the immune system. Mice from congenic lines, genetically almost identical except for known differences at specific loci determining tissue compatibility, tend to have slightly different life-spans.

Immune responses change with advancing age, and changes in the population of lymphocytes from the thymus (T cells) seem to be particularly important in immunological age changes. Diane Popp of the Oak Ridge National Laboratory has used infusions of lymphocytes between congenic mice differing only at the major histocompatibility locus to analyze immunological aging. In her particular stocks, she observed strong immune reactions in young mice, moderate reactions in middle-aged mice, and again strong reactions in old mice. Other investigators, working with different genetic materials, have found weak reactions in old mice. Her convincing interpretation of her data is that the many naive immunocompetent cells in young mice give a strong response. Her middle-aged mice have already used up many of their naive T cells by reacting to agents in the environment, and have had time to build up only relatively few "memory cells"; hence, their total immune response is only moderate. Her old mice have almost no naive T cells, but a much greater supply of descendants of previously exposed "memory" cells, and can again respond very effectively. Use of single-gene differences has thus provided evidence of a very interesting mechanism of aging: Immunological aging may be a natural consequence of living! Some genetic differences in aging may be differences in the balance between immune cell populations.

Single-gene differences affecting blood values can also affect total life-span. Substitution of a particular deleterious allele for its normal counterpart makes the difference between a mouse with normal blood values and one with severe macrocytic anemia. These anemic mice die at considerably earlier ages than do their normal counterparts. Using marrow-cell implants to supplant the abnormally functioning blood-forming tissue of anemic host mice with histocompatible normal blood-forming cells results in a chimeric mouse, most of whose body carries the defective mutant allele, but which also has normally functioning marrow cells carrying the normal gene. This kind of chimeric mouse soon develops normal blood values which it maintains throughout the rest of its life and lives considerably longer than do untreated anemic mice. In fact, the chimeric mice live almost as long as do mice which carry only the gene responsible for normal blood formation in all body tissues. These experiments show that a single gene difference, producing differences within one tissue only, can materially change total life-span.

SUMMARY

The most important lesson in this chapter concerns the existence of genetically controlled differences, as well as environmentally caused differences, in patterns of aging, at least in experimental animal populations. Aging changes are diverse and seem to have a variety of causes.

A second lesson is that some problems, including detection of genetic effects on aging, can often be tackled more effectively by using animal models than by analyzing human populations.

A third lesson is that differences in the processes of aging may be caused by interacting effects of many genes, as in differences observed between inbred strains, or may be caused by effects of genes at one locus only. Single-gene-induced differences can be exploited to analyze specific aspects of the processes of aging.

ACKNOWLEDGMENT. The preparation of this chapter has been supported by U.S. Public Health Service program project AG-00250, from the National Institute on Aging.

BIBLIOGRAPHY

Cutler, R. C. 1975. Evolution of human longevity and the genetic complexity governing aging rate. *Proceedings of the National Academy of Sciences of the USA* 72: 4664–4668.

Hart, R. W., and R. B. Setlow. 1974. Correlation between deoxyribonucleic acid excision-repair and life-span in a number of mammalian species. *Proceedings of the National Academy of Sciences of the USA* 71: 2169–2173.

Russell, E. S. 1966. Lifespan and aging patterns. Chap. 26. Pages 511–519 in E. L. Green, ed. *Biology of the laboratory mouse*, 2nd ed. McGraw-Hill Book Company, New York.

Russell, E. S. 1972. Genetic considerations in the selection of rodent species and strains for research in aging. Pages 55–68 in D. C. Gibson, ed. *Development of the rodent as a model system of aging* [DHEW Publication (NIH)72–121]. Department of Health, Education and Welfare, Washington, D.C.

Russell, E. S. 1978. Analysis of genetic differences as a tool for understanding aging processes. Pages 515–523 in D. Bergsma and D. E. Harrison, eds. *Genetic effects on aging*. Alan R. Liss, Inc., New York.

Smith, G. S., R. L. Walford, and M. R. Mickey. 1973. Lifespan and incidence of cancer and other diseases in selected long-lived inbred mice and their F_1 hybrids. *Journal of the National Cancer Institute* 50: 1195–1213.

15

The Dilemma of Aging Parents: Increased Risk of Genetically Abnormal Offspring

EDWARD L. SCHNEIDER

The increased risk faced by the older parent of having a genetically abnormal child is of great clinical, biological, and sociological importance.

On the clinical level, increased maternal age has been shown to be related to an increased frequency of offspring with chromosomal (multiple-gene) abnormalities, while increased paternal age is now known to be associated with an increased incidence of offspring with single-gene disorders.

The most common parental age-related disorder is the chromosomal condition, the Down syndrome. This genetic disease, first described by Langdon Down in the mid-nineteenth century, is one of the most common causes of mental retardation in children. Down named this condition "mongolism" because of the facial appearance of these infants (Figure 1) and, unfortunately, that term still remains in general use. Since this condition has been found in all countries and among all racial groups, it is more appropriate that we call the condition by the name of the original descriptor. The risk of a mother having a child with the Down syndrome rises from 1 in 2300 at age 20 to 1 in 40 after age 45 (Table 1). It was this high frequency of Down syndrome infants born to older parents that first led to the investigation of parental age effects.

The chromosomal disorders which demonstrate parental age effects account for one in every 200 births. Even this high figure represents only a fraction of the actual chromosomally abnormal human fetuses which

EDWARD L. SCHNEIDER, M.D. • Laboratory of Cellular and Comparative Physiology, Gerontology Research Center, Baltimore City Hospitals, Baltimore, Maryland 21224 of the National Institute on Aging, National Institutes of Health, Bethesda, Maryland 20014

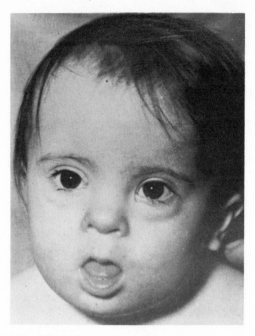

Figure 1. Face of a child with the Down syndrome. (From Smith, *Recognizable patterns of human malformation,* copyright 1970 by W. B. Saunders Company, Philadelphia.)

are conceived. Recent studies indicate that 7% of human fetuses are chromosomally abnormal. For an older mother, this figure might approach 24% or higher.

On a biological level, both the maternal and paternal age effects are of considerable interest. The increased frequency of abnormal offspring associated with advanced parental age will be referred to in this chapter as either maternal age or paternal age effects, depending on whether the mother's or father's age appears to be playing the major role.

Why does maternal age produce one type of disorder and paternal age a second type? To approach this question, one must examine the physiology of their respective germ cells. The oocyte (egg) starts division in fetal life and then remains in that stage for 13 to 50 years until just prior to ovulation. By contrast, sperm are being continuously produced in the male after puberty. How these alternative physiological pathways result in the respective maternal and paternal age effects will be one of the major themes of this chapter.

On a socioeconomic level, the cost of just one parental age-related disorder is staggering. If the frequency of annual Down syndrome births (approximately 8000) is multiplied by the average cost of specialized care

Table 1. *Risk of Having An Offspring with the Down Syndrome**

Maternal age range	Risk
<20	1/2300
20–24	1/1600
25–29	1/1200
30–34	1/870
35–39	1/290
40–44	1/100
>45	1/45

*From R. D. Collman and A. Stoller, "A survey of mongoloid births in Victoria, Australia, 1942–1957," *American Journal of Public Health* 52: 813–829.

for these children ($5000 per year) and the average life expectancy of a Down syndrome newborn (30 years), the total cost is $1.2 billion per year. One must also consider the nonmonetary cost in terms of the hardships that this disease imposes upon the families of Down syndrome patients. These include attention taken from other children, parents forced to stay home to take proper care of these children, and special educational arrangements.

MATERNAL- VS. PATERNAL-AGE EFFECTS

It was previously mentioned that one type of disorder, multigene or chromosomal, was associated with maternal age and a second type, single gene, with paternal age. How were these relationships determined?

Although clinicians had for many years observed that infants with the Down syndrome (mongolism) were more frequently born to older parents, it was not until 1933 that L. S. Penrose, a British geneticist, first concluded that maternal age played the major role and that fathers of Down syndrome children were older than most fathers only because their spouses were older. Many epidemiological studies have confirmed this observation, indicating that with chromosomal disorders there is a strong maternal age effect and that paternal age does not appear to play a major role.

Recently, new cytogenetic techniques have allowed us to gain further insight into the question of maternal versus paternal causes of chromosomal disorders such as the Down syndrome. The human

250   EDWARD L. SCHNEIDER

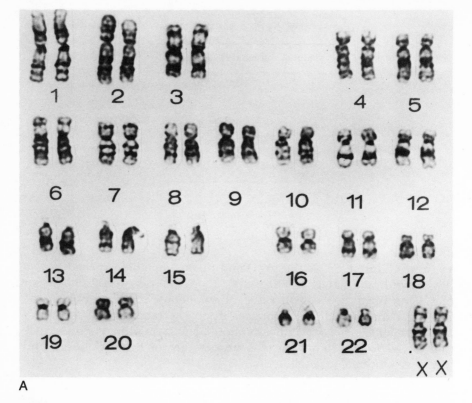

A

Figure 2. (A) Normal female chromosome complement. (B) Normal male chromosome complement. (Courtesy of Helen Lawce, University of California, San Francisco.)

genome is comprised of thousands of individual genes packaged in 22 pairs of autosomal (non-sex-determining) chromosomes, plus either 2X chromosomes in a female (Figure 2A) or 1X and 1Y chromosome in a male (Figure 2B). Most patients with the Down syndrome possess an extra chomosome number 21 in all their cells (Figure 3). With these new cytogenetic techniques, the additional chromosome can sometimes be identified as having come from the mother or father. In most of the informative cases of Down syndrome examined to date, where maternal or paternal chromosomes can be identified, the extra chromosome appears to have come from the mother. Thus, cytogenetic evidence appears to confirm the previous epidemiological studies. In approximately 30% of Down syndrome patients, these same techniques have revealed a paternal origin for the additional 21 chromosome. However, in contrast to the clear elevation in maternal age when the extra chromosome was of maternal origin, the paternal age in these cases was not

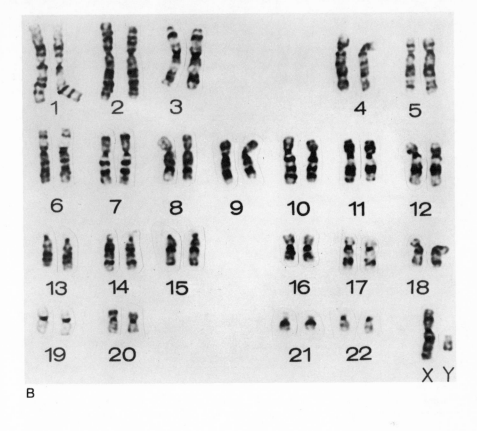

B

significantly increased. Therefore, although maternal aging may not be responsible for all Down syndrome births, it appears to be a contributing factor in the majority.

MATERNAL-AGE-RELATED DISORDERS

The most common maternal-age-related disorder is the Down syndrome. In Table 2, the other disorders that are associated with increased maternal age are listed. As in the Down syndrome, infants with these disorders possess an additional chromosome. Down syndrome involves trisomy (having triple instead of the normal double complement) for chromosome 21, while two other syndromes involve chromosome 13 and chromosome 18, respectively. These disorders which involve autosomal chromosomes are quite severe. Children with the Down syn-

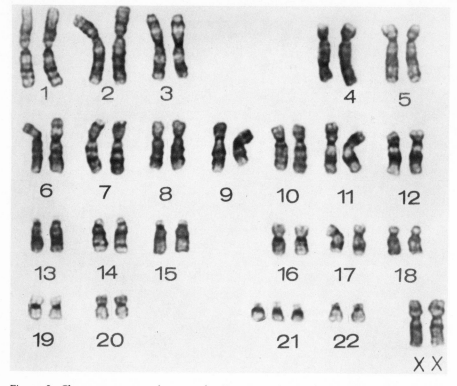

Figure 3. Chromosome complement of a Down syndrome female. Note the extra 21 chromosome. (Courtesy of Helen Lawce, University of California, San Francisco.)

drome, the mildest of the three autosomal trisomies, have severe mental and physical retardation. Children with the other two syndromes, the Edward and the Patau syndromes, rarely survive the first year of life. The other syndromes on the list involve the sex chromosomes: the X or the Y. These syndromes are much less severe than the autosomal conditions and one of them, triplo-X, leads to almost no clinical disabilities. The lack of disability from having an extra X chromosome is a very interesting biological phenomenon. One X chromosome in every female cell is inactivated shortly after conception. This, in effect, gives the female the same equivalent amount of functioning genome as the male, since the Y chromosome is considerably smaller than the X (Figures 2A and 2B) and contains little genetic information other than the genes which determine male sexual development. In conditions such as triplo-X and Klinefelter syndrome, in which the individual has XXX or XXY sex chromosome complements, the extra chromosome is also inac-

Table 2. Genetic Diseases Associated with Advanced Maternal Age

Syndrome	Chromosome complement
Down syndrome	Trisomy-21
Edward syndrome	Trisomy-18
Patau syndrome	Trisomy-13
Klinefelter syndrome	XXY
Triplo-X syndrome	XXX
Double aneuploidy	e.g., Trisomy-21 + XXY

tivated during early development and, therefore, causes little clinical disability.

MECHANISMS OF THE MATERNAL-AGE EFFECT

What is the cause of the increased frequency of chromosomally abnormal offspring born to older mothers? There have been a number of suggested explanations, and these proposed etiological factors can be placed into three categories:

1. Normal aging of the oocyte and/or the female reproductive system.
2. Cumulative exposure of the oocyte and/or reproductive tract to agents capable of inducing chromosomal abnormality.
3. Increased sensitivity of the oocyte and/or reproductive tract to certain agents or environments with increasing age.

These categories are not mutually exclusive, and two or even all three may play a synergistic role.

Normal Aging

The oogonial cell (precursor cell to the egg) must divide twice before the final product, the oocyte, is produced. The first division commences shortly before the birth of the female and then is arrested in this division phase (zygotene stage of the first meiotic prophase) for between 13 and 50 years. Every 28 days between menarche (ages 11 to 15) and menopause (ages 40 to 55), a hormone stimulus from the pituitary gland to the ovary causes one or more oogonia to complete this first division and to start the second meiotic division. Once again, the division cycle is arrested and the oocyte leaves the ovary and enters the fallopian tube where it may be fertilized. At fertilization, the second meiotic division is completed.

The process of cell division is a complicated one. Twenty-three pairs of chromosomes must be equally divided among the daughter oocytes. If one extra chromosome goes to one daughter oocyte, the resultant fertilized egg may produce a child with 47 instead of the normal 46 chromosomes. This child would then have either an autosomal or sex-chromosome trisomy, depending on the nature of the extra chromosome. During this long first division, which can take between 13 and 50 years to complete, the chromosomes remain paired and, at division, they must separate cleanly into daughter cells. Failure of a chromosome pair to separate, called nondisjunction, may well be a time-dependent reaction resulting in production of increasing numbers of chromosomally abnormal oocytes with maternal aging. In fact, the same cytogenetic techniques employed for discriminating between maternal and paternal origin of an additional chromosome have indicated that most cases of maternal chromosomal nondisjunction occur during this first meiotic division.

Cumulative Exposure

If the maternal age effect is related to the action of certain agents or environments, the older mother might have a higher frequency of these abnormalities because of her longer exposure period. An example might be the role of x-irradiation, in which an older mother might be expected to have had a greatei degree of exposure. In addition, the older mother is more likely to have had medical problems such as gallbladder disease or gastrointestinal disorders which result in more than routine radiological exposures.

Increased Sensitivity to Certain Agents

Another possibility is that the tissues or cells of the older mother may be more sensitive to the effects of certain agents such as viruses. Although the viral etiology of the Down syndrome is only a speculation at this time, an older mother's impaired immune competence might predispose her to viral infection, which, in turn, might lead to the production of chromosomal abnormalities.

PROPOSED CAUSES FOR MATERNAL-AGE-RELATED CHROMOSOMAL DISORDERS

Radiation

For many years, epidemiological studies have been conducted on mothers of Down syndrome children to determine if these women had an increased exposure to x-irradiation. These studies have been both prospective, performed before the birth of the affected child, and retrospective, performed after the birth of a child with the Down syndrome. In both types of studies, a higher incidence of x-irradiation exposure was observed in mothers of Down syndrome infants when compared to controls, mothers of chromosomally normal infants.

In a region of Kerala, India, with high background radiation in the soil due to the presence of radioactive minerals, the frequency of Down syndrome births is considerably higher than that seen in other areas of Kerala.

In the laboratory, experimental x-irradiation of normal human cells in tissue culture has produced chromosomal abnormalities similar to those seen with increased maternal age. In addition, x-irradiation of female mice prior to pregnancy resulted in increased frequencies of chromosomally abnormal fetuses.

Together this evidence indicates that radiation is one of the leading factors suspected of playing a role in the maternal age effect.

Chemicals

The last few decades have witnessed an explosion in the number of drugs now available to the public. In addition to prescription and proprietary drugs, our diet is becoming increasingly populated with chemical additives such as food colorings, preservative compounds, and other substances, such as monosodium glutamate.

We know from laboratory experiments that chemicals are quite capable of causing chromosomal abnormalities. Although several drugs, such as thalidomide, have been implicated in birth defects, none has been demonstrated to induce chromosomally abnormal offspring in man. There was some question over whether LSD could lead to chromosomal breakage and malformations; however, evidence for such an effect by LSD has been challenged.

Infectious Agents

One of the most intriguing proposed causes for the maternal-age effect is an infectious agent. Such an agent might have a predilection for older women because of their decreased immune capabilities. If an infectious agent does contribute to the maternal-age effect, it is unlikely that it would produce severe symptoms that would lead a woman to consult a physician. Epidemiological studies should have detected a serious illness related to offspring with chromosomal abnormalities. It is, therefore, much more likely that such an infection would be subclinical, like the common cold.

Two Australian workers, R. D. Stoller and A. Collman, surveyed all the Down syndrome births in Australia and found annual and seasonal variations similar to those produced by infectious diseases. In addition, they noted clusters of Down syndrome births in certain areas of Australia. In fact, they concluded that it was likely that an infectious agent was responsible for the Down syndrome. However, epidemiological studies in other areas of the world have not confirmed their observations. Therefore, despite the convincing work of Stoller and Collman, the infectious origin of the Down syndrome still remains to be proven.

If an infectious agent were involved, viruses and mycoplasmas (unicellular organisms whose size is between bacteria and viruses) would be the most likely candidates for this etiological role. Both viruses and mycoplasmas have been shown to induce chromosomal abnormalities in cultured human cells. Of the two, viruses are of additional interest since they are capable of attaching to chromosomes and even integrating their genomes with the host genome. One could, therefore, envision that a virus attached to a chromosome might lead to nondisjunction of the chromosomal pair during cell division.

Altered Immunity

It has been shown in human beings as well as in experimental animals that immune function declines with age (see Chapter 13). Altered immunity might lead to chromosomal nondisjunction by two mechanisms. First, altered immune competence might predispose older mothers to infectious agents. Second, altered immunity can be manifested by autoimmunity (see Chapter 13). Several investigators have reported that autoantibodies increase as a function of normal aging. Studies of the Down syndrome by Phil Fialkow of the University of Washington, Seattle, and others have revealed an increased frequency of thyroid autoantibodies in mothers of Down syndrome children. Of even greater importance was the finding of L. Dallaire of McGill University, Montreal, Canada, that mothers of Down syndrome children have an increased frequency of antibodies against ovarian tissue. Although it is not clear how autoimmunity might lead to nondisjunction, one might speculate that antibodies reacting with the surface of oocytes could lead to abnormal cell division and, thus, abnormal chromosome segregation.

Delayed Fertilization

Another hypothesized cause for the maternal age effect is delayed fertilization of the oocyte. There is some evidence that in man and other animals there is age-related decline in the ability of the female reproductive system to transport the oocyte to the fallopian site of fertilization. The decreased frequency of sexual intercourse that is observed in older couples might also contribute to delayed fertilization of an oocyte. In experimental animals, it has been demonstrated that delayed fertilization of oocytes can, indeed, lead to the production of chromosomally abnormal fetuses. However, these fetuses are usually triploid (possess three copies of each chromosome instead of the normal two). Chromosome analysis of these fetuses indicates that, most often, this extra chromosome complement comes from abnormal oocyte division rather than from fertilization by two sperm. Since triploidy is extraordinarily rare in human beings, the relationship between delayed fertilization and chromosomal nondisjunction in man is unclear.

Genetic Predisposition

The medical literature contains many reports of several chromosomal disorders occurring in the same family. In lower organisms, there are specific genes which lead to chromosomal nondisjunction. It has, therefore, been suggested that there may be specific genes in man which lead to a predisposition toward nondisjunction. A maternal age effect in this context might be the result of age-related expression of such a gene. In fact, there are many examples of mutant genes in man with delayed expression such as Huntington's chorea, the disorder that afflicted Woody Guthrie. In this condition, the individual remains relatively healthy until the third or fourth decade of life, when the disease manifests itself. However, the bulk of the evidence accumulated to date does not support a genetic predisposition toward chromosomal nondisjunction in man.

APPROACHES TO STUDYING THE MATERNAL-AGE EFFECT

Because of the obvious ethical and practical limitations to human experimentation, the majority of human studies have been epidemiological. This has led our laboratory and others to search for appropriate animal models for studying the maternal-age effect. The mouse was chosen since it has been used to study a number of other human disorders, including diabetes, muscular dystrophy, and albinism. Another important advantage of the mouse is its relatively short lifespan and reproductive period.

Both our laboratory and that of G. Watanabe and M. Yamamoto of Niigata University School of Medicine, Japan, have found the mouse to be a good experimental animal for studying the maternal-age effect. In both studies, a statistically significant increased frequency of chromosomally abnormal fetuses was found in older mice when compared to controls. Watanabe and Yamamoto x-irradiated their group of mice and produced an increased incidence of chromosomally abnormal fetuses in both the young and old group. In our laboratory, Jill Fabricant has assessed the genetic and immunological components of the maternal-age effect in mice and found that neither factor appears to play a crucial role. Certainly, much work is needed before conclusions can be drawn from these experiments on mice. It is evident, however, that the mouse is a good model for the maternal-age effect in man.

Some of the intriguing experiments that should be carried out include transferring old oocytes to young mice recipients and the converse. These experiments may allow us to delineate between the intrinsic (oocyte) and extrinsic (uterine environmental) components of the maternal-age effect.

PATERNAL-AGE EFFECT

In contrast to the maternal-age effect, paternal-age effects have received little attention because of the relatively low frequency of disorders related to increased paternal age (Table 3). Achondroplasia, a form of dwarfism, is the most common of these syndromes and has an incidence of 1 in 50,000 live births, approximately 100 times less frequent than the Down syndrome. However, as with the Down syndrome, the

Table 3. Genetic Diseases Associated
with Advanced Paternal Age

Achondroplasia
Apert syndrome
Myositis ossificans progressiva
Marfan syndrome
Retinoblastoma (bilateral)
Basal cell nevus syndrome
Waardenburg syndrome
Crouzon syndrome
Oculo–dental–digital syndrome
Treacher–Collins syndrome
Progeria
Acrodystosis

risk of having children with achondroplasia increases up to fiftyfold as a function of paternal age. Of particular interest is the fact that this risk estimate increases in a similar exponential fashion, as does the risk figures for the maternal-age-related disorders.

Most of the disorders associated with advanced paternal age are inherited in an autosomal dominant fashion. This type of inheritance implies a mutation in a single gene. Unfortunately, little is known about the causes of the paternal-age effect. Since spermatogonia (sperm cell precursors) are continuously producing sperm in the male, a single mutation in one spermatogonium could lead to the production of many abnormal sperm. As with the maternal-age effect, the paternal-age effect could be due to accumulated exposure to mutagenic agents, increased susceptibility of aged spermatogonia to mutagenic agents, or merely to normal aging's creating errors in DNA replication and/or errors in repair of normal DNA damage. Certainly, we hope that laboratories will tackle this problem as well as the maternal-age effect.

ADVICE TO OLDER PARENTS

Fortunately, major advances in human genetic diagnosis now permit chromosomal analysis of human fetuses in the early part of pregnancy with minimal risk to the mother. This procedure, called amniocentesis, involves the removal of a small amount of fluid from the normal fluid reservoir that surrounds the embryo. This amniotic fluid normally contains cells desquamated from the fetus. These cells are then placed into tissue culture and, after approximately 4 weeks, sufficient cells are accumulated for chromosomal analysis. In this manner, the chromosomal complement of the fetus can be accurately determined.

Amniocentesis can be performed as early as the fourteenth gestational week, and the results can be made available to parents as early as the eighteenth week. Parents can then elect to abort those fetuses with abnormal chromosome complements. It is important that prenatal diagnosis be accompanied by genetic counseling. Parents should be advised of their chances of having an abnormal child as well as the risks of the procedure. Studies of the frequencies of abortions occurring after amniocentesis indicate no significant increase over the number that would be expected at this stage of pregnancy. Therefore, the procedure appears to be reasonably safe for both the mother and her unborn child.

Currently, there are approximately 100 centers in the United States which offer genetic counseling and prenatal diagnosis by amniocentesis. While there is no exact age above which prenatal diagnosis is recommended, most centers utilize 35 or 37 years as their criterion for at risk status. To date, well over 10,000 amniocenteses have been performed for prenatal diagnosis of genetic disorders. The most frequent indication for prenatal diagnosis has been advanced maternal age. Other indications for prenatal diagnosis include family histories of inherited biochemical and sex-linked disorders. Of the 4855 amniocenteses performed for advanced maternal age, 131 (3.7%) resulted in diagnosis of chromosomally abnormal fetuses. With the expansion of these prenatal diagnostic centers and the dissemination of the appropriate information to older mothers, one can anticipate a significant drop in the frequency of maternal-age-related disorders such as the Down syndrome.

Unfortunately, the disorders related to paternal age are difficult to diagnose during early pregnancy. It is hoped that the future may bring improved genetic, biochemical, or radiological techniques which might make early prenatal diagnosis of paternal-age-related disorders possible.

CONCLUSIONS

With maternal and paternal aging, there is a marked increased risk of having genetically abnormal children. Maternal age appears to be the more serious consideration, since the risk of an older mother having a chromosomally abnormal offspring may be as high as 1 in 20. Although there has been considerable speculation as to the etiology of these parental-age effects, the mechanisms remain to be elucidated. The most likely explanation for the difference between maternal and paternal effects is the long (13- to 50-year) period that the oocyte remains in the first meiotic division. However, other factors, such as radiation, infectious agents, altered immunity, delayed fertilization, and genetic predisposition, may also play contributing roles. Progress toward understanding the mechanisms for the maternal-age-related increased incidence of chromosomal disorders in man is being made utilizing the mouse as an experimental model. The goal of these studies is to prevent the tremendous fetal waste that occurs with maternal aging. However, until this goal is achieved, prenatal diagnosis by amniocentesis remains the best option for the older, pregnant mother.

Although the paternal-age-related effects have not received much attention, it is hoped that future research will also be directed at this important area.

BIBLIOGRAPHY

Golbus, M. S. 1977. The prenatal diagnosis of genetic disorders. In *Advances in obstetrics and gynecology*. The Williams & Wilkins Company, Baltimore.

Kram, D., and E. L. Schneider. 1978. Result of reproductive aging: production of genetically abnormal offspring. In E. L. Schneider, ed. *The aging reproductive systems*. Raven Press, New York.

16

Endocrinology and Aging

PAUL J. DAVIS

The inevitability of hormone-mediated senescence in the female reproductive system well in advance of death of the organism has long focused attention on the possibility that hormones mediate the aging process. That aging might reflect, at least in part, deficiency of hormone action raised the attractive prospect of the postponement of aging with hormone replacement therapy. This prospect in large measure explains the inordinate amount of attention which hormonal factors have periodically received in attempts to account for senescence. It also explains the blemishes on nineteenth- and early-twentieth-century medical history of attempts to graft into old men the gonads of goats (and other animals) to effect "rejuvenation." Old scientists were among those who generously offered to be the recipients of such grafts.

After a brief introduction to hormone physiology, this discussion will consider the interrelationships of hormone action and aging from two aspects: (1) Is aging a hormone-modulated process? (2) How does the aging process affect various hormone systems?

ENDOCRINE GLAND PHYSIOLOGY

Hormones are chemicals released into the blood by specialized tissues (endocrine glands). They are composed either of chains of amino acids (polypeptides) or of complex rings of carbon atoms (steroids) (Figure 1). Transported in the blood, hormones subsequently act to regulate selective aspects of metabolism, protein synthesis, or growth in other endocrine target glands — in which case they are termed "trophic" hormones — or in nonendocrine target tissues (liver, muscle). Polypeptide hormones, such as thyrotropin, adrenocorticotrophic hormone (ACTH), and insulin, tend to be membrane-active; that is, they express

PAUL J. DAVIS, M.D. • Endocrinology Division, Department of Medicine, Medical School of the State University of New York at Buffalo, Buffalo, New York 14215

(A)

AA = AMINO ACID

(B)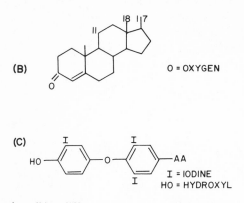

O = OXYGEN

(C)

I = IODINE
HO = HYDROXYL

Figure 1. Schematic structures of types of hormones. (A) Polypeptide hormones. Examples are thyrotropin (TSH), adrenocorticotropic hormone (ACTH), parathyroid hormone (PTH), and insulin. Antidiuretic hormone is a polypeptide consisting of only eight amino acids in a chain, whereas TSH consists of two polypeptide chain subunits, each of which contains more than 100 amino acids. (B) Steroid hormones. The steroid structure consists of three 6-carbon rings and one 5-carbon ring. Examples of such hormones are cortisol, aldosterone, and the sex steroids estrogen and androgen. Binding of hydroxyl (−OH) groups at C11 or C17, or the addition of a 2-carbon chain at C17, result in striking differences in end-organ specificity and function for these steroids. (C) Thyroid hormone. Represented here is triiodothyronine (T_3). Addition of another iodine molecule to the left-hand ring results in the formation of tetraiodothyronine (thyroxine or T_4).

their effects by acting on target-cell outer membrane to influence intracellular content of a "messenger" (e.g., cyclic nucleotides) or of ions such as calcium (Figure 2). In contrast to polypeptide hormones, steroid hormones usually act after crossing the cell membrane and influencing nucleus-directed synthesis of proteins. This action is dependent upon the formation of complexes between the steroid and intracellular soluble proteins ("receptors"). The thyroid gland elaborates a hormone which is neither a polypeptide nor a steroid but consists of two amino acid residues (tyrosine and phenylalanine) linked together by an oxygen molecule (Figure 1). Interestingly, thyroid hormone actions both on cell membrane transport systems and on nucleus-directed protein synthesis (Figure 2) have been postulated.

The release of endocrine gland secretions is controlled by the concentration in the blood of those substances which increase in amount in response to the level of hormone activity. This is a "feedback loop" system (Figure 3). Secretion of trophic hormones, for example, is a function of the pituitary gland; such secretion is regulated by the blood level of target-organ products. Thus, thyrotropin (TSH) is released by the pituitary and is capable of stimulating the thyroid gland, located in the neck, to secrete thyroid hormones [i.e., thyroxine (T_4) and triiodothyronine (T_3)]. TSH secretion is decreased when levels of T_4 and T_3 in blood increase ("negative feedback inhibition"). If circulating levels of T_4 and T_3 fall to abnormal levels, TSH release is increased.

Another layer of hormonal complexity is added to the discussion when we see that, in addition to feedback control, trophic hormone secretion is apparently subject to control from the hypothalamus. The hypothalamus is anatomically continuous with the pituitary gland. The hypothalamus has its own hormone system, the "releasing factors" (Figures 3 and 4). Releasing factors stimulate the release of trophic hormones by the pituitary. For example, thyrotropin (TSH) is secreted when the hypothalamus produces thyrotropin-releasing hormone (TRH) to act on the pituitary.

Endocrine deficiency in the case of the pituitary–thyroid hormone axis can thus be defined at a variety of levels: (1) releasing factor (at the level of the hypothalamus), (2) trophic hormone (at the level of the

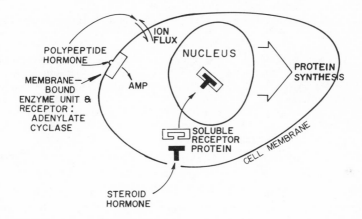

Figure 2. Schematic representation of possible mechanisms of hormone actions. Polypeptide hormones — such as PTH, TSH, or ACTH — activate a cell membrane-bound enzyme system, adenylate cyclase, resulting in an increase in intracellular cyclic AMP. cAMP is a "messenger" thought to trigger hormone-mediated events in the cell. It is also possible that polypeptide hormones may act on membranes to alter transport of ions (such as calcium). Steroid hormones apparently have little interaction with cell membranes, but once inside the cell they are complexed with "receptor" proteins in the cell cytoplasm. The only way in which steroid hormones can gain access to the cell nucleus — where they influence DNA-directed protein synthesis — is in a complex with these receptor proteins. Steroid hormones known to work in this manner are progesterone, androgen, estrogen, and perhaps cortisol. Thyroid hormone has important roles in regulating heat production in higher animals and in modifying rates of protein synthesis within cells. It is possible that thyroid hormone-modulated heat production ("calorigenesis") reflects primarily a membrane action of the hormone — that is, stimulation of sodium transport. Thyroid hormone also is taken up by cell nuclei, but apparently without the need for a cytoplasmic "receptor."

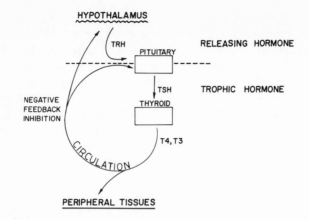

Figure 3. Classic "loop" of hormone physiology: the interaction of the pituitary and thyroid glands. The thyroid gland makes thyroxine (T_4) and triiodothyronine (T_3), hormones which act on peripheral tissues (liver, muscle), as described in the legend to Figure 2. The pituitary gland controls T_4 and T_3 release from the thyroid gland by secreting TSH, a trophic hormone, into the blood. The pituitary release of TSH is sensitive to blood levels of T_4 and T_3. Thus, high levels of T_4 or T_3 in blood block TSH release ("negative feedback inhibition"). Low blood levels of T_4 or T_3 result in the stimulation of TSH secretion by the pituitary gland. In turn, pituitary release of TSH is sensitive to the presence of TRH (thyrotropin-releasing hormone), a substance secreted by the brain (hypothalamus). Thus, neural control of the pituitary–thyroid loop is possible. How important this neuroendocrine input is to hormone physiology is not yet clear. The dashed line represents the boundary of the nervous system. TRH is produced in the brain and released into blood vessels which carry it to the pituitary.

pituitary gland) and (3) at the level of the thyroid gland itself. In addition, inadequate thyroid hormone action will result if there is *resistance* to T_4 or T_3 action at sites in peripheral nonendocrine tissues. Feedback loops, similar to those described in Figures 3 and 4, also relate the adrenal gland cortex and gonads to the hypothalamus and pituitary. Against this background of endocrine gland physiology, we can now examine possible relationships between aging and hormone action.

THYROID FUNCTION

Because of the similarities between the clinical picture of thyroid hormone deficiency — languor, stolid facial expression, skin pallor — and aging, a substantive effort has been made to implicate inadequate thyroid hormone action in the process of "normal aging." It is obviously attractive to conceive of delaying senescence with a simple maneuver such as administering readily available thyroid hormone. In fact, however, it is difficult to implicate the thyroid in the aging process. Levels in blood of thyroid hormones — T_4 and the more metabolically active T_3 — do not appear to decline significantly with age. A number of reports do describe a selective fall in serum T_3 concentrations with aging, but these reports have not carefully eliminated T_3 measurements made in elderly patients with *non*thyroidal illness. Sudden or chronic non-thyroidal illness is usually accompanied by a fall in serum T_3 concentration, regardless of patient age, and it is this phenomenon which is likely to account for the conclusion that a decline in serum T_3 content was "specific" for advancing age. Evidence from various sources indicates that it is unlikely that depressed levels of circulating T_3, were they to be satisfactorily documented, would result in a thyroid-deficiency state when serum T_4 levels remain in the normal range. Depressed concentrations of serum T_3 in patients with nonthyroidal illness probably reflect

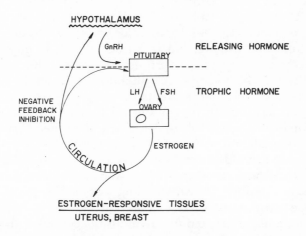

Figure 4. Another "loop" of hormonal feedback: the interaction of the pituitary gland and gonads. Similarities between this loop and the pituitary–thyroid axis, schematically shown in Figure 3, are apparent. GnRH, gonadotropin-releasing hormone, is a hypothalamic product capable of inducing the release of LH (luteinizing hormone) and, to a lesser extent, FSH (follicle-stimulating hormone). FSH and LH synchronize effects in women on the ovary to cause follicle maturation, estrogen production by the follicle, and, cyclically, ovulation. While the feedback circuit is represented here as a negative feedback arrangement, it is important to point out that periodic bursts of estrogen actually provoke LH release (i. e., "positive feedback"). Long-standing high estrogen levels in blood do suppress LH. FSH is not very susceptible to suppression by estrogen. It is the usually monthly burst of LH in the premenopausal female which triggers ovulation. In the menopausal woman, low ovarian estrogen secretion and low blood levels of this hormone result in high FSH and LH production.

impaired conversion of one form of thyroid hormone (T_4) to another (T_3) in tissues outside the thyroid, such as the liver and kidney. As a first step in the analysis of possible interrelationships between thyroid function and aging, we can conclude that effectively normal levels of thyroid gland hormones are found in the circulation, regardless of subject age.

Nor is aging characterized by a decrease in serum concentration of thyrotrophic hormone (TSH). There is, however, an apparent decrease with age in men in the amount of TSH released by the pituitary gland in response to the administration of synthetic hypothalamic releasing hormone (TRH). This interesting observation raises the possibility that neuroendocrine input to this, and perhaps other, endocrine gland systems is faulty as aging progresses. This thesis is discussed in more detail in Chapter 19. It is important not to overinterpret changes in TRH responsiveness with aging, however, because of the wide range found in normal young subjects of responsiveness to this releasing factor. Too, the surprising finding in the rat that immunoassayable levels of TRH in peripheral blood do not change as a function of adequacy of thyroid gland function also must temper efforts which are made to emphasize deficient TRH action as a hallmark of aging.

There are other changes in thyroid hormone physiology which occur as a function of aging, but it is also difficult to implicate these changes mechanistically in the process of aging. Frequently, these changes in thyroid physiology mirror alterations with age in the function of other *non*endocrine organ systems. For example, thyroid hormone synthesis and release by the thyroid gland decrease with age, but this change is apparently secondary to decreased "peripheral" (i.e., outside the thyroid) degradation of thyroid hormone. The fact, therefore, that the thyroid gland in elderly subjects may be "less active" in terms of making thyroid hormone cannot be viewed as evidence of a thyroid deficiency state (hypothyroidism) as long as blood levels of T_4 and T_3 persist in the normal range. The uptake of radioactive iodide by the thyroid gland, a test of thyroid function, also decreases with advancing age. Rather than a herald of thyroid failure, this change is apparently an adjustment made by the gland in response to increased blood levels of iodide which characterize decreasing kidney function in aged subjects. Finally, responsiveness of the thyroid gland to the action of TSH appears to be intact in elderly individuals. To this point, then, we can summarize available data as showing the impact of age on the thyroid hormone axis to be biologically unimportant and as failing to support a role for deficient thyroid hormone action in the normal aging process.

Recently, renewed interest in the thesis that the thyroid axis may be involved in modulating aging has been generated by observations that suggest that senescence is a thyroid hormone-resistant state. Data are available from animal studies (rat) which indicate that with age, thyroid hormone contributes less significantly to metabolic rate (oxygen consumption), despite the persistence of normal concentrations of the hormone in serum. Historically, the major function of thyroid hormone in adult mammals has been visualized to involve maintenance of a normal metabolic rate — that is, appropriate oxygen consumption and heat production. A pituitary gland "factor" has been postulated which may block the end-organ effects of thyroid hormone and thus produce a state of hypothyroidism in aged animals.

These provocative observations in an animal model cannot currently be viewed as relevant to man. In man, there is no evidence that old age is characterized by decreased thyroid hormone action. In fact, thyroid hormone action appears to be normal in senescence or perhaps tissue sensitivity to T_4 and to T_3 is heightened in old age. If end-organ resistance to thyroid hormone occurred with aging, we would expect blood thyroid hormone levels to rise. Resistance to thyroid hormone action is occasionally observed as a genetic disorder and is characterized by extremely high blood concentrations of T_4 and T_3 as an attempt is made by the pituitary–thyroid hormone axis to overcome the resistant state. Finally, if the TRH-responsiveness observations cited earlier are valid, they are contrary to what would be expected in a thyroid-deficiency state. That is, in individuals who are hypothyroid because of thyroid gland destruction, an exaggeration of the pituitary outpouring of TSH in response to TRH has been well documented. We can conclude, then, that the pituitary–thyroid hormone axis does not play a major role in aging in man. The observation of impaired responsiveness of the pituitary gland to thyrotropin-releasing hormone in elderly men suggests that there may be impaired neuroregulation of the pituitary as aging proceeds. The consequences of this are unknown.

ADRENAL GLAND FUNCTION

Like the pituitary–thyroid axis, the pituitary–adrenal gland cortex axis appears to function normally in senescence. The cortex produces cortisol, among other hormones, a steroid which is critical to survival because of its important effects on glucose metabolism; cortisol secretion is controlled from the pituitary gland by ACTH. The degradation of cortisol by nonadrenal tissues and secretion of cortisol by the adrenal cortex do decrease with age, but blood levels of cortisol persist within the normal range throughout senescence. Recent studies buttress the view that the response of adrenocortical cells to ACTH and the capacity of the pituitary gland to release ACTH in response to physiologic stimuli are quite intact in healthy elderly subjects. The adrenal cortex also produces aldosterone, an important salt-retaining hormone, and a number of other steroids. The secretion of aldosterone is only modestly influenced by ACTH but is sensitive to the end product of a complex cascade of chemical reactions which occur in the blood and are initiated by renin, an enzyme primarily of kidney origin. Levels of serum potassium also influence aldosterone secretion.

In the case of the renin–aldosterone axis, it is clear that the "responsiveness" of the system *is* decreased with age and that renin levels in blood, as well as aldosterone secretion, decline as aging proceeds. The term "responsiveness" here relates to the effects on renin activity in blood and on aldosterone release of changes in body posture (upright posture vs. supine posture) and altered oral salt intake on this salt-retaining hormone system. It is unknown what, if any, are the physiologic consequences of these changes.

GONADAL FUNCTION

Follicle-stimulating hormone (FSH) and luteinizing hormone (LH) are pituitary-source polypeptide trophic hormones whose site of action (target) is the gonad. It is this hormonal axis, as pointed out above, which has drawn particular attention to the possibility that hormones modulate the aging process. It is apparent that there is no impairment of the ability of the pituitary gland to secrete FSH and LH in the menopausal and postmenopausal woman; as ovarian sex hormone (estrogen) secretion wanes at the time of menopause, FSH and LH production by the pituitary increases via a feedback loop (Figure 4). Late in the menopause (i.e., eighth decade of life), secretion of these two pituitary hormones may finally decrease. It is generally believed that failure of estrogen secretion by the ovary is the immediate cause of the menopause. It has been suggested, alternatively, that a primary hypothalamic or pituitary process could initiate the menopause, for example through development of an elevated feedback threshold for estrogen. The latter thesis involves aberrant behavior of pituitary or hypothalamic sex hormone receptors which control FSH and LH secretion and failure of these receptors to suppress trophic hormone release in the presence of normal circulating estrogen levels. The inappropriately high FSH and LH levels in blood which then result are regarded, in this thesis, as the immediate cause of ovarian failure (overstimulation → exhaustion).

Although a large number of animal studies have been carried out to weigh the alternatives of primary (that is, within the ovary) and secondary (that is, hypothalamus/pituitary gland-mediated) ovarian failure as the initial step in the menopause, no such studies have been carried out in animals with menstrual rather than estrous cycles. Interesting animal studies do support the possibility that the environment of the ovary, rather than primary changes within the ovary, may condition ovarian failure in estrous animals. For example, ovaries from old rats transplanted to younger animals may resume normal function, and ovaries from normal young rats, transplanted to senescent animals, may abruptly lose function. In women, however, the best evidence to date is that the primary biologic event of the menopause occurs within the ovary. A significantly reduced number of ovarian follicles, the basic organizational units of the ovary, are present in the ovary of the 45 year-old woman compared to that of the adolescent. Many of these residual follicular units are thought to be somewhat insensitive to LH and FSH action. They do not mature properly, are inadequate sources of estrogen, and thus provoke the characteristic increases in LH and FSH secretion as the pituitary attempts to stimulate normal follicular development. The mechanism(s) by which ovarian senescence is programmed, however, is unknown. Also unaccounted for is the decrease in estrogen production from *non*follicular sources in women, a decrease which occurs according to the same timetable as ovarian follicular failure.

There is no question but that the menopause is a dramatic hormonal and gonadal event. It is unjustified, however, to use this model to hypothesize hormone modulation of other inevitable events of senescence. The timetable for ovarian failure cannot at this time be viewed as representative of other hormonal axes. The menopause itself does not directly contribute to the aging process. Loss of estrogen secretion does alter certain risks in women, for example, that of atherosclerosis. Pathologic bone demineralization in women ("postmenopausal osteoporosis") begins long before menopause, and its mechanism is not established; it is not corrected by chronic estrogen administration. The validity of estrogen replacement therapy in postmenopausal women is discussed in Chapter 17. It is clear that such therapy carries material risks (hypertension, hyperlipidemia, carbohydrate intolerance, endometrial carcinoma, thromboembolic events) which must be weighed against possible benefits.

In the case of the human male, it has long been held that sex steroid (androgen: testosterone) concentration in blood was well supported throughout the aging process. More recent statistics indicate rather conclusively, however, that testosterone levels in blood do decline in aging males. These data have been derived from studies which carefully compare subjects of varying ages. Nonetheless, there are enormous individual variations in testicular androgen secretion capability, regardless of age, so that elderly subjects may sometimes have higher testosterone secretory rates than much younger males. Serum LH and FSH concentrations rise, as would be expected, as gonadal hormone production falls. The fact that a decrease in testosterone secretion has been documented statistically as a function of age is no basis upon which to propose an inevitable lack of sex steroid action in aging men and to recommend routine androgen replacement therapy. The loss of sexual potency which occurs coincidentally with advancing years is a highly complex process in which so many psychological and some physical factors play a role that it is unjustified to attribute to decreased testosterone production any kind of major responsibility.

OTHER HORMONAL AXES

The thyroid, adrenal cortex, and gonadal axes represent three endocrine systems under trophic control by the pituitary. A number of other hormones exist which are not directly under pituitary control or are, themselves, pituitary hormones released into the circulation but without specific endocrine targets. Examples of such hormones are antidiuretic hormone (ADH) and growth hormone (GH). ADH is a water-conserving hormone which is stored in the pituitary gland and whose target organ is the kidney. Recent studies of water metabolism in elderly subjects suggests that ADH has quite normal secretory dynamics as aging progresses. It is known that the ability to make a concentrated urine declines with age, however, and it has been postulated that perhaps this phenomenon reflects decreased ADH action; that is, with a decrease in ADH secretion, more water would be excreted by the kidney with a given amount of salt, and a more dilute urine would result. It is now clear that secretion of ADH may, in fact, be exaggerated in aged subjects because the end organ (kidney) is somewhat resistant to hormone action. Receptors in the brain and cardiovascular system which regulate ADH secretion — receptors which are sensitive to the blood osmolality — appear to have heightened sensitivity in old subjects to the administration of ethanol and of sodium chloride, manipulations which turn off and turn on, respectively, the secretion of ADH. As is the case with certain other hormonal systems, then, alterations with age in ADH secretion appear to be secondary to changes with age in other, nonendocrine organs.

Growth hormone (GH) is a pituitary polypeptide which has diffuse peripheral tissue effects and complex feedback mechanisms. GH is, of course, crucial to fetal and childhood development but cannot be described as an important contributor to metabolic processes in aged subjects. Its dynamics have, nonetheless, been studied in elderly individuals, and recent observations involving nonobese subjects indicate that the secretion of GH from the pituitary in response to provocative stimulation — such as lowering of blood sugar concentration — is diminished with aging. Another pituitary polypeptide, prolactin (PRL), is a hormone important in the regulation of events which result in milk production in the postpartum woman. Its physiologic role in men has not been defined. Receptors for PRL do exist in the prostate gland, however, and in liver, so that PRL may have a broader physiologic role than has been previously thought. Although there are sex differences in PRL secretion in response to TRH — a hypothalamic releasing factor which promotes both TSH and PRL secretion — there appear to be no age-related changes in PRL release.

Parathyroid hormone (PTH) is a nonpituitary, calcium-mobilizing polypeptide with important actions on bone, on the kidney, and on metabolism of vitamin D. It is secreted by specialized endocrine tissue — the parathyroid glands — usually embedded within the substance of the thyroid gland. PTH secretion, however, is probably not related to thyroid function. Secretion of PTH is influenced chiefly by the concentration in blood of ionized calcium (Ca^{2+}), rising or falling, respectively, in response to decreases or increases in Ca^{2+}. There has been interest in the dynamics of PTH in aged subjects because of the loss of bone mineral and bone matrix (osteoporosis) which develop as a concomitant of aging. Excessive PTH production in certain pathologic states may certainly result in such demineralization of bone. There is no definitive evidence, however, to implicate PTH in the loss of bone mass which accompanies normal aging. Slightly increased and decreased PTH levels in blood have both been described as concomitants of aging, and it is fair now only to state that no change in PTH dynamics is "characteristic" of aging.

Conclusions which may be made from the surfeit of observations which attempt to relate function of the endocrine glands to aging are the following:

1. There is no satisfactory evidence which substantiates a contribution of endocrine gland function to the process of aging.
2. Aging is characterized by normal circulating levels of thyroid hormone and cortisol, but an appropriately decreased secretion of these substances in response to decreased degradation of hormones in peripheral tissues.
3. Endocrine tissue responsiveness to trophic hormones and stress is intact in the cases of the adrenal cortex and thyroid gland.

4. The menopause is a hormone-mediated event which chronicles but does not regulate aging. It appears to be a process which is primary in the ovary.
5. The ovary is the only endocrine gland whose functional capacity predictably declines with normal aging. Androgen production by the testis tends to fall with age, but wide interindividual variations do not allow us to describe decreased testosterone secretion as an inevitable concomitant of aging.
6. For certain polypeptide hormones, such as parathyroid hormone and prolactin, no impact of aging has been consistently described. Antidiuretic hormone secretion is intact in older individuals, but kidney response to ADH may be impaired.
7. Data which are available regarding neuroregulation of the endocrine axes imply that such input may be decreased with age. This possibility is discussed in Chapter 19.

INSULIN

An additional hormone — insulin — is important to this discussion because its secretion is influenced by aging and because these aging changes have an impact on clinical medicine. Although controversy has persisted regarding the changes in insulin secretion in man which may be characteristic of aging, well-substantiated data drawn from nonobese subjects indicate that advancing age is characterized by decreased insulin release. Here, "decreased insulin release" refers to insulin secretion which occurs in response to an oral carbohydrate (glucose) challenge. As a result of decreased insulin secretion, the rise in blood sugar which occurs after a glucose challenge is higher than in individuals with normal responsiveness. Basal insulin secretion (i.e., insulin release after an overnight fast) is not influenced by age. Insulin turnover — the fate of the hormone once it has been released into the circulation from its site of synthesis in islet cells in the pancreas — is also unchanged with age. Tissue responsiveness (sensitivity) to insulin is also unaffected by normal aging.

It is important to characterize what changes in glucose tolerance are acceptable as part of "normal aging," so that abnormal glucose tolerance can be recognized. "Abnormal glucose tolerance" is a definition of the disease state of diabetes mellitus. The apparent incidence of diabetes mellitus will, therefore, be importantly influenced by how we choose to regard changes in glucose tolerance in older individuals. The changes under discussion, however, are small and relate to blood sugar elevations after a glucose load, not in the fasting state. It is clear that the incidence of overt diabetes mellitus is substantially increased in patients over the age of 35. "Overt" refers to a state of elevation of fasting blood sugar, to glucose tolerance changes beyond those expected with apparently normal aging or to the appearance of complications of diabetes mellitus (in the peripheral nervous system, eye, or kidney). The incidence of diabetes mellitus is at least five times greater in subjects aged 35 years and over, with an incidence peak at age 50 years. The reasons for this increase are not clear.

PERSPECTIVE

Conclusions drawn in this chapter are almost exclusively based on observations of extracellular dynamics: hormone secretion, blood concentration of hormones, and hormone degradation. These observations minimize the role played by hormonal factors in "normal aging" and emphasize that the changes which tend to occur in endocrine function with aging are adjustments to alterations in function of *non*endocrine organs.

Two caveats are appropriate. First, while the impact of endocrine gland changes on the process of aging has been described as minimal in this chapter, the consequences of possible changes in *neuroendocrine* function with aging have not been discussed and are analyzed in detail elsewhere (Chapter 19). Second, few measurements presently exist of the possible impact of aging on the interaction of hormones with peripheral tissue sites (Figure 2) where their action is expressed. Although physiologically trivial hormonal changes are seen with age in terms of release and transport of hormones, substantive changes in the response of tissues to hormones may exist as a function of age. Indeed, as is discussed in Chapter 18, recent evidence suggests that in animals certain kinds of cortisol receptors decline in number with age and that such decline is accompanied by decreased tissue responsiveness to this hormone. On the other hand, cell membrane enzyme systems (such as adenylate cyclase: Figure 2) which are stimulated by polypeptide hormones to initiate intracellular hormonal events may be *more* hormonally responsive with age. Further developments along these lines of investigation are critical to the drawing of a conclusive picture of the interrelationships of aging with functions of the various endocrine glands.

ACKNOWLEDGMENT. This chapter cites recent investigations carried out in the laboratories of the following individuals: R. I. Gregerman (National Institutes of Health), R. Andres (National Institutes of Health), W. D.

Denckla (Roche Institute), R. D. Utiger (University of Pennsylvania),
V. M. Dilman (Leningrad), L. P. Romanoff (Worcester Foundation),
M. G. Crane (Loma Linda University), G. Crepaldi (University of
Padua), and I. S. Edelman (University of California, San Francisco).

BIBLIOGRAPHY

Andres, R., and J. D. Tobin. 1976. Endocrine systems. Chapter 14 in C. E. Finch and
L. Hayflick, eds. *Handbook of the biology of aging.* Van Nostrand Reinhold Company,
New York.
Finch, C. E. 1976. The regulation of physiological changes during mammalian aging.
Quarterly Review of Biology (Mar.) 51: 49.
Leathem, J. H. 1972. Endocrine changes with age. In A. M. Ostfeld, ed. *Epidemiology of aging*
[DHEW Publication (NIH) 75–711]. Department of Health, Education, and Welfare,
Washington, D.C.

17

Hormone "Replacement" in the Aged: Proceed with Care

EDWIN D. BRANSOME, JR.

The concept of "replacing" diminishing hormone production by giving medications as age advances has become quite popular in the last decade, especially as far as estrogens (female hormones) are concerned. Advertisements in medical journals, for example, suggest that all menopausal women should be placed on conjugated estrogens: "Treat her with Premarin."[1] It is also widely (and correctly) assumed that production of testosterone, the only important circulating male hormone, diminishes as the males of our species grow older. Only one other hormone has been shown to diminish with age: 1,25-dihydroxy vitamin D3, the active hormonal form of a sterol which was erroneously thought to be a vitamin until about 9 years ago. "Vitamin" D is not a necessary component of our diet; our bodies can make it if we are exposed to sunlight. In addition, vitamin D must undergo several chemical changes before it can become fully active.

Figures 1 and 3 provide some basic information on the structure and biosynthesis of each of these hormones. There is no doubt that some elderly patients exhibit unquestionable signs of estrogen, or testosterone, or of vitamin D deficiency, and less doubt still about the desirability of treating the patient by replacing the missing hormones. But what about the majority of older people who have symptoms and signs which are too gradual and subtle to be recognized as indicative of a "disease"? Broader questions must be asked:

1. Are these hormone-deficiency states an inevitable consequence of aging?
2. Who has a deficiency? Everybody who is lucky enough to

[1]Premarin® is the Ayerst brand of conjugated estrogens.

EDWIN D. BRANSOME, JR., M.D.● Section of Metabolic and Endocrine Disease, Medical College of Georgia, Augusta, Georgia 30901

survive to some arbitrarily defined "old age," or a select few?

3. Should hormone replacement be prescribed for older people (as most vitamins are) whether or not there is a demonstrable deficiency in the aged in general, or should they be given only after careful evaluation of the individual?

4. If there is a deficiency, what is the balance of the risks vs. the benefits of taking hormones?

Although there are no hard and fast answers to any of these questions, the medical literature of the last several years has provided enough information to allow some tentative conclusions. We know enough to, at least, make recommendations concerning estrogens for the symptoms of menopause, estrogens for prevention of "osteoporosis" (the loss of skeletal bone) after menopause, vitamin D for patients who have developed additional bone disease, and the giving of testosterone to some elderly men. I will take up each of these areas in turn. At the end of this chapter, a list of books and journal articles will provide readers with the opportunity to obtain a broader and deeper understanding of the problems I will discuss.

ESTROGEN AND THE MENOPAUSE

Menopause, marked by the cessation of spontaneous menstrual bleeding, may take place at any time from the early forties to the mid-fifties. In the United States, the average age of menopause is 49.

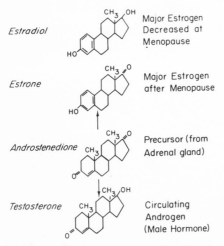

Figure 1. Structures of gonadal hormones and their biosynthetic precursors. Each corner of the two-dimensional ring structures shown represents a carbon atom. Hydrogen molecules bound to the carbon skeletons are not shown. Estradiol and estrone thus are 18-carbon steroids; androstenedione and testosterone have 19 carbons. Estradiol, the major circulating estrogen in premenopausal women—and the most potent—is produced by the ovary. Estrone is the principal circulating estrogen in postmenopausal women and is synthesized outside the ovary from androstenedione, a steroid without significant sex hormone activity which is produced by the adrenal gland. Androstenedione is also the major precursor of the testosterone produced by the testes and by peripheral tissues in both men and women (see the text for further details).

Table 1. Signs and Symptoms of Menopause

Blood levels of estrogen drop.
Blood levels of pituitary gonadotropins (FSH and LH) rise.
Hot flushes
   Periodic diaphoresis (sweating) and dilatation of small blood vessels in the blush area:
      patchy redness.
Libido (sex drive) is usually unaffected or increases as the fear of pregnancy is removed.
Atrophy of glandular breast tissue occurs (variably).
Atrophy of the vaginal mucosa (lining) occurs.
   "Senile vaginitis" is the severe extreme, with dryness, irritability, and a tendency to
      bleed. Sexual intercourse may be quite painful.
Decreased formation of dermal collagen leads to thinned, less elastic skin, especially of the
      face and extremities.
Decreased bone formation (osteopenia, "senile" or "postmenopausal" osteoporosis) with
      loss of skeletal mass, resulting in increased incidence of hip, wrist, and vertebral body
      fractures (variable).
Risk of myocardial infarction (heart attacks) increases.

Changes in the ovaries seem to be the most important: the ovaries become progressively insensitive to stimulation by gonadotropins, hormones from the pituitary gland, which are necessary for the development of ova and the normal production of estrogens and progesterone by the ovary. The blood levels of the gonadotropic hormones increase as the ovary becomes less responsive to their effect. Graafian follicles and ova therefore cease to develop properly. The synthesis of estradiol (Figure 1) decreases; the amount of estrogen circulating in the blood is thus diminished; the remaining estrogen is predominantly estrone, an estrogen weaker than estradiol. The source of estrone is androstenedione, a steroid which comes from the adrenal gland and has no significant hormonal activity itself. Conversion of androstenedione to estrone occurs in tissues other than the ovaries. The amount of estrone formed and the evidence of estrogens being present in postmenopausal women varies, may be increased in obese women, and is rather unpredictable. If cells from the epithelium (lining) of the vagina are examined under a microscope, 30 to 40% of postmenopausal women will show some "estrogenic" effects of estrone production.

The signs and symptoms of menopause are listed in Table 1. Intermittent hot flushes which occur after estrogen production drops are transient, occurring after blood levels of pituitary gonadotropic hormones have increased, and usually disappear after a year or two. Although some women find hot flushes extremely unpleasant, many virtually ignore this sign of vasomotor instability at menopause. Indeed, occasionally, hot flushes may not occur at all.

Other manifestations are present to a variable extent, becoming problems for some women as they grow older, or never being significant to others. Atrophy of the vaginal mucosa can be a severe enough problem to result in symptomatic vaginitis in some women who have no microscopic evidence of endogenous estrogen effect on their vaginal epithelium. The treatment is local (cream) or systemic (pills, injections, or subcutaneous pellets).

POSTMENOPAUSAL OSTEOPOROSIS

Postmenopausal osteoporosis, a disease in which there is an accelerated decrease in the amount of skeletal mass, is the most prevalent bone disorder of men and women. The amount of cortical dense bone of the extremities and the more spongy trabecular bone of the spine, pelvis, and ends of long bones (see Figure 2) gradually diminishes over a period of years, starting at about age 20. Resorption of bone from the inner (endosteal) surfaces exceeds a less-than-normal rate of bone formation. Women deprived of estrogen in early adulthood lose bone at an accelerated rate. Although it is not yet clear whether women who develop osteoporotic bone disease after a natural menopause are those who have the lowest endogenous estrone production, there is now little doubt that giving estrogen replacement will protect women against osteoporosis. A controversy of a few decades has been settled by several studies published within the past few years. Loss of testosterone relatively early in life (from castration or hereditary hypogonadism) will lead to significant osteoporosis in men. Menopausal loss of estrogen leads to significant diminution of bone in about 30% of white women who reach their early to mid-sixties; "senile osteoporosis" is thus more common (at

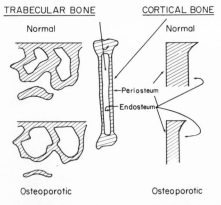

Figure 2. Diagram of the changes of bone occurring in osteoporosis. Trabecular (spongy) bone is present at the ends of long bones and in the axial skeleton (spine, pelvis). Cortical (lamellar) bone is more dense and principally comprises the shafts of long bones. In both types of bone, there is constant remodeling; after young adulthood, resorption exceeds accretion very slightly so that there is a loss of skeletal mass with age. In osteoporosis, loss from bone trabeculae and from endosteal (inner) surfaces of cortical bone is excessive. In senile or postmenopausal osteoporosis, the accretion or replacement of bone seems to be subnormal, in part because of a lack of gonadal hormone — estrogen or testosterone.

Table 2. Predispositions to Osteoporosis

Age: >55 in women; >65 in men
Sex: More common in women; with early castration in either sex
Race: Most common in Caucasians; less in Orientals; still less in Negroes
Slender body habitus (obesity protective)
Cigarette smoking (moderate to heavy)
Inactivity
Inadequate diet: Low calorie intake
                       Low protein intake
                       Low calcium intake
                       High phosphate intake
Diseases: Chronic alcoholism
                Diabetes mellitus
                Intestinal malabsorption
                Postgastrectomy
                Adrenal corticosteroid excess
                Hyperthyroidism
                Uremia (chronic kidney failure)
                "Collagen" disease (e.g., rheumatoid arthritis)

least three to four times) in elderly white women aged 65 and over than in white men.

As bone mass progressively diminishes, bones become more prone to fracture on weight bearing and stress. The incidence of fractures of the distal radius (wrist fractures) suffered during falls, and of vertebral body fractures (leading to back pain and to progressive spinal deformity), is increased. Most serious of all is the increased incidence of fractures of the "hip," (of the femoral neck), inasmuch as complications of immobilization after a "hip fracture" are a major cause of death in the elderly. "Relative vitamin D deficiency" may be an additional contributing factor in many femoral neck fractures (see below).

Predispositions to osteoporosis are summarized in Table 2. The racial tendency is probably attributable, as is the sexual dichotomy, to differences in the amount of bone mass formed by early adulthood. For the rest of the list, either bone formation is decreased (in females, with sex-hormone deficiency, dietary deficiencies, diabetes mellitus), or the normal process of bone resorption is accelerated.

What about elderly patients who already have significant osteopenia or even symptomatic osteoporosis? Giving estrogens to such women often results in rapid improvement of back pain. Daily administration of relatively low doses of estrogens (conjugated mares' estrogens, estradiol, diethylstilbestrol, mestranol, or ethinylestradiol) will not restore much, if any, of the bone that has been lost, but will certainly slow down the rate of bone loss in the future by raising the depressed rate of bone formation. Even though the bone mass does not appear to be appreciably increased on x-ray examination, the incidence of fractures seems to diminish even where estrogens are given relatively "late in the game."

I know of no reliable data concerning whether testosterone (or other androgen) administration to elderly men with osteopenia is of benefit. My personal impression, based on attempts to treat only a very few such patients, is that testosterone administration is not of much use in elderly men who already have bone disease (more on this below). I can find no evidence that giving testosterone (in addition to estrogens) to postmenopausal women results in any more protection against bone disease than is given by estrogen alone. What testosterone will do is stimulate the inappropriate growth of facial hair and bring about other signs of masculinization in women, side effects which are usually considered undesirable.

Although there are no current treatments which have been proven unequivocally effective in restoring measurable amounts of bone to osteoporotic patients, there are a few experimental therapies which have not yet been adequately tested. Giving calcium, phosphate, vitamin D, and sodium fluoride has been enthusiastically advocated by one group of investigators, but trials by other groups of doctors have not been encouraging. A new and very interesting approach has been reported by a team of British and American investigators: weekly injections of a synthetic (the N-terminal amino acids 1–34) fragment of human parathyroid hormone. A preliminary report has suggested that this new approach is really able to reverse osteopenia in the aged.

THE PROS AND CONS OF ESTROGEN REPLACEMENT

Some of the evidence that estrogen replacement may be indicated in postmenopausal women has already been discussed. Table 3 summarizes the benefits of giving estrogens. Since the vasomotor symptoms of estrogen decrease are themselves transitory, they do not justify continued administration. Stopping the medication may itself result in a brief period of hot flushes, but the complaints are much milder than

Table 3. Benefits of Estrogens in the Menopause

Relief of menopausal vascular symptoms (hot flushes/flashes).
Reversal of atrophic vaginitis.
Protection against bone loss (senile osteoporosis or osteopenia).
Protection against benign and malignant breast disease.

Table 4. Effects of Estrogens on Clinical Laboratory Determinations

---

Increased serum triglycerides and β-lipoproteins.

Decreased serum Xa factor (activated factor X clotting inhibitor).

Elevations of serum enzymes [especially hepatic lactic acid dehydrogenase (LDH) and alkaline phosphatase levels].

Increased plasma renin substrate levels.

Decreased serum inorganic phosphate levels.

Suppression of phytohemagglutinin (PHA)-induced lymphocyte transformation (re: the immune system).

Reduced serum levels of vitamins, minerals, etc.
   Zinc
   Tryptophan (amino acid)
   Ascorbic acid (vitamin C)
   Pyridoxine (vitamin B_6)
   Folic acid
   Retinol (vitamin A)
   Riboflavin (vitamin B_2)

---

those which led to estrogen therapy in the first place. Treatment of atrophy of the vaginal epithelium may be effective with a topical or local application of a cream containing estrogen, but if "senile vaginitis" is present, systemic estrogens (pills, injections, or pellets) will be necessary. Although prophylactic treatment with estrogens is of proven efficiency in protecting groups of women against severe osteopenia after menopause, we should remember that no more than one out of three white women in their sixties develop bone disease. We do not yet know how to identify specific individuals at risk, even with statistical data on who might be predisposed.

The statistical evidence of protection against breast disease is based primarily on the history of large groups of women who have received estrogen-containing oral contraceptives, despite the fact that most breast cancers occurring before menopause are estrogen-dependent.

Many women and their physicians are also of the opinion that estrogen replacement is an antidote for depression and other emotional difficulties in the period around the menopause. Unfortunately, this purported benefit has thus far been virtually impossible to prove: methods of measuring the complaints as well as the improvements remain elusive.

The changes in the results of laboratory tests listed in Table 4 are common manifestations of the use of estrogen-containing oral contraceptives. To my knowledge, there have been no comparable studies of large groups of older women on estrogen replacement therapy. In most instances, the abnormalities shown in laboratory tests do not predict clinically significant side effects of estrogens. There will be serious

Table 5. Possible Side Effects of Estrogens

Thromboembolic disease
  Venous thrombosis of the legs
  Cerebrovascular accidents (especially in moderate to heavy smokers)
Blood pressure elevation*
  Increased diastolic pressure (common; reversible)
  Severe hypertension (uncommon)
Cancer
  Possible increased incidence of uterine (endometrial) cancer
  Possible increased incidence of breast cancer after prolonged (>10 yr) use*
Gallbladder disease
  Increased tendency to form gallstones
  Cholecystitis (inflammation) more frequent
Susceptibility to infections*
  Increased incidence of chicken pox, herpes simplex, rubella (German measles)
  Increased incidence of urinary tract infections
Rare reactions
  Chorea (involuntary movements)
  Jaundice (cholestatic hepatitis)
  Exacerbation of hyperlipoproteinemia (Type V)

---

*Based on the effects of estrogen-containing oral contraceptives.

side effects of estrogen administration in a few women, however (see Table 5).

Thromboembolic disease, inflammation of the veins of the legs or pelvis (phlebitis), may be increased and may occasionally result in blood clots (thrombi) being carried to the lungs, where they lead to a pneumonia-like response (pulmonary embolization). The risk of cerebrovascular accidents ("strokes") from thrombi (clots) in medium-sized arteries of the brain is significant and is increased further in women who are moderate to heavy smokers. This side effect of taking estrogen may be "dose-related." Because of the many reports that women taking oral contraceptives containing more than 50 micrograms of estrogen (mestranol or ethinylestradiol) developed thromboembolic disease, the U.S. Food and Drug Administration several years ago compelled the removal of many "sequential" oral-contraceptive preparations from the market. Although it has been known for some time that estrogens may alter the biochemical mechanism of blood clotting, it is only quite recently that the precise abnormality — low levels of factor Xa, a component of Antithrombin III (an inhibitor of clotting), was identified. This new knowledge has already proven to be of value for patients on estrogen therapy who undergo surgery; administration of small doses of an anticoagulant drug (heparin) will apparently remedy the factor Xa deficiency, and has been found to protect patients against thromboembolic disease.

To return to the question of estrogen "replacement" in the post-menopausal woman, my statements above should suggest the wisdom of giving the *lowest effective* dose to control symptoms. For protection against osteoporosis, it appears that 1.25 mg of conjugated mares' estrogens a day or the equivalent will suffice.

Mild elevation in diastolic blood pressure has been found when large groups of women on or off estrogen-containing oral contraceptives were compared, but does not seem to be of much clinical significance. Increased activity of a hormonal system (the renin–angiotensin system) maintaining blood pressure cannot usually be documented, even though the amount of one element of the system, circulating "renin substrate" is increased. A rare patient will, however, develop severe hypertension, which, if neglected, may become irreversible and fatal. The precaution against this uncommon side effect of estrogen (we do not know whether the problem is significant in older women) is to make sure that blood pressure is monitored within the first 6 to 12 months of starting on estrogen. If a patient is already mildly to moderately hypertensive, giving estrogens might have no effect at all, but the patient should be observed very closely by her physician. Estrogens should be discontinued if blood pressures rise and stay in the hypertensive range (diastolic pressures of 95 mm of mercury or above). Hypertensive effects of chronic estrogen therapy may take as long as several months to wear off.

In December 1975, two papers were published in the *New England Journal of Medicine* recounting an increased risk of endometrial (uterine) cancer in women who had been taking conjugated mares' estrogens as estrogen "replacement." Several additional articles offered corroboration. All the reports have been retrospective (after the fact) and have been flawed by a lack of agreement between pathologists as to what the precise microscopic criteria of the diagnosis of endometrial cancer are, but it has been known for a long time that high doses of estrogen may lead to uterine cancer in animals and in humans. There is, also, some evidence that the incidence of this common cancer might be increased in obese, diabetic, or hypertensive women, or in women of low parity (few or no children).

It may be some time before the controversy is resolved over whether low doses of "replacement" estrogens are harmful. Only women who have not had hysterectomies seem to be at risk. A reasonably safe policy is to have the patient "cycle" — taking the drug only 3 weeks out of the month. Intermittent estrogen will usually result in sloughing of the tissue (and bleeding) each month or every few months. Additional administration of a progestational agent every 6 months will reduce bleeding in women who have built up a significant amount of endometrial tissue, and who have not bled after estrogen cycles.

If women have had endometrial cancer previously, or have a history of treated cancer of the uterine cervix or breast, they should not receive "replacement estrogens." Although they very well might have no problems with recurrent cancer, the risks to such patients seem, to me at least, to outweigh any potential benefits of taking estrogens.

A tendency to gallbladder disease is attributable to estrogens making bile more "lithogenic" — it tends to form more stones. In the presence of symptomatic gallbladder or biliary tract disease, estrogen replacement should probably be stopped.

Increased susceptibility to infections has been related by one British study of young women taking estrogen-containing oral contraceptives. This statistical association may or may not apply to older women on "replacement" estrogens. I have listed a few reactions (Table 5) to warn that such uncommon problems might be caused by estrogens.

VITAMIN D AND OSTEOMALACIA IN THE AGED

It has become apparent in the last several years that some of the metabolic bone disease seen in the aged may not be solely a result of the decreased bone formation in osteoporosis. Another sort of abnormality, osteomalacia, the adult equivalent of rickets, has been frequently involved in the hip (femoral neck) fractures in elderly Britons and Scandinavians. Osteomalacia is a bone disease in which resorption of bone (rather than just formation as with osteoporosis) is increased, and bone that is formed is of poor quality. What seems to be responsible is impaired formation of the active hormonal form of vitamin D probably because of decrease in the activity of an enzyme protein in the kidney which affixes an hydroxyl ($-OH$) group to the C1 atom of the sterol (see Figure 3). As a result, there is impairment both of calcium absorption from food reaching the small intestine and of bone formation as well.

For a physician to determine whether osteomalacia, as well as osteoporosis, is present in a specific patient is at present quite difficult. The special laboratory tests needed and the equipment for needle biopsies of pelvic (iliac crest) bone are available at only a few university medical centers. At present, there simply is not any convenient method to determine who has osteomalacia or who should be given a hormonal form of vitamin D to treat the disease. Relatively weak vitamin D and calcium supplements are often given to patients with hip fractures and osteopenia on an empirical basis, the patients then being closely observed for improvement in their bone disease or for their development of an elevated blood calcium level, an indication of vitamin D toxicity.

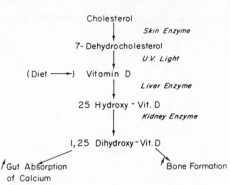

Figure 3. Synthesis of the hormonal form of vitamin D. Vitamin D may be derived from the diet or made by the skin under the influence of sunlight. The active hormonal form of vitamin D is necessary for the absorbtion of calcium from the diet and for the formation (accretion) of bone. If the biochemical steps shown do not occur properly in the liver and the kidney, there will be an increase in the amount of parathyroid hormone produced by parathyroid glands in the neck. The concentration of calcium in the blood will, then, be maintained solely through removal of calcium from the bone. Bone will thus be lost, and new bone that is synthesized will be abnormal and deficient in calcium. The resulting bone disease is called osteomalacia in adults or rickets in children.

TESTOSTERONE AND THE AGING MALE

In men, there is no equivalent of the relatively abrupt menopause seen in women. The gradual decline of serum testosterone measurements in men from their fifties on is to some extent responsible for the loss of libido and in muscle and skeletal mass inevitably seen in elderly men. The effects of testosterone "replacement" are frequently disappointing, because a decrease in the output of the circulating species of "male hormone" is not the sole factor involved in declining sexual performance. Indeed, there is some evidence that, with age, tissues also become less responsive to testosterone.

The principal complication of giving testosterone is that the drug will usually increase the extent of prostate gland hypertrophy (enlargement), sometimes to the point of symptomatic partial obstruction of the elimination of urine through the urethra. Benign prostatic hypertrophy is quite common, having been estimated to be present in 60 to 70% of males in their sixties or older. Occult undiscovered cancer of the prostate is also a relatively common disease and may be made much worse by increased male hormone.

Testosterone may thus be given if the serum levels of the hormone are low; if they are not, it is not worth implementing. If a man complains of impotence (albeit most impotence at any age has a significant psychogenic component) or has osteopenia as well as testosterone deficiency, he should receive replacement therapy. Oral preparations of drugs related to testosterone are usually ineffective; testosterone works best if administered by periodic intramuscular injection. Any such patients should be followed by a physician for the possible development of disease of the prostate.

CONCLUSIONS

I have reviewed the several situations in which hormone "replacement" in the aged is apropos. The most important and most common decision to be made is whether to prescribe long-term estrogen therapy for postmenopausal women. The need for such replacement has only very recently been shown to be justified for prevention of clinically significant loss of skeletal mass. Although no more than one out of three women will develop postmenopausal osteoporosis, we cannot really say who will do so before the fact. Categorical administration of estrogens to all older women is clearly unwarranted because of the undesirable side effects which the therapy risks. It should be obvious that until our knowledge improves a great deal, replacement of estrogens should be decided on an individual basis and should be supervised by a physician.

Administration of testosterone to older men is justified only if the specific individual is obviously deficient in the hormone. Only a relatively small number will qualify.

Giving testosterone to postmenopausal women is, in my opinion, difficult, if not impossible, to justify.

A deficiency in the metabolism of vitamin D to its natural hormonal form may be involved in severe bone loss in the elderly and can be corrected by giving vitamin D and calcium. Assessment of this problem is technically difficult at present and calls for equipment and skills which, as mentioned above, are available at only a few places in the United States.

I hope that I have been successful in explaining the nature of these problems to readers heretofore unfamiliar with them, while suggesting that the question of "replacement" is specific to each area and complex even within that area. We should, therefore, not only "proceed with care," as the title of this chapter suggests, but should proceed with *medical care* when hormone replacement is considered.

BIBLIOGRAPHY

The Menopause (General)

Greenblatt, R. B., V. B. Mahesh, and P. G. McDonough, eds. 1974. *The menopausal syndrome*. Medcom Press, New York. [This book contains many pertinent articles.]
Weg, R. B. 1975. Sexual inadequacy in the elderly. In R. Goldman, M. Rockstein, and M. L. Sussman, eds. *The physiology and pathology of human aging*. Academic Press, Inc., New York, p. 203.

Postmenopausal Osteoporosis

Garn, S. M. 1975. Bone-loss and aging. In R. Goldman, M. Rockstein, and M. L. Sussman, eds. *The physiology and pathology of human aging.* Academic Press, Inc., New York, p. 39.

Meema, S., and H. E. Meema. 1976. Menopausal bone loss and estrogen replacement. *Israel Journal of Medical Sciences* 12: 601.

Thomson, D. L., and B. Frame. 1976. Involutional osteopenia: current concepts. *Annals of Internal Medicine* 85: 789.

Side Effects of Estrogen Replacement

Fregly, M. J., and M. S. Fregly, eds. 1974. *Oral contraceptives and high blood pressure.* Dolphin Press, Gainesville, Fla.

Hoover, R., L. A. Gray, P. Cole, and B. MacMahon. 1976. Menopausal estrogens and breast cancer. *New England Journal of Medicine* 295: 401.

Kistner, R. W. 1976. Estrogens and endometrial cancer. *Obstetrics and Gynecology* 48: 479

McDonald, P. C., and P. K. Siiteri. 1974. The relationship between the extra-glandular production of estrone and the occurrence of endometrial neoplasia. *Gynecologic Oncology* 2: 259.

Rosenberg, L., B. Armstrong, and H. Jick. 1976. Myocardial infarction and estrogen therapy in post-menopausal women. *New England Journal of Medicine* 294: 1256

Royal College of General Practitioners. 1974. *Oral contraceptives and health.* Plenum Press, New York. [An interim report of a 20-year (1958–1978) study of the side effects of estrogen-containing oral contraceptives.]

Salhanick, H. A., D. M. Kipnis, and R. L. Vandewiele, eds. 1969. *Metabolic effects of gonadal hormones and contraceptive steroids.* Plenum Press, New York.

Tritapepe, R., C. DiPadova, M. Zuin, M. Bellami, and M. Podda. 1976. Lithogenic bile after conjugated estrogen. *New England Journal of Medicine* 295: 961.

Wessler, S., S. N. Gitel, L. S. Wan, and B. S. Pasterack. 1976. Estrogen-containing oral contraceptive agents. A basis for their thrombogenicity. *Journal of the American Medical Association* 236: 2176.

Wynn, V. 1975. Vitamins and oral contraceptive use. *Lancet* i: 561.

Osteomalacia in the Elderly

Aaron, J. E., J. C. Gallagher, J. Anderson, L. Stasiak, E. B. Longton, B. E. C. Nordin, and M. Nicholson. 1974. Frequency of osteomalacia and osteoporosis in fractures of the proximal femur. *Lancet* i: 229.

Alevizaki, C. C., D. G. Ikkos, and P. Singhelakis. 1973. Progressive decrease of true intestinal calcium absorption with age in normal man. *Journal of Nuclear Medicine* 14: 760.

Testosterone and Elderly Men

Huggins, C. 1947. The etiology of benign prostatic hyperplasia. *Bulletin of the New York Academy of Medicine* 23: 696.

Rubens, R., M. Dhondt, and A. Vermeulen. 1974. Further studies in Leydig cell function in old age. *Journal of Clinical Endocrinology and Metabolism* 39: 40.

Vermeulen, A., R. Rubens, and L. Verdonck. 1972. Testosterone secretion and metabolism in male senescence. *Journal of Clinical Endocrinology and Metabolism* 34: 730.

Whitmore, W. F., Jr. 1973. The natural history of prostatic cancer. *Cancer* 32: 1104.

18

Altered Biochemical Responsiveness and Hormone Receptor Changes during Aging

GEORGE S. ROTH

Loss of vitality is one of the most obvious and important manifestations of aging. This decline in performance is particularly noticable under conditions of stress or exposure to various physical and chemical agents. Cases in point range from reduced ability of senescent individuals to avoid oncoming traffic, to decreased resistance to bacterial and viral infection, to less efficient metabolism of ingested fatty foods. In such situations, the ability of senescent organisms to make appropriate physiological and biochemical responses is markedly altered. Muscle contraction and relaxation, cell division, nutrient utilization, and numerous other biological phenomena are under rigid control, and impaired ability to properly regulate these activities in response to internal and external challenge may have disastrous consequences for the aged organism. Almost all such responses on the part of an organism either directly involve or are somehow modulated by the action of hormones. In fact, the markedly altered ability of aged organisms, tissues, and cells to respond biochemically to hormonal signals has been established by numerous laboratory documentations, some of which will be discussed below. Consequently, it is of utmost importance to gerontologists (those who study aging) to determine exactly how this process influences the various mechanisms of hormone action. Only after establishing this can reasonable attempts to restore such mechanisms to a youthful state be initiated.

GEORGE S. ROTH, Ph.D. • Endocrinology Section, Clinical Physiology Branch, Gerontology Research Center, Baltimore City Hospitals, Baltimore, Maryland 21224 of the National Institute on Aging, National Institutes of Health, Bethesda, Maryland 20014

AGING AND THE MECHANISMS OF HORMONE ACTION

Hormones are chemical messengers which circulate throughout an organism, whose function is to signal specific tissues and cells to perform certain biochemical and physiological functions at appropriate times. We are fortunate in that endocrinologists (scientists who study hormones, their production and effects) have recently shed much light on the complex biochemical mechanisms by which these messengers signal cells to carry out these important life processes. Hormones can be arbitrarily grouped into two classes; those whose actions are initiated on the cell surface, and those which act more directly on the cell genetic machinery. This is not to say that hormones of the former category do not ultimately affect the genome in some cases; it serves only as a working classification according to initial sites of action. The surface-active hormones include insulin, epinephrine (adrenaline), and glucagon (an important regulator of carbohydrate metabolism); the gene-active group includes the steroids (sex hormones and antiinflammatory agents such as cortisone) and thyroid hormones. One property which all these hormones share is the requirement that they attach to specific cellular "receptors" prior to initiating their biochemical responses. The surface-active hormones bind to receptors on the cell membrane, while the gene-active hormones interact with receptors in the cytoplasm, the fluid material inside the cell (Figure 1).

This fact was discovered by administering the different classes of hormones, in radioactive form, to experimental animals such as rats and mice. Various tissues were removed, sliced into very thin sections, and examined by special photographic techniques in which radioactive compounds could be localized. Those hormones classified as "surface-active" were found to be situated primarily on the outer membrane of cells while the "gene-active" hormones accumulated mainly in the cytoplasm and nucleus, where the cellular information transfer systems are located. More sophisticated procedures, by which cells could be disrupted and separated into their component parts, using high-speed centrifugation, later confirmed these findings. Parallel chemical analyses showed that most hormone receptors are protein molecules.

Following hormone binding to these receptor proteins, a multitude of biochemical events occur, and specific patterns are characteristic of particular classes of hormones. Almost all of these biochemical steps are candidates for age-dependent modification, and their number probably includes many events which contribute in part to altered hormone action during senescence. However, one must keep in mind that the initial required step common to the action of essentially all known hormones is receptor binding. Furthermore, in many instances the degree of re-

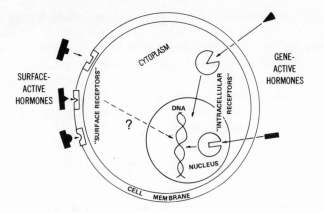

Figure 1. Subcellular location of receptors for different hormone classes. Cells consist of an outer surface membrane, enclosing the soluble cytoplasm and the DNA-containing nucleus. Hormones whose receptors are located on the membrane include insulin, epinephrine, and glucagon. Receptors for steroids and thyroid hormones are found inside the cell, in the cytoplasm and nucleus. Hormones of this latter category act more or less directly on the genome. Surface-active hormones may in some cases indirectly act on the genome, although their initial actions are on the cell membrane.

sponse is proportional to the amount of receptor-hormone complexes which form. In fact, during a number of developmental, disease and altered metabolic states, such as obesity and diabetes, as well as diurnal and ovarian cycles, there occur changes in hormone receptor concentrations which result in proportionate alterations in biochemical responsiveness. Several years ago, this information spurred the hypothesis that changes in receptor levels might account, at least partially, for some of the altered biochemical responsiveness to hormones which occurs during aging. Since that time, receptor studies have occupied a small but ever-increasing portion of the research on age changes in hormone action. The following sections of this article will attempt to summarize the major findings in this area and to place them in perspective with the overall decline in biochemical vitality characteristic of senescence.

CHANGES IN HORMONE RECEPTORS AND RESPONSIVENESS DURING AGING

Different cell types respond to different hormones in different ways. Generally, a cell is thought to be a "target" for particular hormones if it possesses specific receptors for these hormones. One group of hormones for which almost all cells are targets are the steroids. These iinclude certain sex hormones as well as some antiinflammatory agents, such as cortisone and related compounds. Because of their general presence in mammalian tissues, steroid receptors were the first to be studied during aging. Investigations have focused mainly on measuring the concentrations of receptors present as well as the "affinities" or tightness with which they bind steroid hormones. Extracts were prepared from various tissues known to physiologically respond to steroids. So far, tissues from rats, mice, rabbits, chickens, and human beings have been examined. Radioactive steroids were added to these preparations. Hormone which did not bind to receptors was removed by adsorption to activated charcoal, special filtration, and other physical separation procedures. In this way, the remaining "receptor-bound" hormone could be measured using instruments which quantitatively detect radioactivity. By repeating this basic experiment using various levels of hormone and constant portions of cellular extract, both the receptor concentration and affinity could be evaluated. Initial findings revealed that steroid hormone receptor concentrations appear to be reduced during aging in a number of tissues, including brain, fat, liver, muscle, and prostate gland. However, the ability to tightly bind steroids is not changed in those receptors which are detectable. Thus, the aging effect seems to depend more on the ability of tissues to maintain normal receptor concentrations than on the functional capacity of those receptors which are present (Figure 2).

However, not all hormone receptors have been found to change during senescence. The insulin receptor of liver and other tissues is a case in point. Nevertheless, many initial studies did demonstrate that certain hormone receptor concentrations are reduced during aging. They did not attempt to establish whether receptor loss actually results in decreased responsiveness to hormones within the same tissues and cells. Since the ultimate goal of these investigations was an understanding of the causes of altered hormonal responsiveness during aging, a subsequent group of studies have been conducted in an attempt to elucidate the role of receptor changes in this process. In this work, receptor levels were determined in ways similar to that mentioned above, but in addition, various physiological and biochemical effects of the hormones were measured in cells and tissues of animals of different ages. Reductions in various types of biochemical responses to hormones, which appear to be closely related to loss of hormone receptors during aging, have now been observed (Figure 3).

One group of investigations using rats has revealed that the ability of a class of steroid hormones, the glucocorticoids, to control nutrient transport and metabolism is reduced some 60% between maturity and senescence. The rate of nutrient entry into cells and subsequent breakdown into energy and by-products can be measured both by analytical

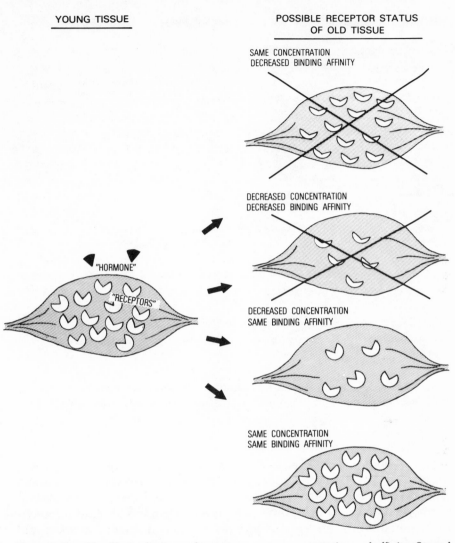

Figure 2. Possible effects of aging on hormone receptor concentration and affinity. Several explanations are possible for decreased hormone binding in aged tissues: unaltered receptor concentration but decreased binding affinity, decreased concentration and decreased binding affinity, and decreased concentration but unaltered binding affinity. The last of these seems to be the general case in which receptor changes have been noted during aging. It should be emphasized, however, that not all receptors are necessarily subject to such age alterations, as depicted on the lower right.

biochemical techniques and use of radioactively labeled compounds. It has been shown that the glucocorticoids rigidly control these processes. This loss of control, observed thus far in aged white blood cells and fat cells, is closely paralleled by 60% reductions in glucocorticoid receptor

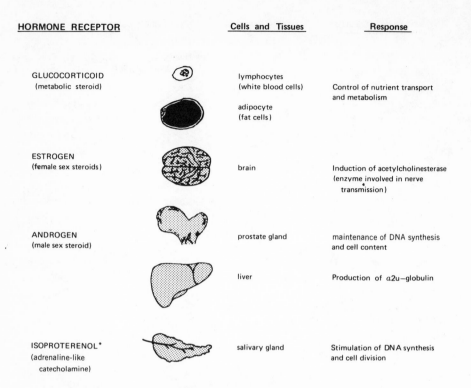

HORMONE RECEPTOR	Cells and Tissues	Response
GLUCOCORTICOID (metabolic steroid)	lymphocytes (white blood cells) adipocyte (fat cells)	Control of nutrient transport and metabolism
ESTROGEN (female sex steroids)	brain	Induction of acetylcholinesterase (enzyme involved in nerve transmission)
ANDROGEN (male sex steroid)	prostate gland liver	maintenance of DNA synthesis and cell content Production of α2u—globulin
ISOPROTERENOL* (adrenaline-like catecholamine)	salivary gland	Stimulation of DNA synthesis and cell division

Figure 3. Correlations between reduced biochemical responsiveness and hormone receptor concentrations during aging. *Age reduction in entry into tissue, receptor binding not measured.

concentrations. It has also been shown that control of nutrient metabolism and hormone binding to receptors are maximal at the same hormone levels. Moreover, if receptors are blocked with inactive steroid hormone analogs, no response can occur. Hormone binding to receptors, therefore, seems to be required to control nutrient transport and metabolism. Consequently, loss of receptors during aging appears to result in reduction of this response.

Another set of studies with rats have reported that the ability of the female sex steroids, the estrogens, to induce certain enzymes required for the proper transmission of nerve impulses in the brain is progressively reduced during aging. Enzymes are proteins which perform various cellular functions. Production of certain of these proteins is stimulated by various hormones. A concomitant age-related decrease in estrogen receptor concentrations in the brain was also observed. Thus, as with the nutrient metabolism control response mentioned above, loss of steroid receptors during aging seems closely, if not causally, related to decreased biochemical responsiveness in various tissues and cells.

Finally, the ability of a number of hormones and various agents to stimulate DNA synthesis (replication of the cellular genetic information) and cell division has been found to be markedly impaired during aging. One case involves reduced stimulation of this response by adrenaline-like hormones or "catecholamines" (in particular, one called isoproterenol) in rodent salivary glands. Unlike steroid hormones, whose receptors are intracellular, catecholamine receptors are located on the cell surface. Unfortunately, until recently, technical problems have made it impossible to make true measurements of these membrane receptors. However, earlier studies indicated that the entry of these hormones into the salivary gland is reduced with increasing age, suggesting possible decrease in receptor levels. Future work, employing the latest methodology, should determine whether or not receptors are actually lost. In any case, reduced efficacy of the male sex hormones, the androgens, in controlling DNA and cell content of the prostate gland has also been noted during aging. Very precise measurements have revealed that the receptors for these hormones, which are intracellular, are indeed reduced in concentration. A similar loss of androgen receptors, which seems responsible for an impaired ability to produce a particular globulin type protein in the livers of senescent rats, has also been reported.

In essence, reductions in a number of biochemical responses to hormones in the cells of a wide variety of tissues seem to be at least partially due to loss of hormone receptors during aging.

MECHANISMS OF HORMONE RECEPTOR LOSS DURING AGING: TOPIC FOR PRESENT AND FUTURE INVESTIGATIONS

Since hormone receptor changes seem to play an important role in at least some biochemical responses altered by aging, it is now desirable to elucidate those mechanisms by which receptor changes occur. Such mechanisms can be envisioned as operating on at least three possible levels: molecular (intrinsic to particular target tissue cells), cellular (changes in cell populations within particular tissues) and neuroendocrine–systemic (alterations in the internal environment of the organism).

Since some initial studies of receptor changes during aging employed tissues, such as liver and brain, which contain many complex cell types, it could not be determined whether receptors were lost from individual cells, or receptor-containing cells were lost from the whole tissues. This problem was overcome by techniques which allowed the

CONTROL BY HORMONES AND/OR OTHER BIOCHEMICAL AGENTS

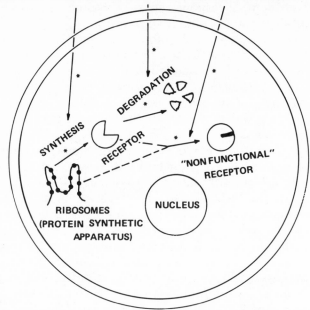

Figure 4. Possible mechanisms of hormone receptor loss from target cells during aging. Molecular processes which might be altered during aging to result in decreased cellular concentrations of functional hormone receptors include synthesis (assembly), degradation (disassembly), possible presence of receptors in a nonfunctional state, and the effect of hormones and other biochemical agents on these processes. *Possible alterations during aging.

isolation of defined cell populations from animals of different ages. Classes of cells based on size, density, resistance to certain enzymatic treatments, and other distinguishing properties can be separated by various physical and chemical manipulations. In addition, certain classes of nerve and fat cells, which do not divide or decrease in number during most of the life-span, were studied. Essentially, all these cells are the same age as the organism. Both of these cell populations exhibit receptor loss during aging. Thus, receptor loss from individual cells does occur, suggesting a "molecular," rather than a strictly "cellular," mechanism.

Having established this point, a number of aspects of the molecular biology of hormone receptors can now justifiably be studied during the aging process (Figure 4). For example, decreased receptor levels in aged cells could be due to reduced rates of synthesis on ribosomes, the cellular protein factories, or to increased rates of degradation by intracellular enzymes (in other words, changes in the processes by which receptors and other proteins are assembled and disassembled in the normal course of cellular metabolism). Such possibilities should certainly be explored. It is also possible that apparently decreased receptor concentrations may be due to the presence of nonfunctional receptors which cannot be measured by present binding assays. Certain populations of enzymes exhibiting age effects similar to those of hormone receptors (i.e., decreased apparent concentrations but comparable binding properties) have already been reported to contain some nonfunctional molecules. In these analyses, the enzymes in question are separated from other cellular proteins on the basis of distinguishing physical and chemical properties, such as size, electrical charge, and solubility. The purified proteins are then injected in animals (usually rabbits or goats), which make an immune response to the foreign substances and produce molecules called antibodies which can specifically attach to these proteins. These antibodies are then used for identification and analyses of the original enzymes in various preparations in the laboratory. Techniques such as these could also be applied to the receptor problem. Nonfunctional receptors, which cannot be measured by present hormone binding assays, might be detected by such "immunochemical" methods.

Finally, more emphasis should be placed on examining those factors which control hormone receptor quantity and characteristics under "normal" conditions, independent of the aging process. It has already been established that certain hormones and biochemical agents control the levels of receptors for other hormones. For example, during the ovarian cycle, fluctuations in circulating estrogen levels cause changes in the concentration of receptors for, and responsiveness to, progesterone, another female hormone. Similar observations have been made regarding control by other hormones of receptors for male sex hormones — catecholamines, growth hormone, and a number of others. Obviously, a whole new regulatory level for hormonal responsiveness has been uncovered. Work in this area may eventually pinpoint age changes in specific control steps required for maintenance of normal receptor levels. Even more intriguing is the possibility of restoring biochemical responsiveness lost during aging by appropriate biochemical manipulation of hormone receptors.

BIBLIOGRAPHY

General

Cuatrecasas, P. 1974. Membrane receptors. *Annual Review of Biochemistry* 43: 169–214.
King, R. J. B., and W. I. P. Mainwaring. 1974. *Steroid cell interactions.* University Park Press, Baltimore.
O'Malley, B. W., and A. R. Means, eds. 1973. *Receptors for reproductive hormones.* Plenum Press, New York.

Technical

James, T. C., and M. S. Kanungo. 1976. Alterations in atropine sites of the brain of rats as a function of age. *Biochemical and Biophysical Research Communications* 72: 170–175.

Kanungo, M. S., S. K. Patnaik, and O. Koul. 1975. Decrease in 17β-estradiol receptor in brain of aging rats. *Nature* 253: 366–367.

Roth, G. S. 1975. Changes in hormone binding and responsiveness in target cells and tissues during aging. Pages 195–208 in V. J. Cristofalo, J. Roberts, and R. C. Adelman, eds. *Explorations in aging.* Plenum Press, New York.

Roth, G. S., and R. C. Adelman. 1975. Age related changes in hormone binding by target cells and tissues; possible role in altered adaptive responsiveness. *Experimental Gerontology* 10: 1–11.

Roth, G. S., and J. N. Livingston. 1976. Reductions in glucocorticoid inhibition of glucose oxidation and presumptive glucocorticoid receptor content in rat adipocytes during aging. *Endocrinology* 99: 831–839.

Roy, A. K., B. S. Milin, and D. M. McMinn. 1974. Androgen receptor in rat liver: hormonal and developmental regulation of the cytoplasmic receptor and its correlation with the androgen dependent synthesis of α_2u-globulin. *Biochimica et Biophysica Acta* 354: 213–232.

Schocken, D. D., and G. S. Roth. 1977. Reduced β-adrenergic receptor concentrations in aging man. *Nature* 267: 856–858.

Shain, S. A., R. W. Boesel, and L. R. Axelrod. 1973. Aging in the rat prostate, reduction in detectable ventral prostate androgen receptor content. *Archives of Biochemistry and Biophysics* 167: 247–263.

19

The Brain and Aging

CALEB E. FINCH

The brain controls or influences most organ and cell functions in the body, either directly by contact through nerves, or indirectly through hormones. Immediately subservient to the brain is the pituitary, our master endocrine gland. Neurohormones from the brain, called releasing factors, control each of the various pituitary hormones: adrenal cortical trophic hormone (ACTH), thyroid-stimulating hormone (TSH), gonadotrophins (LH, FSH), and so on. In turn, these pituitary hormones control the adrenal cortex, thyroid, gonads, and so on, and influence many other endocrine functions (Figure 1). How aging alters the functions of the brain and pituitary is of obvious interest, and yet surprisingly little is known.

There is no agreement on even the basic characteristics of aging, which indicates the difficulties of understanding biological aging. The parable of the blind man trying to describe an elephant also illustrates the problems of researchers in aging. Some of our dimness of sight comes from confusion between the effects of aging and diseases of aging and from the absence of reliable data about how aging alters many basic body processes.

The interrelations of aging and disease need special comment. As an organism becomes older, the likelihood of a specific disease or degenerative lesion increases. For many degenerative conditions, the increase is an exponential function of age which proceeds in parallel with the increased mortality rate. A majority of researchers are using "animal models" to study aging, since the laboratory mouse and rat undergo many of the same changes in metabolism, reproduction, and neural functions as those in human beings. The high degree of inbreeding and control of diet and environment possible in the laboratory clearly en-

---

CALEB E. FINCH, Ph.D.● The Andrus Gerontology Center, University of Southern California, Los Angeles, California 90007. This chapter is contribution no. 31 from the Neurobiology Laboratory, Andrus Gerontology Center.

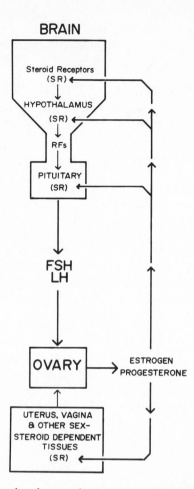

Figure 1. Interactions of the female reproductive neuroendocrine axis, showing that the brain and pituitary receive hormonal signals (estrogen, progesterone) as well as send out hormones. FSH: Follicle-stimulating hormone; LH: luteinizing hormone; SR: steroid receptors, found in all sex-steroid-responsive tissues, including the brain.

hances uniformity in the pattern of aging. Even so, not all individuals have the same lesions. For example, my students James Nelson and Keith Latham found that aging inbred mice with lymphatic tumors or lung disease have lower blood testosterone and testes weight, whereas aging mice without these lesions showed no change. Possibly the normal levels of testosterone in some older men are a consequence of their relatively fortunate health. The decline of testosterone levels in any average group of men may then be the effect of disease rather than of aging per se. Because age-related diseases can influence most endocrine glands, including the adrenal, thyroid, gonads, and pituitary, it may be some years before the facts of hormone changes during aging are agreed upon.

The possibility that aging changes in the brain contribute to some hormonal shifts is being explored in a number of laboratories. The loss of regular estrous cycles in aging rats is the best-studied case so far. There is a great deal of excitement among neurobiologists about the reactivation of estrous cycles in middle-aged (20-month old) laboratory rats by certain drugs which influence certain nerve endings in the brain. L-Dopa and some other drugs are able to enter the brain from the blood and increase the levels of catecholamines (substances like adrenaline) in nerve cell endings (Figure 2). L-Dopa is normally derived from dietary tyrosine and is the precursor of catecholamines in the brain and elsewhere in the body. It is generally believed that nerve endings of the hypothalamus (a region at the base of the brain) (Figure 1) require particular levels of catecholamines for regular female reproductive cycles. In many parts of the nervous system, catecholamines are thought to act as

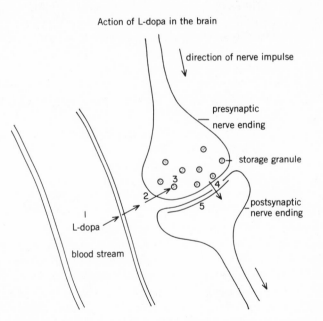

Action of L-dopa in the brain

Figure 2. Entry of L-dopa into nerve endings. (1) crossing the blood capillary wall and probable decarboxylation; (2) entry to the presynaptic nerve ending; (3) entry to the storage granule after conversion to dopamine; (4) release into the synapse by nerve impulse; (5) action on the postsynaptic receptors. Metabolic sequence: L-dopa → dopamine → noradrenaline.

neurotransmitters; that is, after firing of a nerve cell, they diffuse across the synapse and act on receptors in the postsynaptic cell. The ability of L-dopa to reactivate estrous cycles in old rats thus implies that metabolic deficiencies in hypothalamic nerve cells or in their receptors arise during aging. As will be described, L-dopa is effective in the treatment of Parkinson's disease because it is converted directly to catecholamines. Thus, it can compensate for major losses of catecholamines in some parts of the brain. There is as yet no evidence that postmenopausal women taking L-dopa for Parkinson's disease regain normal menstrual cycles.

The loss of female reproductive cycles during aging most likely involves more than one type of aging process. For example, the number of egg cells (oocytes) and ovarian follicles which can make sex steroids is fixed at birth in mammals, and declines irreversibly thereafter. Human beings are born with about 500,000 egg cells and, by the age of 50, may have less than 10,000. It was widely believed, until the L-dopa experiments with rats, that this loss of egg cells and hormone-producing follicles was the *only* cause of female reproductive aging. Now it is possible to speculate that aging in *both* ovaries and brain cells contributes to the loss of regular cycling and fertility. It may even be possible that aging changes in the ovary cause aging changes in some parts of the brain. In any case, the aging of reproductive functions illustrates the complexity of aging processes and warns against the idea that aging processes have a simple, single cause which could yield to a simple, single treatment. It seems most unlikely that a universal cure for aging changes will be found.

The possibility that brain catecholamine deficiencies arise during aging is being studied by our laboratory as well as others. We have used healthy aging male mice for our studies because they do not show changes in testicular function which could influence the brain's metabolism. We and others have found that aging does indeed impair catecholamine metabolism in the hypothalamus and other parts of the brain. Both the levels and the rate of metabolism (turnover) of catecholamines are reduced. These findings directly support the hypothesis that L-dopa reactivates estrous cycles in old female rats by temporarily compensating for impaired hypothalamic catecholamine metabolism. Our results on male mice now allow us to study the more complex aspects of female aging and to distinguish changes due to gonadal failure during aging from those of other origins. It may be some time before the exact role of L-dopa is known, because there are at least four different types of catecholamine-containing nerve cells in the hypothalamus. To understand aging in the brain will require exacting biochemical measurements of minute groups of brain cells, whose func-

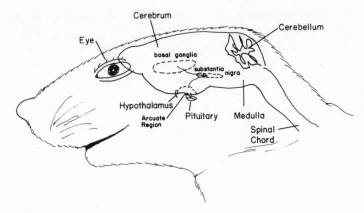

Figure 3. Major regions of the rodent brain. The dopamine-containing neurons of the substantia nigra may project to the putamen and hypothalamus (Kizer *et al.*, 1976).

tions are very obscure in young animals. We have recently improved the sensitivity of our assays for catecholamines and now can detect 5 trillionths of a gram (5 picograms) of the catecholamines: dopamine, noradrenaline, or adrenaline. With these techniques, we hope to map the details of how aging alters key neural pathways.

In addition to impairments in the hypothalamus, we also found 25% reductions of dopamine in a nearby area of the brain called the basal ganglia (Figure 3). The basal ganglia are also involved in Parkinson's disease and become more than 80% depleted of dopamine. This causes involuntary tremors of the hands and rigidity of the postural muscles. The depletion of dopamine which we observe in normal aging mice was recently found similar to that in the *normal* human brain during aging. Although mice and men age at vastly different rates, the extent of change is remarkably proportionate to the length of life-span. Even though the depletions of catecholamines are not as severe as those occurring during Parkinson's disease, they suggest a reason why Parkinson's disease is rarely observed in the young. We hypothesize that losses of catecholamines are a normal aspect of aging and that some individuals have greater depletion of dopamine than others; such individuals may be at high risk for Parkinson's disease. On the average, about 1% of the adult population in the United States ultimately show Parkinson symptoms. Perhaps Parkinson's disease is aging gone from bad to worse. One type of Parkinson's disease is associated with a viral encephalitis epidemic ("encephalitis lethargica"), which swept this country in the early 1920s. This type of encephalitis is known to cause damage to dopamine-containing neurones in the basal ganglia. Many of the survivors recovered to live normal lives, but 20 or more years later developed Parkinson's symptoms. We speculate that the onset of symptoms was delayed so long because the viral damage to the basal ganglia was insufficient to cause Parkinson symptoms until decades later, when the increasing dopamine depletion of normal aging was combined with the initial deficiency.

Another facet of the trend for impaired function of dopamine neurones during aging is revealed by the effects of certain antipsychotic drugs, such as phenothiazines (e.g., thorazine) which acts by impairing dopaminergic functions. These can produce side effects similar to the dopamine deficiency of Parkinson's disease. In very high doses, phenothiazines can temporarily induce Parkinson's symptoms in normal young humans; in older ones, Parkinson symptoms commonly appear at lower doses. The greater sensitivity of older humans to the Parkinsonian effect of phenothiazines may thus result from the addition of the drug effect to the underlying age-related impairment in dopamine functions.

Recently, direct neural pathways between the basal ganglia and the hypothalamus were discovered by neuroanatomists. The neurons of the substantia nigra in the basal ganglia which influence muscle-movement control centers also may send axons to the endocrine centers of the hypothalamus (Figure 3). Thus, age changes in the dopamine neurones of the substantia nigra may be implicated in the apparently different aging phenomena of movement and endocrine disorders.

A different and striking effect of L-dopa was found by the late George Cotzias, who introduced the treatment of Parkinson's disease by this drug. In high doses, L-dopa markedly increases the average longevity in one type of laboratory mice (Figure 4) and even reduces the incidence of tumors. These provocative results may mean that deficiencies of catecholamines underlie a number of life-shortening conditions. Careful research to understand these results is being continued by Cotzias' associates at the Sloane–Kettering Foundation in New York City.

What other phenomena of aging could be viewed as consequences of brain catecholamine impairments? The list could include almost all organ systems, and physiologic and mental processes, since catecholamines influence many body functions via endocrine and neural controls. Too little is known about other aspects of the brain to speculate intelligently.

A wide variety of endocrine changes have been reported during aging besides those in reproductive function. These include shifts in blood levels of hormones from the adrenal, thyroid, and pancreas and in the threshold for secretion of hormones. It seems remarkable that each

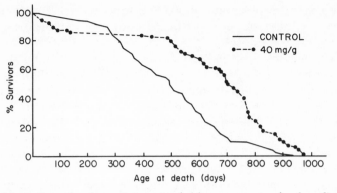

Figure 4. Longevity and L-dopa. Swiss mice fed large amounts of L-dopa lived about 4 months longer than the controls. Because this colony of mice had below-average longevity, further studies are being done to determine if longevity extension by L-dopa also applies to longer-lived mouse strains (Cotzias *et al.*, 1977).

age group from puberty onward has a characteristic pattern of metabolism which is constantly and slowly changing. There does not appear to be any final "steady state" which characterizes adult life. Since the hypothalamus either directly or indirectly controls these glands, the influence of aging in the brain on endocrine regulation must be seriously considered. For example, in old rodents, in contrast to the young, adrenal steroids are not secreted during fasting. Since the adrenals respond just as well to ACTH injected by the experimenter in separate studies, an age-related impairment is being sought at higher levels at the pituitary and hypothalamus. It is even possible that immunological changes of age may be controlled by hormones or other factors ultimately regulated by the brain since the thymus gland is influenced by the pituitary and since many white blood cells contain receptors for catecholamines. The complex relationships by means of which the various endocrine glands control the brain, and vice versa, will make it very difficult to sort out cause and effect. It is even possible that the study of aging will lead to the discovery of new hormones, as suggested by Donner Denckla's research, and yet unknown physiological mechanisms.

I have proposed the view that the two-way interrelationships between the brain and the hormones and tissues it controls could be the basis for the absence of a "steady state" in metabolism during adult life. As changes occur, the brain may respond, promoting new changes. Such a cascade of neural, endocrine, and metabolic events has also been proposed to control the timing of puberty. Such mechanisms may be the underlying force in postmaturational aging as well. If this is true, one would predict that true arrest of maturation would prevent aging and greatly prolong the life-span.

The changes of sleep pattern during aging give a glimpse of other possibilities relating to neurochemical and behavioral changes of age. During aging, the pattern of sleep shows predictable changes in normal humans. The number of spontaneous awakenings during the night increases with age, while the time spent in deep sleep and rapid-eye-movement (REM) sleep decreases. Apparently, a longer stay in bed is required as one grows older to achieve the same total amount of rest. These quantitative studies confirm the general observation that with aging, sleep becomes less intense. Since the same trends are also observed in laboratory rodents, organic causes or causes intrinsic to the brain appear to be involved rather than "psychologic" factors. The "organic" component of sleep changes during aging may result from changes in amine-containing neurones. The catecholamine- and 5-hydroxytryptamine-containing nerve tracts of the brain stem (medulla and pons) are known to have a key role in regulating sleep. For example, destruction of the locus coeruleus in young rats causes a disappearance of REM sleep. Since the locus coeruleus is a major source of the noradrenaline-containing neurones, sleep, too, may change during aging because of impairments of catecholamines. Harold Brody's recent observation that there are 20 to 30% losses of locus coeruleus neurones in normal humans during aging is thus of great interest.

In some individuals, the loss of normal sleep may impair mental functions. According to one study, the number of spontaneous awakenings was inversely correlated with performance on an intelligence test, whereas the amount of REM sleep was directly proportional to performance. In humans, it is clear that subtle age changes in sleep associated with cell loss in a minute brain region could have important psychological and sociological consequences. It is striking how few neurons the locus coeruleus contains — only 40,000 of the 200 trillion nerve cells in the human brain or less than one-millionth of all brain cells. Each of the neurons from the locus coeruleus has a vast number of terminals, 10,000 or more, which, in turn, could spread their influence widely.

The loss of neurons from the locus coeruleus during aging in humans must be considered in the perspective that many other groups of neurons do not show loss during normal aging. For example, in the inferior olive, neurons persist in full number even to advanced ages beyond 90 years. The often-repeated belief in a "loss of 100,000 neurons each day" is probably a gross exaggeration. The best studies show a high degree of selectivity about which neurons are lost and at what ages the loss occurs.

In conclusion, the demonstration of selective aging changes in the normal brain of short- and long-lived mammals has recently opened the way for serious study of cause-and-effect relationships in physiological, behavioral, and mental changes of aging. More basic information must be gathered, however, before critical and testable hypotheses can be formulated. The ultimate goal of such research is to evaluate how the various changes of aging determine the susceptibility to diseases and disorders of aging and limit the healthy life-span.

ACKNOWLEDGMENTS. This research was supported by grants to C.E.F. from the NIH (AG-00117), the NSF (PCM76-02168), by General Research Support Grant S05 RR07012-07, by the Glenn Foundation for Medical Research (Manhasset, N.Y.), and the Orentreich Foundation for the Advancement of Science (New York, N.Y.).

BIBLIOGRAPHY

General

Cooper, J. R., F. E. Bloom, and R. H. Roth. 1976. *Biochemical basis of neuropharmacology*, 2nd ed. Oxford University Press, New York.

Finch, C. E., and L. Hayflick, eds. 1977. *Handbook of the biology of aging*. Van Nostrand Reinhold Co., New York.

Greengard, P. 1977. Second messengers in the brain. *Scientific American* 237: 108–119.

Technical

Cotzias, G. C., S. T. Miller, L. C. Tang, P. S. Papavasiliou, and Y. Y. Wang. 1977. Levodopa, fertility, and longevity. *Science* 196: 549–551.

Everitt, A. V., and J. A. Burgess, eds. 1976. *Hypothalamus, pituitary, and aging.* Charles C Thomas, Springfield, Ill. [See chapters by D. Denckla (pituitary–thyroid) and P. Ascheim (hypothalamic–pituitary–ovary).]

Finch, C. E. 1975. Neuroendocrinology of aging: a view of an emerging area. *BioScience* 25: 645–650.

Finch, C. E. 1976. The regulation of physiological changes during mammalian aging. *Quarterly Review of Biology* 51: 49–83.

Kizer, J. S., M. Palkovits, and M. J. Brownstein. 1976. The projections of the A8, A9, and A10 dopaminergic cell bodies: evidence for a nigral hypothalamic-median eminence dopaminergic pathway. *Brain Research* 108: 363–370.

Terry, R. D., and S. Gershon, eds. *Neurobiology of aging*. Raven Press, New York. [See chapters by J. Feinburg (sleep) and H. Brody (neuron number).]

20

Evolutionary Biology of Senescence

RICHARD G. CUTLER

The ancient covenant is in pieces; man knows at last he is alone in the universe's unfeeling immensity, out of which he emerged only by chance. His destiny is nowhere spelled out, nor is his duty. The kingdom above or the darkness below: it is for him to choose.

Chance and Necessity — Jacques Monod, 1971

BIOSENESCENCE: CONCEPTS AND MECHANISMS

Man's self-awareness has proven to be uncomfortable to him; he realizes that his life-span is finite, and knows of nothing that will prevent his physical and mental health from slowly declining to the point where death will be unavoidable. Naturally man, with his characteristic curiosity and love of life, wonders why this is so, how it happens, and what can be done, or should be done, about it. Is our life-span sufficient? If we should live any longer, would it be unbearable? An important difference exists between simply living out one's natural life-span and living the same number of years in an optimum state of vigor and health. Man's optimum period is from about 14 to 30 years of age, or only about one-fifth of his life-span.

To describe the processes causing the decline in vigor and health, I use the term biosenescence rather than aging, as aging can have non-biological connotations. "Chronoage" refers to chronological age, whereas "bioage" refers to biological age or degree of biosenescence.

What causes biosenescence? The most popular idea is the "wear-

RICHARD G. CUTLER, Ph.D. • Gerontology Research Center, Baltimore City Hospitals, Baltimore, Maryland 21224 of the National Institute on Aging, National Institutes of Health, Bethesda, Maryland 20014

and-tear" concept. But since living systems can repair and rebuild themselves, why aren't these capabilities used fully to fix up the wear and tear? Another difficulty with the wear-and-tear hypothesis is how to explain the wide range of life-spans characteristic of different mammalian species. Different species have varying maximum life-span potentials (MLPs[1]) because of inherently different biosenescent rates. It is not really known how complex biosenescent processes are, or whether something can be done to lengthen useful life-spans. If biosenescence is really so complicated, how did closely related species obtain vastly different life-spans when they are physiologically and molecularly very similar? To investigate this problem, I have chosen an evolutionary approach to study longevity and biosenescence.

EVOLUTION OF LIFE: WHAT CAN IT TELL US ABOUT BIOSENESCENCE?

Evolutionary studies show that man and other species represent a recorded history of their own evolutionary past, from the morphological to the biochemical levels. The power of the evolutionary approach to aid in understanding the biochemical nature of life is now being appreciated. Morphological records of our ancestors can be found in our bodies. For example, we have vestigial tissues which can best be explained on an evolutionary basis. Relics of our evolutionary past have been found at the biochemical level in the amino acid sequences of proteins, the nucleic-acid-base composition of chromosomes, and in the design of metabolic pathways. These studies indicate that man was not constructed perfectly but by chance-and-necessity or opportunistic processes that occurred over millions of years by random mutations and natural selection. Biosenescence may be an unavoidable, natural by-product of evolution; and, if biosenescence is a universal phenomenon present in all forms of life, it may have played an important role in shaping the basic life processes themselves.

Given the importance biosenescent processes may have played in the evolution of life, it is strange that this subject has not been examined more thoroughly in the past. Discussions of biosenescence or longevity in books on biology or evolution are not easy to find. I believe this has been a serious oversight. In my studies in this area, I have found explanations for the very nature of life itself through a search to understand the nature of biosenescence. I have been led to the conclusion that life and biosenescent processes coevolved from the very origin of life itself.

[1]See Glossary for definitions of terms and abbreviations.

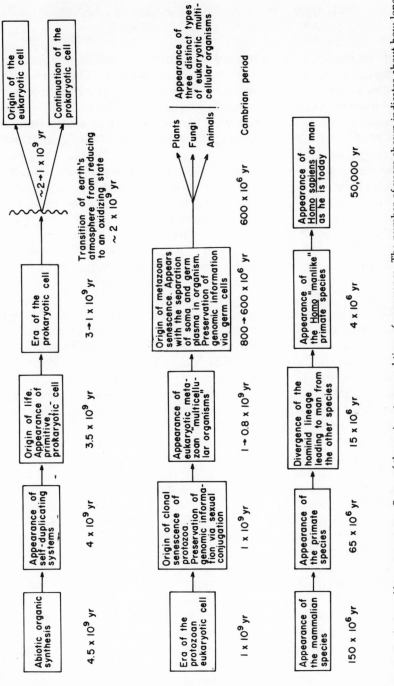

Figure 1. Evolution of human systems. Some of the major stages and time of occurrence. The number of years shown indicates about how long ago that stage appeared.

ORIGIN OF INTRINSIC BIOSENESCENT AND
ANTIBIOSENESCENT PROCESSES

Life is thought to have formed spontaneously millions of years ago from the same basic constituents that are now found in all its forms. The major steps are shown diagrammatically in Figure 1. The components of life are nucleic acid bases, amino acids, and carbohydrates or sugars. These substances are thought to have been formed by an abiotic process, beginning about 4.5×10^9 years ago when the earth's atmosphere was in a reducing state. This atmosphere consisted mainly of N_2, CH_3, H_2O, and H_2. From these gases, and an energy source such as lightning, ultraviolet light, or volcanic heat, the nucleic acid bases and amino acids were formed spontaneously. Proof that such reactions can take place has been shown by actually producing them in the laboratory. Particularly interesting is that, of the countless different molecules that could have been produced experimentally, a significant fraction of the molecules formed are those actually found in life. The basic building blocks for life originated abiotically, but why and how did the first living systems emerge, and what physical forms did they take?

The genetic system forms the unique characteristics of life. Thus, to explain the origin of life, one must explain the origin of the genetic system. I suggest that life, past and present, is an antibiosenescent process that preserves information, which continues to preserve itself. This constitutes the unique feedback enabling the living system to develop through natural selection. Genetic information, which is nonphysical, is what is preserved. These ideas are understood by examining the major steps leading to self-duplicating systems.

We begin with the genetic system, which contains information stored in an ordered sequence of four nucleic acid bases forming the polymer DNA. This informational DNA is transcribed and then translated into an ordered sequence of amino acids, which, in turn, forms a protein. The proteins do the work of the cell, and the genetic system simply stores the information that defines the amino acid sequence. Different proteins have different amino acid sequences capable of structural or enzymatic functions. DNA also contains regulatory information which determines how, when, and where total information (regulatory, enzymatic, and structural) will be read.

About 4×10^9 years ago, a mixture of nucleic acid bases and amino acids, and later polymers of nucleic acid bases and amino acids, were spontaneously formed in a primeval soup. Both the nucleic acid and amino acid polymers had short life-spans, due to the environmental conditions that led to their formation. These polymers probably underwent some type of postsynthesis modification, but were always soon degraded back to the nucleic acid bases or amino acids. The origin of the self-duplicating system rose from two unique properties of the nucleic acid and amino polymers.

The first was the ability of a nucleic acid polymer to act as a template to guide or arrange free nucleic acid bases along its length in a complementary manner of base pairing. This complementary nucleic acid polymer could also act as a template, and the resultant polymer would be a copy of the original parent molecule. The duplication process was not precise, and the mechanism of how the original duplication process exactly occurred is now known. But undoubtedly such a process occurred with the very important result that the life-span of the base sequence of the original parent polymer could be greatly extended by the duplication process, despite the subsequent destruction of the parent strand. The second unique property of the nucleic acid polymer was its ability to influence the amino acid sequence of proteins during their spontaneous polymerization. Therefore, duplication may have been helped by the presence of proteins whose amino acid sequences were influenced by the nucleic acid polymer.

The critical step occurred when DNA randomly evolved a nucleic acid base sequence that could direct the synthesis of a protein which helped the DNA to duplicate itself, at sufficient accuracy and speed to prevent the loss of this information. This information, formed by chance, escaped death even though the physical system that preserved it, the nucleic acid and amino acid polymers, had a finite life-span.

The preservational process of self-duplication gave rise to a new and far-reaching property of the DNA. The information it contained had the capability not only to be preserved, but also to grow in a selective manner. The sequence of the DNA was constantly altered by a random addition of bases that increased its length, and by the innate inaccuracy of its duplication. This latter process is called mutation. As the new nucleic acid base sequences were acquired, they were read out through the template-guidance reaction of protein synthesis. The entire polymer would be preserved through natural selective processes in accordance with its value in preserving this new genetic information, plus the original information it contained.

The original biosenescent processes were the natural environmental hazards of the self-duplicating system, and the original antibiosenescent processes were the duplication process and the associated proteins which were coded for by the preserved information. Hence, intrinsic biosenescent and antibiosenescent processes orginated. From the antibiosenescent processes emerged a stable life process.

Since the origin of life nothing basically has changed. The continued evolution of life to higher forms represents a coevolution of antibiosenescent and new biosenescent processes, which gave rise to new antibiosenescent processes, and so on. From this analysis arises the concept that all life systems represent, in both their morphology and biochemistry, this entangled coevolutionary history.

In summary: (1) Life originated when the capacity evolved to preserve information that could preserve itself; (2) this preservation required the ability to duplicate the information within a critical period of time; and (3) the origin and perpetuation of life represents a triumph of antibiosenescent over biosenescent processes.

INTRINSIC BIOSENESCENT AND ANTIBIOSENESCENT PROCESSES OF THE LIVING CELL

About 3×10^9 years ago, the first forms of life appeared, generally classified as prokaryotic cells. The prokaryotic cell was essentially the only form of life for over 2×10^9 years, or about 50% of the total time life has existed. The prokaryotic cell evolved due to several important events, such as the gradually diminishing abiotic production of basic raw materials necessary for life. This resulted from an increasing ozone layer in the atmosphere that absorbed an increasing amount of ultraviolet light, which gave rise to the selection of these forms of life that were more capable of coping with a decreasing availability of ultraviolet-producing nutrients. The increasing variety of living forms added new selective pressures, which favored a living system that could best compete for a limited supply of nutrients. A harmful internal environment, which was a by-product of otherwise beneficial reactions, also contributed to the evolution of the prokaryotic cell.

This last effect is an intrinsic pleiotropic biosenescent process and will prove to be extremely important in understanding how and why we grow old. The presence of pleiotropic processes — having both harmful and useful effects — is well recognized by geneticists. The effects of pleiotropy include the effects of the primary process itself and secondary branching effects. The epigenetic nature of development implies secondary pleiotropy as an almost universal principle. Even persistent vestigial tissues or other useless characteristics might be present because of the pleiotropic effect.

How pleiotropic biosenescent processes might have developed is as follows. Proteins evolved by a chance-and-necessity operation. By chance, different types of proteins occurred, and by necessity, new genetic information became fixed as its advantage proved greater than its disadvantage. Enzymes are selected on the basis of their ability to act on a certain substrate. This specificity can never be expected to be perfect. Enzymes would be expected to act on other substrates that would not be beneficial to the organism, and many reactions are probably harmful. The greater the number of different enzymes and substrates existing within a cell, the higher the probability would be of side reactions that were not originally selected. A similar problem existed for the enzyme products. These products were made for their beneficial effect, but they may also have some type of interference or competition effect with other cellular functions. Thus, as the cell became more complex, internal environmental hazards increased and harmful pleiotropic intrinsic reactions occurred.

I see no way for any living system to avoid the presence of such pleiotropic genes whose fixation in the genetic system depends on chance-and-necessity factors operating over a limited time interval. Therefore, through necessity of survival, other processes evolved at a later time that were not concerned directly with the vital functions of the cell but developed solely to reduce the harmful effects of these pleiotropic reactions. These are intrinsic antibiosenescent processes.

Three major classes of environmental hazards act toward destroying the information content of the genetic system: (1) external hazards, (2) natural instability of the biological constituents of the living system, and (3) harmful intrinsic pleiotropic effects of otherwise beneficial reactions. The maintenance of the living system requires protective and repair processes directed against all three of these hazards. Because these environmental hazards qualitatively and quantitatively change in time, protective and repair processes also had to change.

The prokaryotic cell, during the period of 1 to 3×10^9 years ago, represented a very complex living system that evolved much of the original genetic information coding for regulatory processes and the structure of proteins necessary for life. This information is found in the higher eukaryotic multicellular organisms existing today. Prokaryotic cells contained many different types of intrinsic cellular pleiotropic biosenescent processes that were counteracted, in turn, by many different types of intrinsic antibiosenescent processes.

A dramatic change occurred in the earth's atmosphere about 2×10^9 years ago that had profound consequences on all forms of life, and probably resulted in the almost complete elimination of all living systems. The earth's atmosphere changed from a reduced to an oxidative state as the result of oxygen released from the photo-breakdown of water, and later by living organisms.

Oxygen was harmful to almost all constituents of the cell, and protective processes evolved. An advantageous use of oxygen developed by the production of energy. Oxygen is still toxic to many biological constituents, and important protective processes against oxygen can be found in the cell. When oxygen was used to produce energy by oxidative phosphorylation, it became pleiotropic, with both beneficial and harmful effects. Oxygen is a major intrinsic pleiotropic biosenescent agent. The processes that evolved to reduce its harmful side effects are major antibiosenescent processes that can now be found in all aerobic cells.

The preservation of living systems throughout their evolution was a delicate matter. Antibiosenescent processes never evolved in great excess, and at no time did cells exist that were completely free from the effects of biosenescent processes.

A most remarkable event of the oxygen-crisis period was the origin of the eukaryotic cell, which contained a true nucleus. Because the eukaryotic cell has many properties similar to the prokaryotic cell, a popular explanation for the appearance of this new type of cell is that it developed by a symbiotic process between previously existing prokaryotic cells. Features of the eukaryotic cell act as antibiosenescent structures that enable the eukaryotic organism to have a better capacity for survival in an oxidative environment and better protection against the harmful pleiotropic factors inherited from prokaryotic cells. The genetic apparatus is enveloped by basic proteins which may be involved in regulating the flow of genetic information, but they could also act as a protective shield for the DNA against intrinsic damaging agents. The nuclear membrane may have originated to maintain the reducing environment from which the cell evolved and to further protect it from harmful pleiotropic agents in the cytoplasm. The compartmentalization of the oxidative-phosphorylation reaction by a mitochondrial organelle, which utilizes oxygen and gives off many harmful by-products, such as free radicals, could have been an important means for isolating such reactions from the rest of the cell. A similar reason for compartmentalization would apply to lysosomes and peroxisomes. These examples support the proposition that much of the basic morphological, as well as biochemical, aspects of the cell were molded by antibiosenescent processes acting against intrinsic pleiotropic biosenescent processes.

Another important antibiosenescent process of the cell is its capacity to degrade and resynthesize many of its constituents. This process, turnover, removes defective components. Even incorrect information in the DNA can be corrected by a turnover process — DNA repair — using the DNA complementary strand. If complete turnover is possible, then a cell does not need to divide, but complete turnover appears possible only by cell division. If complete turnover does not occur at a rate at least equal to the rate of accumulation of defective components, death of the cell is inevitable. Some populations of eukaryotic cells accumulate biosenescent damage in spite of division. With the loss of the genetic information of these cells, death can only be prevented by genetic recombination before a critical number of division cycles elapse.

The presence of some strains that do not require genetic recombination tells us that the difference in genetic information is slight between the delicate balance of biosenescent and antibiosenescent processes. Why accumulation of biosenescent damage evolved when it could have been easily prevented may be, as T. M. Sonneborn suggested: to enforce recombination for the long-term genetic advantage to the species. Whatever the reason, one can be sure it was for long-term preservation of the genetic information representing the species. Thus, there is basically nothing different between cells that do or do not undergo a finite number of divisions in terms of the presence of biosenescent processes or the requirement for antibiosenescent processes.

ORIGIN OF BIOSENESCENCE IN THE MULTICELLULAR ORGANISM

The free-living eukaryotic cell is thought to have evolved by 10^9 years ago all the metabolic pathways, synthesis capabilities, general cellular morphology, and features of the genetic apparatus that exist today. Thus, when the eukaryotic cell began to colonize into multicellular organisms, these organisms inherited all the intrinsic cellular biosenescent and antibiosenescent processes originally present in the free-living cells. The sea anemone, which has complete cell turnover similar to a clone of free-living eukaryotic cells, has been used mistakenly as an example that immortality can exist in multicellular organisms. Each cell of the sea anemone has intrinsic biosenescent processes which require antibiosenescent processes, such as cell division, to preserve the genetic information of the organism. What is immortal is the population of cells, which in fact simply represents the preservation of the genetic information that must be immortal in all living systems.

The multicellular organism offered many advantages, but disadvantages also arose, related to the delicate balance between biosenescent and antibiosenescent processes. As the metazoan organisms became more complicated by processes of differentiation, many types of cellular-differentiated states, such as muscle and neuron cells, limited the freedom of cell division. The neuroendocrine system evolved to communicate regulatory messages governing homeostatic and developmental processes of the entire metazoan. Like all other products, the hormonal factors were pleiotropic, as they also had side effects that were detrimental to the organism. A reduction of the biosenescent processes

,associated with differentiation and developmental processes was necessary; otherwise, the cells making up the organism would lose their capacity to continue preserving their genetic information.

The solution to this problem involved only a slight change in the makeup of the multicellular organism. The organism was separated into soma and germ plasm cells. Therefore, it was no longer necessary to preserve the genetic information of the entire organism. The soma, or body, part of the organism could be allowed to undergo biosenescence, and the genetic information in the germ plasm cell was preserved. Thus, the major problem of how to prevent accumulation of biosenescent processes in multicellular organisms was really not solved, but avoided, by allowing the soma to undergo biosenescence and preserving the germ plasm. This much easier task, achieved by chance-and-necessity evolutionary processes, was the breakthrough required for the future evolution of complex metazoan organisms. How was the genetic information in the germ plasm kept intact? I believe this was done through utilizing the same antibiosenescent processes that were inherent in the primitive free-living eukaryotic cell, plus a few additional processes. Some of the antibiosenescent processes that may be used to preserve the germ plasm are as follows.

For the ovary: (1) A highly condensed genetic apparatus involving special histone proteins giving maximum protection to the DNA; (2) a minimum readout of information from the genetic apparatus, further protecting the DNA; (3) low specific metabolic rate of the cell, which would reduce the rate of production of pleiotropic biosenescent agents; and (4) the presence of synapses between homologous chromosomes during meiosis to permit additional postreplication repair capacity, suggested by Rolf Martin of Brooklyn College, New York.

For the gonads: Processes similar for the ovaries but additionally: (1) the presence of an active state of division, spermatogenesis, which acts to remove slowly dividing cells that are likely to have genetic defects; and (2) selection of the most viable cells by their migratory pathway to fertilize an ovum.

Additional processes that select only those germ cells with complete information are the abortion of abnormal fetuses, and the decreased ability of defective organisms to reproduce. However, even these processes do not completely eliminate the loss of information in some of the germ plasm cells, as evidenced by the rapid increase of genetically abnormal fetuses, not always aborted, that occur as a function of increased chronoage.

In summary, the separation of the metazoan organism into soma and germ cell components represents an important means of avoiding intrinsic cellular biosenescent processes. This separation led to the evolution of the higher metazoan organisms, and to the origin of biosenescent processes for the soma, which can die with no serious consequence to the species. The biosenescence of the soma was merely the outcome of mechanisms used to ensure the continued preservation of the genetic information that can preserve itself. The separation of soma and germ cells might not have even taken place if intrinsic cellular biosenescent processes were not important or were a trivial matter to overcome.

The separation of soma and germ plasm was quickly followed by the appearance of many different types of metazoan organisms during the Cambrian period, which began about 600×10^6 years ago (Figure 1). Three major types of metazoan systems appeared: plants, fungi, and animals. These biological systems proved the most successful on a long-term basis within the metazoan class and were able to continue to preserve their lineage of genetic information.

Each organism has an optimum life-span to ensure survival of germ cells. Much evidence indicates a process of genetically programmed death in plants, where there would be definite advantages to allow the soma to die. The means to achieve a finite life-span in plants is not genetically programmed biosenescence but genetically programmed death, which can occur without the selection of special death factors or hormones. Merely turn off an antibiosenescent process or a necessary vital process, and all the cells of the organism quickly die. Evolution of life represents a history of the means to preserve life, not the means of how to limit it.

Trees, such as the giant sequoia or bristlecone pine, which reach ages of 3000 or 4000 years, are not really long-lived, as commonly thought, but undergo continual somatic death. The living part of the tree soma consists of a thin layer of dividing cells located in the cambium layer, just inside the bark. There are no long-lived postmitotic cells in trees, and most of the tree that is over 1 year old is not even alive but only represents the walls of dead cells; therefore, trees can be very old chronologically but not biologically. The long life-span of the continuously turning over population of cells making up the cambium layer is similar to bacteria, *Paramecium*, or germ cells maintaining their viability.

REPRODUCTIVE VALUE AS A FUNCTION OF CHRONOLOGICAL AGE

Metazoan development is usually completed shortly after sexual maturation, after which the organism probably will be killed due to environmental hazards. If not killed, the organism undergoes biosenescence and eventually dies. To ensure maximum reproductive probability, soma–germ organisms are under positive selective pressure to reach maturity as soon as possible to prevent the accumulation of biosenescent processes from reaching harmful levels before sexual maturity.

Consider a population in equilibrium with its environmental hazards. Death rate equals birth rate, and the total number of individuals in the population does not change with time. Assume that no member of the population suffers from biosenescence and that all members have an equal probability of being killed regardless of chronoage. Such a population would consist of more young than chronologically old individuals simply because, regardless of the individual's age at death, it is always replaced by a newborn individual. Thus, in all dividing populations, even in the absence of biosenescence, there are always more young than chronologically old individuals. The number of individuals at a given chronoage decreases as the chronoage increases.

With increasing chronoage, an individual's reproductive probability decreases, illustrated in Figure 2a. An exponential decrease in the number of individuals is shown, as would be expected even if they were immortal, and their corresponding reproductive capacity is a function of increasing chronoage. Figure 2b shows the resultant reproductive value of this population as a function of chronoage.

The concept of declining reproductive value with increasing chronoage was first recognized in 1910 by the famous geneticist Ronald A. Fisher, and later in the 1930s by J. Haldane of University College, London. In 1952, Peter B. Medawar of University College, London, and in 1957, George C. Williams of Michigan State University utilized this concept to propose a hypothesis for the evolution of biosenescence in soma–germ-cell organisms. I believe the hypothesis presented by Medawar and Williams is basically correct and represents the first major contribution made toward our understanding of the nature of biosenescent processes in man.

Medawar and Williams suggested that biosenescent processes are pleiotropic in nature, having beneficial effects at one stage of life but harmful effects later. If true, there would be positive selective pressure to prevent the expression of the harmful aspects of the corresponding gene(s) governing the processes until after sexual maturation. The intensity of the selective pressure acting against such genes would follow a continuous function, proportional to the reproductive value curve, as shown in Figure 2c. Hence, only a few biosenescent processes would be expressed up to the age of sexual maturation; then their expression would gradually increase in number and/or intensity in a continuous manner as a function of increasing chronoage.

The postponement of biosenescence would depend on the reproductive value curve, which in turn is dependent upon two parameters; (1) how fast the members are lost from the population, or the intensity of the environmental hazards of the ecological niche of the organism; and (2) reproductive capacity. Organisms would be selected to mature and reproduce as quickly as possible and would finally be killed by natural environmental hazards only after completing a given amount of reproduction and before significant decline of physiological functions through intrinsic biosenescent processes.

The hypothesis of Medawar and Williams suggests a new type of biosenescent process expressed as a developmentally linked, pleiotropic, intrinsic biosenescent effect. This biosenescent effect is minimal up to sexual maturation, but afterward shows a sharp increase in expression, illustrated in Figure 2d. The developmentally linked biosenescent process gives no advantage to the organism. Therefore, selective pressure exists to reduce biosenescent effects and foment preservation of the germ plasm. Hence, the declining slope of the reproductive value

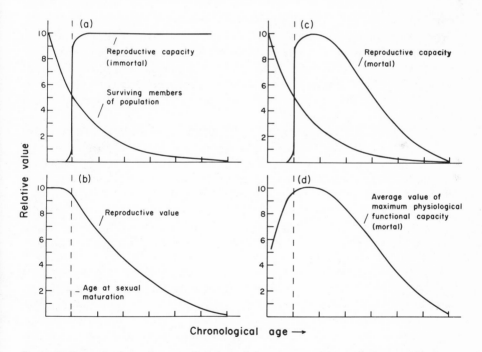

Figure 2. Reproductive value as a function of chronoage and its evolutionary effects on reproductive capacity and physiological functional capacity. (a) to (d) illustrate how the rate of loss of members of a population affect the reproductive value of its members and thus the evolutionary pressure to maintain reproductive capacity and physiological functional capacity.

Table 1. Correlation between Mortality Rate in the Wild and Maximum Life-Span Potential for Some Birds*

Common name	Average annual adult mortality (fraction that is killed)	Maximum life-span potential (yr)
Blue tit	0.72	9
European robin	0.62	12
Lapwing	0.34	16
Common swift	0.18	21
Sooty shearwater	0.07	27
Herring gull	0.04	36
Royal albatross	0.03	45

*Data from D. Botkin and R. Miller, Mortality rates and survival of birds, *American Naturalist* 108: 181–192, 1974.

curve evolved so that the organism did not live as long as possible, to avoid undergoing serious biosenescence. Thus, loss of function due to the accumulation of the developmentally linked biosenescent process is rarely seen and occurs only when natural hazards suddenly diminish. This occurs when animals are placed in a zoo and occurred when man reduced his environmental hazards faster than his longevity could evolve.

Many data exist supporting a chronological age-dependent decrease in reproductive value. For example, animals in the wild rarely live to a chronoage where they would suffer a functional decline. An excellent correlation is found between the intensity of the hazards of the natural ecological niche and the animal's period of optimum biological function, or MLP. Table 1 shows the excellent inverse correlation between the mortality rate for birds in their natural environment and their maximum life-span potentials in captivity. Table 2 shows a similar inverse correlation for some rodent species.

Since man reaches sexual maturity at about 14 to 18 years and full growth at 20 years, serious symptoms of biosenescence should appear soon after the age of 20. We would also expect his average survival time to be somewhere between 30 and 40 years in the natural ecological niche where he evolved, which is exactly what we find. Figure 3 shows a linear general physiological decline, and an increase in biosenescence occurs beginning at about 20 years. The data agree with man's mortality rate being lowest around the age of sexual maturation, and rising at an exponential rate with increasing chronoage.

Table 3 shows that man's average life-span expectancy was only 30 to 40 years until A.D. 1400. When the *Homo sapiens* species first emerged, apparently the decline in physiological and mental processes after 30 years of age was not a problem because man did not normally live past 30. Only in the past 200 years have biosenescent processes become common.

In summary, two important events occurred after the appearance of the soma–germ organism: (1) a new type of pleiotropic biosenescent process linked to development appeared, and (2) the harmful effects of the new biosenescent process became a function of the species' environmental hazards. The intensity and time of appearance of the effects caused by the developmentally linked biosenescent processes are a function of reproductive value. Intensity of continuously acting biosenescent and antibiosenescent processes is also selected according to a species' reproductive value. We now have a rational basis to answer why different animals have different MLPs, why animals grow old, and what biological processes govern the period of optimum vigor and health.

I believe Medawar and Williams were mistaken that biosenescent rather than antibiosenescent processes evolved. For example, they assume that all cells were originally intrinsically immortal, such as bacteria or germ cell populations, and that for the cells of soma–germ-cell organisms, biosenescence was a new development. In 1891, August Weismann made a similar mistake, proposing an evolutionary theory of biosenescence based on the idea that germ cells were intrinsically immortal. However, Weismann did recognize the important role played by

Table 2. Mortality Rate of Some Rodent Species in Their Natural Ecological Environment*

Species	Average mortality rate	Maximum life-span potential	Maximum calorie consumption (kcal/g/MLP)
Peromyscus maniculatus (deer mouse)	63–94% in 1 yr (rarely live over 2 yr)	3040 days	703
Peromyscus leucopus (white-footed mouse)	Rarely live over 2–3 yr	3300 days	672
Mus musculus (field mouse)	99% in 1 yr (rarely live over 1 yr)	1200 days	250
Tamias striatus (Eastern chipmunk)	50% in 1.5 yr	12 yr	561

*Data from A. King, *Biology of peromyscus*, American Society of Mammalogists, 1968, and C. Tryon and O. Snyder, Biology of the Eastern chipmunk, *Journal of Mammalogy* 54: 145–168, 1973.

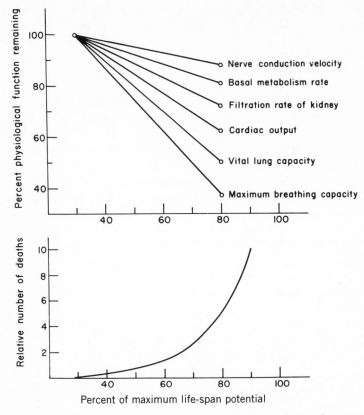

Figure 3. Average loss of physiological functional capacities and mortality rate, as seen taking a cross section of the human population at different ages.

the germ cells in preserving the genetic information of the soma–germ-cell organism. I also disagree with Williams that little can be done to extend life-span because of the great number of harmful pleiotropic effects.

I propose that cells are intrinsically mortal, that both biosenescent and antibiosenescent processes developed simultaneously, and that for the mammalian species only antibiosenescent processes or longevity evolved. I propose that this was achieved in part by slowing down the rate of development. Hence developmentally linked biosenescent processes, as proposed by Medawar and Williams, were effectively eliminated by postponement of the expression of their corresponding genes.

Table 3. Average Mortality Rate of Present-Day Man in His Past and Present Environment*

Time period	Average chronoage at 50% survival (yr)	Maximum life-span potential (yr)
Würm	29.4	69–77
(about 70,000–30,000 yr ago)		
Upper Paleolithic	32.4	95
(about 30,000–12,000 yr ago)		
Mesolithic	31.5	95
(about 12,000–10,000 yr ago)		
Neolithic Anatolian	38.2	95
(about 10,000–8000 yr ago)		
Classic Greece	35	95
(1100 B.C.–A.D. 1)		
Classic Rome	32	95
(753 B.C.–A.D. 476)		
England	48	95
(1276)		
England	38	95
(1376–1400)		
United States	61.5	95
(1900–1902)		
United States	70.0	95
(1950)		
United States	72.5	95
(1970)		

*Some of the data are from E. S. Deevey, The human population, *Scientific American* 203: 195–204, 1960.

COMPARATIVE BIOLOGY OF THE MAMMALIAN SPECIES

Mammals are thought to have evolved from one or a few common ancestors about 150×10^6 years ago. Detailed studies at the physiological, biochemical, and molecular biological levels have shown that mammalian species represent a remarkably closely related group. They have remarkably similar problems of dysfunction, disease, and undergo qualitatively similar processes of biosenescence, yet they do not have the same MLPs. MLP varies from about 1.5 years for a shrew, 3 years for a mouse, to 70 years for an elephant, and 100 years for man. This difference in MLP evolved according to the differences in intensity of environmental hazards, reproductive value as a function of chronoage.

If the biology of the different mammalian species is so similar, what governs their aging rate? I believe part of the answer can be found in the studies of Max Rubner in the early 1900s. He found that a great number

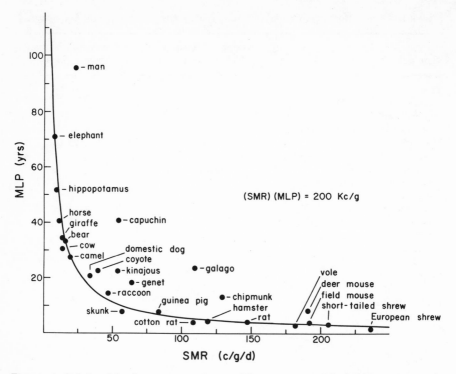

Figure 4. Relationship between maximum life-span potential (MLP) and specific metabolic rate (SMR) for some common mammalian species. SMR is in calories per gram per day.

of mammalian species metabolize about the same amount of calories per gram of tissue per MLP. This remarkable finding is shown in Figure 4, where I have reexamined this concept using recent data on MLP and specific metabolic rates for different mammals. Maximum life-span-potential calorie consumption (MCC) is about 200 kcal/g/MLP. Rubner's finding suggests that metabolism, particularly when used to generate heat, was harmful to the animal. If pleiotropic biosenescent processes exist as a by-product of metabolism, their rate of accumulation should indeed be related to metabolic rate.

Early mammals were much smaller and, therefore, had to have a higher rate of metabolism to maintain their body temperature at about 37°C. These animals also had shorter MLPs than the average of those today, which suggests that a simple and rather rapid increase in longevity may have evolved by simply lowering the rate of production of body heat. Increasing the animal's size or hibernating led to lower heat requirements and so reduced specific metabolic rate. An interesting advantage to these possibilities is that, not only can they evolve quickly, as few genes are likely to be involved, but they can reduce the animal's environmental hazards.

However, there are outstanding exceptions to Max Rubner's rule. For example, some of the living fossil-like animals have an unusually low metabolic capacity (MCC) of about 150 kcal/g/MLP. The most remarkable exceptions are man and a few primate and rodent species, which have an MCC of about 800 kcal/g/MLP. These animals apparently have achieved much more life in terms of MCC than all other mammalian species. It appears that total MCC may be just as good a measurement of an animal's life capacity as MLP, the chronoage parameter.

Most mammalian species vary from about 150 to 800 kcal/g/MLP, and can be divided into three main classes: 150 to 250, 300 to 400, and 700 to 800 kcal/g/MLP. Exceptionally high values of MCC are found for the deer mouse, chipmunk, capuchin, and man, as seen in Figure 4. Figure 5 shows the MCC curves for some of the most common primate species. Primates appear generally to have the highest MCC values. But

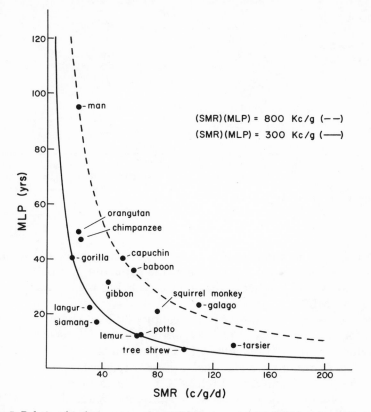

Figure 5. Relationship between maximum life-span potential (MLP) and specific metabolic rate (SMR) for some common primate species. SMR is in calories per gram per day.

Table 4. *Prediction of Maximum Life-Span Potential on Basis of Body and Brain Weight for Some Common Mammalian Species*

Common name	Cranial capacity(cm³)	Body weight (g)	Maximum life-span potential (yr)	
			Observed	Predicted*
Nonprimate species				
Pigmy shrew	0.11	5.3	1.5	1.8
Field mouse	0.45	22.6	3.5	3.2
Opossum	7.65	5000	7.0	5.8
Mongolian horse	587	260	46	38
Camel	570	450	30	33
Cow	423	465	30	27
Giraffe	680	529	34	35
Elephant (India)	5045	2347	70	89
Mountain lion	154	54	19	23
Domestic dog	79	13.4	20	21
Primate species				
Tree shrew	4.3	275	7	7.7
Marmoset	9.8	4.3	15	12
Squirrel monkey	24.8	630	21	20
Rhesus monkey	106	8719	29	27
Baboon	179	16,000	36	33
Gibbon	104	5500	32	30
Orangutan	420	69,000	50	41
Gorilla	550	140,000	40	42
Chimpanzee	410	49,000	45	43
Man	1446	65,000	95	92

*The equation used to predict MLP is MLP = (brain wt, g)$^{0.636}$ × (body wt, g)$^{-0.225}$ and is taken from a paper by G. A. Sacher, Relation of lifespan to brain and body weight in mammals, pages 115–133 in *Ciba foundation colloquia on ageing*, Vol. 5: *The lifespan of animals* (G. E. W. Wolstenholme and M. O'Connor, eds.), J. & S. Churchill Ltd., London, 1957.

the species with the highest MCCs do not necessarily have an unusually high MLP, or vice versa. For example, the capuchin and baboon have MCC values over twice that of the chimpanzee, but the chimpanzee still has a higher MLP.

These data strongly support the relation of the rate of metabolism to the rate of biosenescence. But how can some species have such high MCC values? A discovery made by H. Friedenthal in 1910, and repeated by G. Sacher of Argonne National Laboratory 50 years later, showed that species with an unusually high MCC value also have, in proportion, an unusually large ratio of brain to body weight, where body weight is inversely related to specific metabolic rate. Extra-long-lived mammals

Table 5. Longevity Parameters for Some Domestic Dog Breeds

Breed	Body weight (kg)* (at sexual maturity)	Brain weight (g)*	Observed* MLP (yr)	Calculated* MLP (yr)	Calculated specific metabolic rate (Cal/g/day)	Calculated maximum calorie consumption (kCal/g/MLP)
Pekinese	5.6	58.7	20	25.1	44	403
Dachshund	8.2	70.9	19	21.4	40	312
Fox terrier	7.8	67.9	16	20.4	40	298
Mastiff	42.2	116.5	14	20.0	26	189
Leonberger	47.6	113.0	14	19.5	25	178
St. Bernard	47.0	113.7	14	20.0	25	182

*Data from A. Comfort, Ageing, the biology of senescence, Holt, Rinehart and Winston, Inc., New York, 1964.

seem to have an extra amount of brain for a given body size, compared to their shorter-lived relatives. The extra amount of brain can be calculated by an encephalization index, which can, in turn, be related to learned behavior and cognitive abilities. By comparing body and brain weight, MLP for most mammalian species can be predicted within ±10% of known value. Table 4 shows the accuracy of this formula for a number of mammalian species, and Table 5 for some domestic dogs.

To explain the correlation of the extra amount of MLP with extra brain size, George Sacher suggested in 1957 that the extra brain size is related to a qualitatively improved homeostatic ability for control of the animal, resulting in longer MLP. He emphasized systemic processes, rather than molecular differences, as the basis of the different MLPs of mammals. Another explanation was my own in 1972 that extra brain size does not play a direct role in lowering biosenescent rate but reflects evolutionary selective pressure that gave rise to additional antibiosenescent processes of a molecular and not of a systemic nature.

In summary, all mammalian species have common physiological, biochemical, and molecular biological characteristics and, as a result, have common types of body dysfunctions, pathology, and biosenescent processes. However, they have different MLPs, hence undergo biosenescence the same way qualitatively but at different rates. What governs rate is a key question in gerontological research and perhaps is even more important, and solvable, than what the actual biosenescent processes are per se.

MCC values are often the same despite different MLPs, which shows that although metabolic rate appears to govern the intrinsic rate of biosenescence, some important exceptions to the constant MCC rule exist, including man. I suggest that the MLP of a mammal may be related to its ability to learn from experience, as compared to instinctive abilities. To take full advantage of the ability to learn, and to teach what is learned, more time is necessary than in animals solely dependent upon instinctive behavior. Thus, a mutually enhancing process of increased brain size and longevity may have occurred during the evolution of the mammalian species.

GENETIC COMPLEXITY OF THE PROCESSES GOVERNING LONGEVITY OF THE MAMMALIAN SPECIES

The greatest longevity possible for the soma has almost always proven to be advantageous, and the evolution of mammals followed a general trend of ever-increasing MLPs. Apparently, longevity or increased MLP evolved and not biosenescence or shorter MLPs. A comparison of the MLPs of existing mammalian species and their evolution-ary relationships to one another reveals MLP increased throughout the 150×10^6-year evolutionary history of mammals. Analysis of MLP for extinct species using estimates of body and brain weights from their fossils can be criticized, because a formula that works for living mam-mals may not apply to extinct species. I calculated the MLP for living-fossil species that are considered unchanged for millions of years and found the predicted MLP to be as accurate as for more progressive species, such as pigmy shrew, oppossum, and tree shrew, shown in Table 4, which are considered living-fossil species.

I calculated MLPs for almost every extinct mammalian species where the evolutionary time of appearance and body and brain size are known — about 150 species. MLP increased steadily to the present time for all species whose evolutionary lines did not become extinct. Extinct species apparently did not undergo an increase of MLP for some time prior to their extinction.

The evolutionary or phylogenetic tree for the primate species in Figure 6 shows how MLP increased at a remarkably high rate. The closer the primate species appears to be related to man, the faster the rate of increase in MLP, and man has the highest rate of increase.

To determine how fast longevity evolved along the ancestral de-scendant sequence leading to man, the MLP was calculated using data from more than 100 different hominid fossils. Figure 7 shows MLP in-creased in an exponential manner. About 100,000 years ago the increase suddenly stopped, leaving man's MLP essentially unchanged ever since.

I estimated the complexity of the genetic factors involved in in-creasing MLP from a knowledge of how fast new genes or adaptive mutations can be fixed in a population, comparing this to how fast MLP increased over a certain period of time. Although MLP increased, it may have been accomplished in different species by different mechanisms. I argue that this is unlikely, for all mammalian species, particularly the primate species, are remarkably similar, right down to the molecular level, and have common biosenescent processes. Therefore, the extinct hominid ancestors of man may also have suffered from the same com-mon dysfunctional processes that we and the other primate species have today. Essentially all the biosenescent processes in prehistoric man were reduced in their rate of expression by an equal amount. Such a uniform reduction implies that (1) either a great many genes were changed, if all the different biosenescent processes were independent; or (2) a few primary antibiosenescent processes existed and only a few genes were changed.

Figure 8 shows the results of the calculations of MLP for the

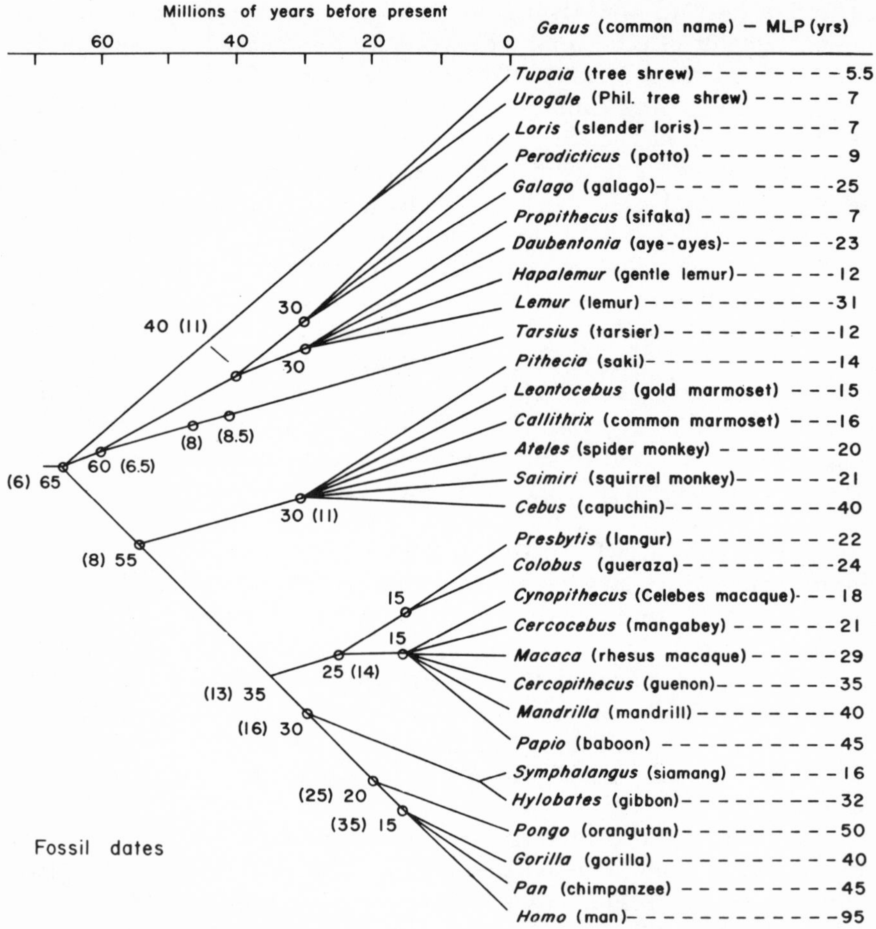

Figure 6. Evolutionary path and relationship of the primates. Numbers in parentheses are estimates of maximum life-span potential of the species that existed at that time. Its neighboring number represents millions of years ago according to geological dating.

hominids. About 100,000 years ago the hominids reached a maximum rate of MLP increase of about 14 years per 100,000 years. In this time, only about one amino acid change in about 250 genes could have occurred out of a total estimate of 40,000 genes per haploid genome. This supports the concept that only a few primary antibiosenescent processes exist for the hominid species, and probably for all mammalian species.

The idea that substantial biological changes occurred during evolution without corresponding changes in the number of structural genes is

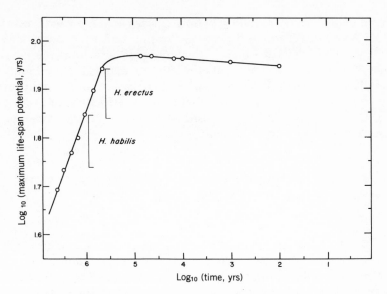

Figure 7. Evolution of increased maximum life-span potential for the hominid species leading to man.

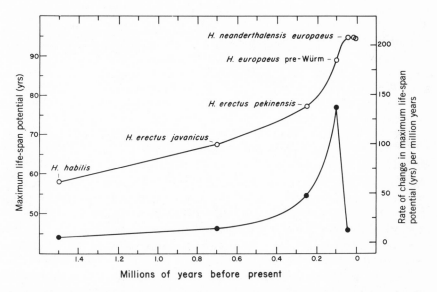

Figure 8. Recent evolution of maximum life-span potential of the hominid species leading to man. Open circles, maximum life-span potential; closed circles, rate of change of maximum life-span potential.

consistent with current thinking of how species in general evolved at the gene level. Allan Wilson and his associates at the University of California suggest, based on a substantial amount of data, that the essential change was in the rearrangement of regulatory genes, and that mammalian evolution can be accounted for on that basis, not by point-mutational changes occurring in genes that code for structural or nonregulatory proteins. *I suggest that increased MLP and extra brain capacity also resulted from such regulatory changes.*

EVOLUTIONARY EVIDENCE FOR CONTINUOUSLY ACTING, DEVELOPMENTALLY LINKED BIOSENESCENT AND ANTIBIOSENESCENT PROCESSES

Intrinsic pleiotropic cellular biosenescent and antibiosenescent processes originated in the free-living eukaryotic cell. These processes exist in today's mammalian organisms, and are of a continuously acting nature because they are involved with everyday pleiotropic effects of cellular metabolism. Another type of pleiotropic biosenescent process exists that does not act continuously but is developmentally linked in its expression. Evolutionary selective pressure operated to retard harmful effects of biosenescent processes beyond maturation by retardation of developmental rate. One can predict, therefore, from the evolutionary data, the existence of developmentally or chronoage-linked biosenescent and antibiosenescent processes.

BIOCHEMICAL EVIDENCE FOR CONTINUOUSLY ACTING BIOSENESCENT AND ANTIBIOSENESCENT PROCESSES

A problem exists in identifying biosenescent processes intrinsic to a cell and those caused by possible harmful effects due to the cell's external environment. One might conclude that intrinsic cellular biosenescence does not exist, and that all changes in a cell with increasing chronoage result from extracellular humoral factors. This possibility cannot be definitely ruled out, but it appears unlikely, according to previous arguments. Maybe antibiosenescent processes are in excess of biosenescent processes, and so do not play an important role in limiting life-span. But this is also unlikely, for no evolutionary selective pressure would exist to select for a characteristic beyond what is actually needed. We find an excellent correlation between environmental hazards, reproductive value, and general rate of physiological decline with chronoage. Nothing indicates that MLP evolved in excess over what was necessary for an animal's genetic information to survive.

Chronoage-Dependent Intrinsic Accumulation of Damage

Many biological data indicate that intrinsic cellular biosenescent processes do exist, as follows.

1. *Chromosomal aberrations.* A progressive increase of aberrations is found throughout the life-span of mouse, guinea pig, and beagle dog, reaching values of over 50% of all cells undergoing mitosis toward the end of the life-span. The increase is continuous throughout the life-span, beginning at sexual maturation and increasing in an exponential manner that might be indicative of positive feedback. An accumulation of damage to internal cellular constituents prohibits successful mitosis. It has never been suggested that the aberrations per se cause biosenescent processes. Also, the lack of a correlation between the accelerated rate of aging, as caused by ionizing radiation and the resultant increase in chromosomal aberration frequency, cannot be used as an argument against the significance of this data, unless it is assumed that radiation-accelerated aging exactly duplicates normal aging processes, which it does not. A chronoage-dependent increase exists in the frequency of chromosomal aberrations in leukocytes of mammalian species and in genetically abnormal germ cells.

2. *Alterations of chromatin.* Alterations were found in liver and brain tissues from the mouse and rat. With increased chronoage we find (a) an increase in binding strength of chromatin proteins to DNA; (b) indirect evidence for the accumulation of DNA–protein and protein–protein covalent cross-links; (c) an increase in the number of DNA breaks or nicks; (d) modification of chromatin protein, such as phosphorylation and acetylation; (e) decrease in ability of ribosomal RNA to hybridize to purified DNA preparations; and (f) a decrease in transcription ability of the chromatin, shown by *in vivo* and *in vitro* tests.

Another indication that chromatin accumulates damage with chronoage is the apparent misregulation of enzyme systems, evidenced by the apparent loss in control of enzymatic-specific activity. Some enzymes show a decrease, others an increase, and others no change in their specific activity with increasing chronoage. Richard Adelman of the Fels Research Institute, Pennsylvania, found an increasing lag time for an indirect induction of specific types of liver enzymes, but no difference if induced by direct hormone stimulation. However, the ability of a cell to respond normally to enzyme induction, direct or indirect, does not necessarily imply that the cell is normal. An example of this is that yeast cells killed by irradiation, as they can no longer divide and soon die, still retain their normal capacity for maximum enzymatic-induction ability.

Extracellular humoral factors can also be involved in altering chromatin, such as the methylation of the DNA of the Pacific salmon, related to the change in hormonal levels.

3. *Cytoplasmic proteins.* Abnormal proteins have been found within the cell but probably result from posttranslational modification. Some changes may be due to spontaneous denaturation and deamination reactions, surprisingly high at 37°C. However, a high percentage of abnormal proteins are not likely to accumulate as a result of mutations in the genes of proteins that make other proteins, as in the protein-error catastrophe, suggested by L. Orgel, Salk Institute, California. Theoretical arguments show this model to be untenable, and most mutations are expected to involve regulatory genes, which represent a larger target for mutagenic agents than structural genes. Mutations in regulatory genes would cause a loss of homeostasis in the cell. I have previously suggested that regulatory abnormal proteins via mutation or posttranslational changes have the potential for positive feedback, resulting in a protein-regulatory catastrophe. No major increase of abnormal proteins or a significant decline in the vital functions of the cell would be expected according to this model, only a change in levels of proteins, their disappearance, or the appearance of new ones.

4. *Age pigments.* Fluorescent pigments accumulate with increasing chronoage in a number of postmitotic cells, such as muscle and neuron. They may result from the side effects of oxygen, which produce different types of free radicals and cross-linking agents. A. Tappel and his colleagues at the University of California have shown that lipid peroxidation is a key reaction, producing malonaldehyde, which can cross-link with the primary amine group of proteins, nucleic acids, or phospholipids. Age pigments reflect the presence of intrinsic pleiotropic biosenescent processes that have the potential to do serious cellular damage. The age pigments per se are not likely to cause alteration or dysfunction.

5. *Finite dividing potential of cells in vivo and in vitro.* Mouse lymphocytes, alveolar duct cells of the mouse mammary gland, and mouse bone marrow cells show a finite division potential in serial transplantation experiments. Human fibroblast skin cells, cultured *in vitro,* have a finite division capacity which, on the average, decreases as a function of increased chronoage of the donor.

6. *Loss of cells.* This occurs with increasing chronoage in many different tissues, such as kidney and liver, and particularly glia and neurons in certain areas of the human brain. The size and number of dendrites in neurons decrease with increasing age. Total cell-cycle time of some types of dividing cells increases with chronoage.

7. *Liver regeneration.* Nancy L. R. Buchler of the University of Leeds, England, found that with increasing chronoage in mouse liver, (a) longer lag time in initiating DNA synthesis occurs, (b) replacement of lost hepatic cells is slower, (c) greater synchrony exists among cells

entering the mitotic cycle, and (d) the older mouse is less tolerant to liver deprivation.

Biochemical evidence of chronoage-dependent accumulation of damage at the extracellular level is as follows:

a. *Mineralization of tissue.* Many tissues and blood vessels accumulate a number of minerals, particularly calcium, reaching levels likely to interfere with normal function. Progeria victims, who probably have only one or a few abnormal genes, show an accelerated mineralization of tissues, in addition to other accelerated aging processes, indicating that a simple regulatory process controls the rate of mineral deposition.

b. *Collagen cross-linking.* Cross-linking of collagen, the most abundant protein in the body, steadily increases in many tissues with age. This process could seriously affect the operation of heart valves, the elasticity of blood vessels, and diffusion of nutrients to cells from blood vessels.

Repair and Protective Processes

Repair and protection act as antibiosenescent processes to reduce the rate of accumulation of cellular damage, implicating intrinsic pleiotropic biosenescence present at the cellular level.

Repair. Only DNA can be repaired; other damaged constituents can only be replaced. Ultraviolet or ionizing-radiation DNA-repair systems are present in mammalian tissues never exposed to such radiation. Therefore, they exist probably to repair damage produced by intrinsic biosenescent processes.

Increased repair efficiency to UV damage was found with increased MLP for different mammalian species in 1974 by Ronald Hart of Ohio State University and Richard Setlow of Brookhaven National Laboratory. This indicates that a longer MLP necessitates an increased capacity to repair DNA, and was predicted according to a hypothesis of biosenescence I presented in 1972 (see the Bibliography).

Protection. The cell has many defenses to slow the rate of accumulation of damage.

1. Skin cells of mice show a chronoage-dependent increase in sensitivity to carcinogenic agents. P. Ebbesen of the Institute of Medical Microbiology, Copenhagen, Denmark, found that sensitivity of skin transplanted from old animals to young or vice versa is not affected by these agents. Sensitivity of the skin to carcinogenic agents seems independent of humoral factors and depends only on intrinsic cellular biosenescent processes. Increase in sensitivity could be from loss in repair ability or a decrease in protective capacity against mutagenic agents.

2. Most mammalian cells have protective systems against toxic agents, such as microsomal mixed-functional oxidative systems. These systems are particularly effective in liver cells. A harmful pleiotropic effect of these systems is that a number of harmless hydrocarbons are changed into very active carcinogens. Arthur Schwartz of Temple University Medical School, Pennsylvania, found that this capacity to produce carcinogens from an inactive hydrocarbon decreased with increasing MLP for different mammalian species.

3. Most cellular and extracellular constituents show a high rate of turnover. Turnover in mammals is extraordinarily high, yet no satisfactory reason for its existence has been given. Liver cells, essentially postmitotic, turn over about half their proteins every 5 days. Albert L. Goldberg of Harvard Medical School has shown that a special class of proteolytic enzymes exists that has high selectivity toward degrading abnormal proteins. Such proteolytic enzymes would not likely exist if abnormal proteins were not a problem. Turnover is, apparently, a protective or an antibiosenescent process that reduces the rate of accumulation of defective cellular constituents.

4. Free-radical scavangers and antioxidants exist in the cell, such as vitamins E and C, selenium, and glutathione. Superoxide dismutases remove the O_2^- radical, and glutathione peroxidase removes lipid peroxidases. The presence of these protective processes reflects the existence of intrinsic biosenescence processes.

Presence of Pleiotropic-Damaging Agents

Metabolic processes are potentially pleiotropic biosenescent processes, supported by the correlation of MLP with SMR (as shown in Figures 4 and 5). The more enzymes and heir products that coexist in a cell, the higher the probability of a harmful side reaction or pleiotropic effect. Data of Horton A. Johnson of Tulane University School of Medicine, Louisiana, indicates that rats living at 9°C undergo accelerated biosenescence. Mean life-span decreased from 700 days to 450 days in controls. Food consumption increased 60% and oxygen consumption 40% over controls living at 25°C.

Normal temperatures of about 37°C are harmful to the cell but represent a trade-off for increased efficiency of enzymatic reactions. Temperature probably increases rate of thermal denaturation of cellular components and decreases specificity of enzyme reactions.

Other potential pleiotropic-damaging agents, or by-products of useful metabolic processes, found in all mammalian cells are aldehydes, oxidizing agents such as O_2 and H_2O_2, and free radicals such as O_2^- and $HO\cdot$.

Correlation of Accumulation of the Damage, Intensity of Damaging Agents, and Protective and Repair Processes with MLP

Most present-day antibiosenescent processes probably did not play an important role in recent evolution of increased life-span, but are, nevertheless, important to survival. These antibiosenescent processes would be at a constant level of intensity in all mammalian species regardless of MLP. Changes that are functions only of chronoage and not of bioage probably also occur in mammals; therefore, they are not true biosenescent processes that limit life-span. To determine the important antibiosenescent and potential biosenescent processes that played an important role in the recent evolution of longevity in mammals, a comparative analysis must be made between different mammalian species having different MLPs. In 1972, I predicted that the antibiosenescent process or rate of change of the potential biosenescent process should be a function of MLP of each mammalian species. For longer-lived species, one would expect higher levels of antibiosenescent processes and a lower rate of accumulated damage due to biosenescent processes. Antibiosenescent processes that do not follow this relation may still be important in determining the MLP of the species but not in their recent evolution of increased MLP. Potential biosenescent processes not showing a correlation with MLP are probably irrelevant to determining MLP and are not true biosenescent processes, although they may show a chronoage-dependent increase of damage.

Amount of the damaging agent, or the efficiency of the protective or repair process, need not change with chronoage or bioage. The constitutive levels of these parameters throughout the life-span determine the rate of accumulation of damage and therefore the rate of biosenescence. One should not necessarily expect a loss of repair capacity or an increase in concentration of a damaging agent as a function of chronoage.

Some parameters which have shown a positive correlation with MLP of different mammalian species are (a) chromosomal aberration frequencies, (b) rate of accumulation of age pigments, and (c) rate of mineralization of tissues. For protection and repair, there are (a) ultraviolet repair capacities, (b) capacities to activate precarcinogens, (c) B/A ratios of lactic dehydrogenase (a higher ratio is found in many tissues with increasing MLP for the primates), (d) selective removal of abnormal proteins, and (e) developmental rates. For damaging agents, specific metabolic rate is inversely proportional to MLP.

GENERAL BIOLOGICAL EVIDENCE FOR DEVELOPMENTALLY LINKED BIOSENESCENT AND ANTIBIOSENESCENT PROCESSES

The primary argument that chronoage and/or developmentally linked pleiotropic biosenescent processes exist is the continuous decrease with chronoage of an individual's reproductive value. Younger individuals would be favored by having the highest levels of antibiosenescent processes or the lowest degrees of expression of continuously acting or developmentally linked pleiotropic biosenescent processes.

A hypothetical example of an intrinsic developmentally linked pleiotropic biosenescent process is a gene that acts to give superior bone strength at a young chronoage, but later results in calcification of connective tissues in the arteries. Developmentally linked biosenescent processes probably exist related to the humoral factors controlling postmaturational growth and development. Various humoral factors involved in the development and differentiation of the soma are undoubtedly potent pleiotropic biosenescent processes that became genetically fixed before the appearance of any mammalian species.

Presence of Neuroendocrine Pleiotropic Humoral Factors

Arthur V. Everitt of the University of Sydney, Australia, suggested that many hormones appear to have both biosenescent and antibiosenescent effects. For example, adrenocorticol hormones appear to inhibit skeletal biosenescence but accelerate the cross-linking of tail tendons. However, growth hormone inhibits cross-linkage of tail collagen but accelerates skeletal aging, promotes cardiovascular and renal disease, and increases neoplasia. Thyroid hormone also accelerates collagen, skeletal, and renal aging, but appears to inhibit atherosclerosis and neoplasia. Most hormones of the anterior pituitary, thyroid, adrenal cortex, and gonads increase the biosenescence of the kidneys and cardiovascular system. But it is difficult to determine if the decrease in synthesis of pituitary growth hormone, thyroxine, cortisol, estradiol, and testosterone, which occurs with increasing chronoage, is a result of biosenescence, a cause of biosenescence, or a protective process to limit harmful effects.

Vladimir M. Dilman of the Petrov Research Institute of Oncology, Leningrad, USSR, also presented evidence for the pleiotropic effects of hormones. He suggested that an intrinsic, continuous, chronoage-dependent increase occurs in the threshold sensitivity of the hypothalamus to inhibitory regulatory factors. This increase may be the basic timing device governing the rate of presexual maturation development, sexual maturation, and menopause. But the ever-increasing threshold level of the hypothalamus to feedback inhibition results in an imbalance of hormonal factors, causing homeostatic failure and diseases of compensational processes.

If Dilman's hypothesis is true, it is a perfect example of a developmentally linked pleiotropic biosenescent process. This process would be predicted from the steadily declining reproductive value curve, particularly after sexual maturation, where reproductive value rapidly declines. The disadvantages of a rising threshold to feedback inhibition governing developmental rate would not have been serious in the historical past, for man and other mammals in the wild would rarely reach the chronoage where this would result in biosenescence.

A developmentally linked biosenescent process is also consistent with data and ideas presented by Caleb E. Finch and associates at the University of Southern California. They suggested that changes in hormone levels at a prematurational age induce alterations of central hormones, such as the catacholamines. Hormone output of the pituitary is altered and, by a feedback process, further changes in neurohormones can occur. The resultant cascading effect, reflecting an inherent instability of the neuroendocrine regulatory system, would progressively limit the organism's homeostatic abilities during adult life.

Thus, control systems that govern developmental processes are probably not stable, closed-looped feedback systems, but open-ended systems that run straight through a program with no stable state at the end of the program. If the animal lives long enough, the evolutionarily selected, beneficial part of the program is played out, as expected on the basis of the decline in reproductive value for the species. Sufficient selective pressure to evolve a stable neuroendocrine regulatory system after the developmental program ended would not exist, simply because too few animals survived past this age for this system to become feasible.

Another example of the open-ended nature of developmental programs comes from the work of O. H. Robertson of Stanford University, California, in the early 1960s. He demonstrated that the Pacific salmon's high rate of biosenescence when it spawns was due to adrenal cortical hyperplasia (an unusually high production of adrenal hormones, the corticosteroids). Salmon prevented from reaching sexual maturity do not undergo the high rate of biosenescence and live a few more years. The sudden biosenescence of the salmon has been thought the result of a genetic program of biosenescence or death that occurs for the good of the species. But it probably results from the salmon's unusually short reproductive value after spawning. I predict that the probability of survival is extremely low for a salmon not undergoing biosenescence, if it reaches the spawning grounds, returns to the sea, and then returns in a few years to the spawning grounds for another round of reproduction. Sufficient selective pressures probably did not exist to evolve an antibiosenescent process to counteract the effects of the first sexual maturation process.

This biosenescent process is genetically programmed because it is developmentally linked. But the adrenal hormones are certainly not "death hormones" that evolved for the sole purpose of limiting the life-span of the salmon for its own good. Rather they are developmentally linked intrinsic pleiotropic biosenescent processes.

Retardation of Development Rate

How did antibiosenescent processes evolve to counteract developmentally linked biosenescent processes? A simple and probable occurrence was a decrease in the rate of development, thereby effecting postponement of the necessary but harmful developmentally linked biosenescent processes. The classical studies of Clive McCay of Cornell University, Ithaca, N.Y., in the 1930s showed that a restricted calorie diet can lengthen the life-span of a rat to about twice its normal value. Longevity was directly related to the extent prematuration development rate was slowed. A restricted calorie diet resulted in an extension of MLP if given after the animal reached sexual maturity, but not nearly as much. I believe this small increase in MLP is a result of another mechanism, related possibly to the stimulation of protein turnover. For example, MLP can also be increased by a periodic fast, which does not affect developmental rate. I found that such periodic fasting accelerates the removal of abnormal proteins, probably mimicking natural living conditions of scant food.

P. S. Timiras and co-workers at the University of California have also shown increased MLP in rats fed a diet deficient in the amino acid tryptophan. Tryptophan is a precursor to serotonin, an important humoral factor of the neuroendocrine system. The tryptophan-deficient diet lowers the functional level of the pituitary — essentially a nutritional hypophysectomy that lowers the levels of pleiotropic hormones. This might also happen in the rats on McCay's restricted calorie diet.

These data are consistent with the idea that the control of growth and development lies in the hypothalamus–pituitary axis, and that diets which slow down the activity of this axis will slow developmental rate. The overall effect is increased chronoage when the effects of the pleiotropic biosenescent processes of development appear.

More evidence that the pituitary may secrete pleiotropic hormones which accelerate biosenescence comes from the studies by W. D. Denckla of the Harvard Medical School. He found that surgical removal of the pituitary, followed by constant administration of thyroid and growth hormones, resulted in rejuvenation, after a 6-month period, in several physiological parameters and in the recovery of chromatin-transcription activity. Denckla concluded that the pituitary may secrete a factor (death hormone) which prevents cells from responding to thyroid hormone. Removal of the source of this factor would result in a rejuvenation of the organism if the inhibitory process is reversible. An exciting aspect of these data is that developmentally linked pleiotropic processes may indeed be reversible.

If developmentally linked humoral factors are pleiotropic and give both beneficial and harmful effects, wouldn't their removal cause more harm than good? Possibly a slowdown in the rate of biosenescence can be effected by the removal of these humoral factors, without serious harmful side effects, if done after their function is completed; or if the adult requires these humoral factors, reduce them to levels found in the young, sexually mature adult.

But will the postmaturational elimination of these humoral factors significantly extend life-span? Probably not, for this would have been the way longevity evolved in the first place. Instead, longevity appears to have solved the problem of developmentally linked biosenescent processes by simply postponing the time of their appearance.

An examination of the rate of development during the evolution of longevity for mammals reveals a good correlation between the age at sexual maturation and MLP, shown in Figure 9. Scattering is probably because age at sexual maturation is not a good parameter to represent the average rate of development for many species. The time span over which sexual maturation occurs is probably also important, short for some species, and occurring over years in others such as man. Growth-limiting factors, which may also have a potentially harmful pleiotropic effect, may not be expressed until a time much after sexual maturation, important for species which continue to grow after maturity. Examining genetically closely related groups such as the primate species, including man, in Figure 10 reveals an excellent correlation between the age at sexual maturation and MLP.

Continuously acting or intrinsic biosenescent processes not linked developmentally are present; hence, reducing the rate of development is insufficient to lower the biosenescent rate of all physiological processes. Clive McCay found that the longer the component of life-span gained by calorie restriction in the presexual maturational period, the shorter the gain in life-span postmaturationally, illustrated in Figure 11. Examination of the animals that had just died revealed that many types of biosenescent processes continued on at the same rate, such as the rate of deterioration of bone structure and the onset frequency of some types of

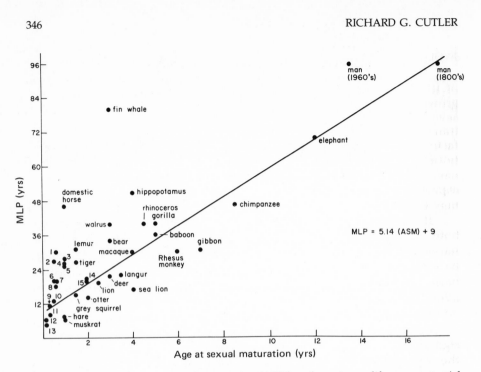

Figure 9. Relation of age at sexual maturation (ASM) and maximum life-span potential (MLP) for some common mammalian species. Names of species not given in the figure are: 1, domestic cow; 2, domestic hog; 3, cat; 4, galago; 5, llama; 6, sheep; 7, dog; 8, goat; 9, chinchilla; 10, aardvark; 11, chipmunk; 12, tree shrew; 13, rat; 14, coyote; 15, beaver.

cancers. Similarly, the rejuvenation experiments of W. D. Denckla did not show a 100% recovery in function of some parameters measured. This might reflect the amount of damage left in the organism caused by the intrinsic biosenescent process of the continuously acting, and not the developmentally linked, type.

Developmentally linked expression of humoral factors, as compared with the continuously acting type, may not be equally important in all mammalian species. The average ratio of MLP to the age of sexual maturation (ASM) for the primate, and many different mammalian species, is 5. This is not true for some rodent species, which have a ratio of about 40, as shown in Figure 12. Figure 13 illustrates the difference by comparing man and the deer mouse on a relative MLP scale. The relative time that these rodents live past their sexual maturation age is about eightfold greater than for the primate species. Developmentally linked biosenescent processes are not too harmful, as they get so much more life-span past their sexual maturation age.

I propose that the resultant biosenescent rate of a mammal is the

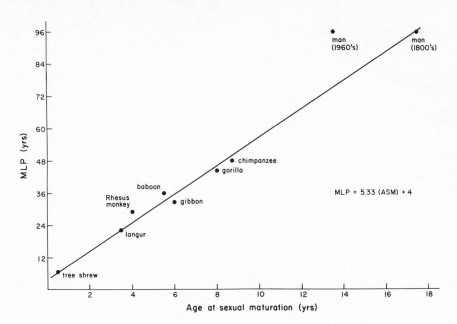

Figure 10. Relation of age at sexual maturation (ASM) and maximum life-span potential (MLP) for some common primate species.

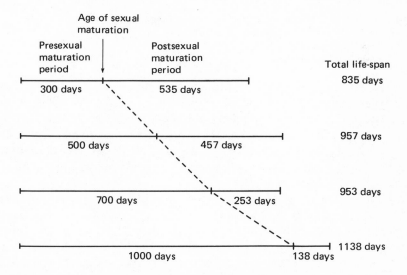

Figure 11. Relation between the amount of increased life-span gained in rats on a restricted calorie diet in the presexual maturation period and the postsexual maturation period.

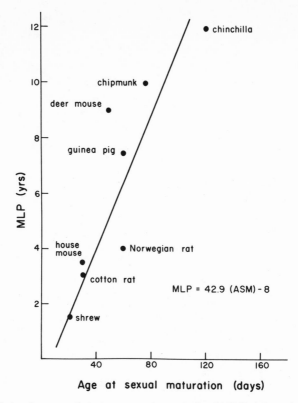

Figure 12. Relation between the age at sexual maturation (ASM) and maximum life-span potential (MLP) for some common rodent species.

sum of two processes: the continuously acting biosenescent process, which is proportional to specific metabolic rate, and the developmentally linked biosenescent process, which is proportional to the time tissues are exposed to humoral factors. Thus, for shorter-lived species such as deer mouse, with a total MLP of 8 years, biosenescence is predominantly due to continuously acting, and not to developmentally linked, biosenescent processes.

This explains why McCay's results (Figure 11) show a progressively decreasing length in postmaturational life-span when prematurational life-span is extended by a calorie-restricted diet. The continuously acting biosenescent processes are not developmentally linked and were not slowed down. The same experiment on a primate species, and not on rats or mice where the ratio of MLP to ASM is 5 and not 40, would bring a much greater increase in postmaturational life-span for the same degree of calorie restriction, or retardation of developmental rate, predicted by this hypothesis.

This hypothesis is consistent with the correlation previously noted between the amount of ultraviolet repair efficiency and MLP for a number of mammalian species found by Ronald Hart and Richard Setlow. The proportion found was logarithmic; when MLP doubled, repair capacity increased by a constant factor. An increase of life-span from 3 to 6 years, as for mouse and chipmunk, is equivalent to the absolute amount of increase in repair, as from 50 to 100 years for chimpanzee and man. Thus, an increase in repair appears more essential to increase the MLP for animals that live a short time and have a higher SMR than for longer-lived animals with much lower SMRs.

Some data do not support the idea that developmentally linked biosenescent processes exist and that a means by which longevity evolved was reduction in the rate of development: for example, the occurrence of precocious sexual maturation, where human females have become sexually mature as early as 1 year of age and have given birth to a normal child at the age of 5 years. Although often a result of an abnormality, such as a brain tumor, many cases are apparently normal in all respects except for an unusually fast rate of sexual development. One person became sexually mature at the age of 2.5 years, gave birth to a normal child, and lived at least long enough to reach menopause at the age of 55 years. Men castrated at an early age, about 8 to 12 years, experienced an extraordinarily high level of gonadotropins throughout their entire life-span, yet records indicate that they could live at least up

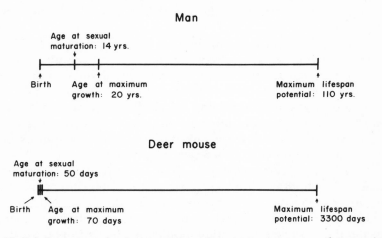

Figure 13. Relative age at sexual maturation and age at maximum growth to maximum life-span potential for man and the deer mouse.

to 63 years of age. But we do not know if the 55-year-old woman or the 63-year-old man were indeed completely normal for their chronoage, or were actually biologically equivalent to a much higher chronoage. If they truly did not experience any accelerated biosenescence whatsoever, we would have to conclude that at least some of the humoral factors governing sexual maturation are not pleiotropic biosenescent processes.

A WORKING HYPOTHESIS OF BIOSENESCENT AND ANTIBIOSENESCENT PROCESSES

My working hypothesis is in two parts, one on all living systems and the other on the mammalian species.

A General Hypothesis of Biosenescent and Antibiosenescent Processes

1. Normal beneficial biological processes within all living systems are postulated to exist, but also act destructively. These are biosenescent processes of two major types: (a) thermodynamic processes that increase the entropy and/or the spontaneous destruction of the physical constituents of the living system, and (b) intrinsic pleiotropic processes having both beneficial and harmful effects, which are by-products of the beneficial processes of the organism. As living systems become more complicated, intrinsic pleiotropic biosenescent processes are more numerous.

2. Special antibiosenescent processes exist within all living systems to counteract the destructive effects of the intrinsic pleiotropic biosenescent processes.

3. Living systems exist on the basis of one principal requirement: the capacity to preserve required information. The variety of life represents the many ways that have evolved which successfully met this preservational criterion.

4. The length of time a living system has been able to preserve its evolutionary line, or set of information, determines its success, not its reproductive capacity or the number of individuals within its population, or other parameters sometimes suggested.

5. The living system exists because of its capacity to preserve information indefinitely within a physical form of finite life-span. Information can grow by mutation and natural selection, and is preserved by duplication. Life exists because of the presence of antibiosenescent processes, but the origin and continuation of life depend upon the duplica-

tion of the genetic information, before duplication capacity is lost to biosenescent processes.

6. Living systems have never been immortal or potentially immortal. The only immortal aspect of life is the continuity of genetic information.

7. The life-span of cell components depends on the balance between the biosenescent and antibiosenescent processes, with biosenescent processes always predominating. The level of damage due to biosenescent processes is never allowed to accumulate to a critical level in the dividing population. Biosenescent components sometimes accumulate to an unusually high level, where sexual recombination is required to preserve the genetic information of the species.

8. Biosenescence of soma cells leads to eventual death of the organism. Biosenescence is tolerable because of the separation of soma from germ cells, where the soma cells are eventually lost, but information in the germ cells is preserved by reproduction. Reproduction must occur before information is lost due to biosenescent processes in the germ cells.

9. The balance of biosenescent and antibiosenescent processes determines the period of maximum functional capacity of the soma. The balance is determined by the species' reproductive value as a function of chronoage, dependent on the natural environmental hazards of the species' ecological niche.

10. Antibiosenescent processes in both protozoan and metazoan organisms never exist in excess, only to the extent necessary to ensure the maximum survival for the organism's genetic information.

11. Protozoan and metazoan organisms show common biosenescent and antibiosenescent processes, based on the common physical nature of genetic material and metabolic processes. However, a great diversity exists in other metabolic processes that are pleiotropic in nature. Therefore, a universal set of primary biosenescent processes and their antibiosenescent processes in all living systems appears unlikely.

12. Much of the morphological and biochemical nature of all organisms living today represents a history of many different biosenescent and antibiosenescent processes that became genetically fixed along their evolutionary pathway. Thus, an understanding of the state of living systems cannot be answered without a knowledge of the coevolutionary history of biosenescent and antibiosenescent processes. A study of processes which determine the length of time a soma will function at its optimum level represents a fundamental inquiry of the living system.

Thus, the biological science of gerontology is clearly not a subdivision of other fields of biology, but a basic science in its own right.

*A Special Unified Hypothesis of Biosenescent and Antibiosenescent
Processes for the Mammalian Species*

1. All mammalian species are unified on the basis of common biosenescent and antibiosenescent processes at the cellular, extracellular, multicellular, and organismic levels.

2. All mammalian species can be unified on the basis of common regulatory mechanisms for increased longevity and cognitive abilities.

3. Essential differences between mammalian species lie in genetic regulatory information. The extremely rapid evolution of increased longevity and extra brain size suggests that genetic regulatory differences are not great, and that perhaps only a few primary key regulatory processes effected these changes.

4. Longevity increased in the mammalian species by two basic operations, both of a genetic regulatory nature: (a) reducing rate of expression of continuously acting intrinsic cellular and extracellular biosenescent processes, accomplished by lowering specific metabolic rate (increasing body size) and increasing expression of antibiosenescent processes, leaving the nature of the processes unchanged; and (b) retarding the rate of expression of the entire genetic program of development leading to the sexual maturation and cessation of growth in the adult, which postpones the genetic expression of developmentally linked pleiotropic biosenescent processes. The qualitative nature of these biosenescent processes remains unchanged in the different mammalian species.

Operations (a) and (b) were both necessary during the evolution of increased longevity to achieve a balanced and uniform decrease in the overall rate of expression of the biosenescent processes of the mammalian organism. The resultant rate of biosenescence is determined by selective pressure of environmental hazards, as reflected by the reproductive value curve.

Some Predictions of the General and Special Unified Hypotheses

1. Intrinsic cellular and developmentally linked biosenescence decreases the functional potential of the genetic information in the soma and germ plasm cells.

2. Many morphological characteristics, metabolic pathways, and some enzymatic reactions will be found necessary for the survival and proper functioning of the cell, not as vital processes but as antibiosenescent processes.

3. Many types of intrinsic, pleiotropic, biosenescent processes will be identified at both the cellular and extracellular levels, expressed both continuously and as functions of development.

4. A genetic program of biosenescence or death, where special factors evolve specifically to shorten the life-span, will not be found for mammalian species.

5. Intrinsic cellular biosenescent processes exist and help determine the overall rate of biosenescence in the mammalian organism. The various cells and tissues are expected to undergo biosenescence at about the same rate relative to the overall rate of biosenescence of the organism. No outstanding weak link or particular clock exists where the rate of biosenescence would be significantly greater than the rest of the organism, to govern, as a "pacesetter," the rate of biosenescence of the organism. The only clock governing MLP of the mammalian species is the rate of specific metabolism, the level of continuously acting anti-biosenescent processes, and the rate of development.

6. The genetic program of development is open-ended, with no inherent design to maintain a stable state in the adult animal. This will be found both at the genetic and systemic levels of regulatory control.

7. Developmentally linked biosenescent processes will be dominant in nature and will increase in intensity (particularly after sexual maturation) with increased chronoage. Humoral factors present in mature animals will accelerate biosenescence. Therefore, in the design of transplantation experiments to test for the presence of intrinsic cellular biosenescent processes, the prediction should seriously be considered in choosing the chronoage of the recipient animal and in interpreting the results.

8. All processes that only retard the rate of development will lengthen life-span.

9. No inherent biological limitation for further increase in longevity in any mammalian species is evident. If selective pressure exists to evolve a longer life-span, it occurs.

10. The comparative approach, using closely related mammalian species of substantially different MLPs, or maximum life-span-potential consumption of calories (MCC), should prove to be a valuable means of studying this special hypothesis and determining differences between the mammalian species that account for their different rates of biosenescence. The primate species represents the ideal animal, but rodents and domestic dogs will prove useful pragmatically.

A MODEL FOR THE FUTURE EVOLUTION OF MAN

From the arguments presented, longer life-span would be evolutionarily consistent for a given species if it furthered probability for survival. Increased longevity provided the means to fully develop cognitive abilities. The coevolution of cognitive ability and longevity automatically resulted in a lowered intensity of environmental hazards, thus providing additional positive selection for increase. Thus, man today has the longevity and cognitive abilities that best suited his situation 10,000 years ago.

Our cognitive abilities, knowledge that death is inevitable, and our unwillingness to grow senile can all be explained by our means to ensure the preservation and continued propagation of the information content of our germ plasm, information which can continue to ensure this preservation and propagation. The ability to be conscious of oneself may not have been selected for but could exist as an unavoidable by-product (pleiotropic?) achieved by a given degree of cognitive ability. Should we increase longevity without also increasing our cognitive abilities? George Simpson, a highly respected biologist and evolutionist, said that only a very stupid person can believe that man is already intelligent enough for his own good. Would that person also believe that man already has sufficient longevity for his own good?

A fascinating aspect, I have found, in studying the evolution of longevity is how one change proves to have multiple advantages. For example, increased longevity permits more time for learning, and for teaching offspring, permitted by a retardation in development rate. Various stages of development were retarded in equal proportions, resulting in an increased juvenile period, which proved advantageous by an extended teaching period.

Changes in genetic regulatory mechanisms which led to retarded development may have similarly occurred to evolve other functions. The fetalization or neonatalization hypothesis put forth in the 1920s by Louis Bolk, a Dutch anatomist, states that stages of development correspond to juvenile or even fetal stages of a related animal or extinct ancestor. Hence, man is fetalized with regard to the great apes, evidenced by 20 traits found in the human adult and also found in the fetal, but not adult, ape.

Developmental control lies at the· hypothalamus–pituitary level. Stephen J. Gould of Harvard University has indicated that only a few genetic changes are necessary to substantially change some of the major morphological features of the body, such as body size, or brain-to-body size ratio (as found in aleleotic and proportional dwarfs). This shows not only how the simultaneous evolution of longer MLP, extra brain size, longer developmental period, plus other human morphological characteristics occurred, but also how they were able to evolve so quickly. The hypothalamus–pituitary axis probably is involved as an important generator of biosenescent processes, and played a key role in the evolution of longevity.

Evolution of longer life-span for man was obviously possible and

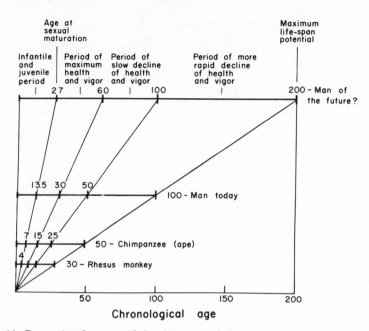

Figure 14. Proportional stages of development of the primate species relative to one another and the projection of these stages of development to a man with a maximum life-span potential of 200 years.

desirable in the past. With the optimistic view that longer life-span is still advantageous, let us see where future man would be if evolution continued along the same biological lines. Figure 14 shows the stages of development for the rhesus monkey, chimpanzee, and man, and their projections in proportion to a man with a MLP of 200 years. By doubling the MLP for man, we clearly double all other stages of life, including the biosenescent stage.

Would the proportional increase of optimum health and vigor be worth the increase of time spent in declining health and vigor? My guess is that it would. The individual would not be a liability, but his increased time of physical and mental health would act synergistically in terms of wisdom, earning power, and creativity and would represent valuable contributions far offsetting the social cost of a longer period of declining health.

The biological characteristics of man with a MLP of 200 years are shown in Figure 15 compared with man today with a 100-year MLP. The additional period spent at the prematurational stage of life actually results in a somewhat less proportional amount of time spent in the declining years for the man with a 200-year MLP.

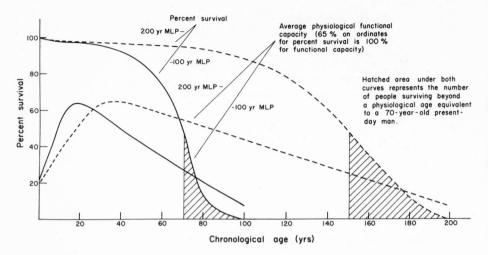

Figure 15. Percent survival and the average physiological functional capacity of man as a function of chronoage and how these parameters would appear if they were proportionally increased to a man having a maximum life-span potential of 200 years.

The 200-year MLP percent survival curve is based on percent survival proportional to physiological function, as for the 100-year MLP survival and physiological function curves. The probability of death, not percent survival, might be proportional to physiological function. In this case, the probability of death would be at a higher value for a longer period of time, resulting in a much faster rate of decrease in percent survival for the older segment of the population. Thus, as MLP increases, the fraction of the older population surviving past a given age would be much less (the percent survival curve would have an increased "exponential rate of loss" component). The two models cannot yet be distinguished, but clearly the latter model would be more desirable, as a much smaller fraction of the population would be in declining health after extending MLP to 200 years as compared to today — not a larger fraction.

How would future *Homo sapiens* appear? Perhaps doubled brain size, although not likely to improve on the quality of the brain, might provide greater redundant capacity for neurons and their supporting cells, thereby postponing the onset of senility. By reducing the developmental rate by one-half, and by doubling the time for growth and maturation of the brain (its growth rate would be the same as it is now, and only one extra division of cells is needed to double its size), a person at the adult stage of life would look much like a 6-year-old in terms of body and brain size proportion; but this 6-year-old-looking individual would be sexually mature and fully grown. A comparison of body and brain proportions of man today and possible changes with further fetalization and retardation of development is shown in Figure 16.

To extend MLP, we need to learn what changed during the natural evolutionary extension of life-span. By the special working hypothesis, changes are required in certain antibiosenescent processes, concerning the intrinsic continuously acting and developmentally linked pleiotropic biosenescent processes. We have the structural genes for these processes, and need only a change, or increase, in their temporal expression, similar to what occurred naturally in the evolutionary increase of life-span.

To counteract continuously acting biosenescent processes, increase (1) DNA-repair capacities, (2) selective removal of abnormal proteins, and (3) superoxide dismutases, antioxidants, and free-radical scavengers. These changes might be achieved by inducing the production of the enzymes involved to higher levels by analog stimulants, by direct enzyme addition therapy, or by genetic engineering.

Because of the high rate of evolution of longevity, not many

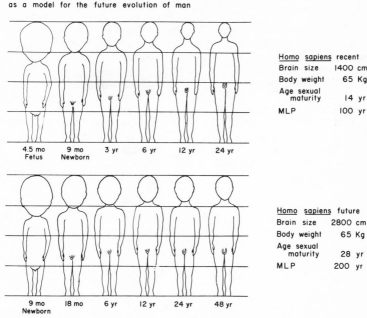

Further retardation of development and fetalization of characteristics as a model for the future evolution of man

Homo sapiens recent
Brain size   1400 cm
Body weight   65 Kg
Age sexual
    maturity   14 yr
MLP   100 yr

4.5 mo   9 mo   3 yr   6 yr   12 yr   24 yr
Fetus   Newborn

Homo sapiens future
Brain size   2800 cm
Body weight   65 Kg
Age sexual
    maturity   28 yr
MLP   200 yr

9 mo   18 mo   6 yr   12 yr   24 yr   48 yr
Newborn

Figure 16. Proportional changes in body and brain size during the developmental stages of man and how these stages might appear in a man whose rate of development is slowed down by a factor of 2 and whose brain size is allowed to double.

changes are likely to be required, and many of the structural genes coding for the antibiosenescent processes may be linked under some common regulatory process, which could be changed. Intrinsic biosenescent changes are not necessarily irreversible and may be subject to some degree of rejuvenation.

Reduction of the developmentally linked biosenescent process might be achieved by further reducing the rate of development. This might be accomplished by slowing the genetic program of the neuroendocrine system by direct biochemical manipulation. Partial rejuvenation by removal of some of the developmentally linked humoral factors, or by restoring levels to a younger chronoage level, should prove possible.

In conclusion, it appears that man's distinguishing characteristics, an unusually long period of physical and mental health, and superior cognitive ability, although extremely complicated phenotypically, are nevertheless controlled by simple genetic regulatory processes. I am optimistic that both of these characteristics can and should be enhanced.

BIBLIOGRAPHY

Cutler, R. G. 1975. Evolution of human longevity and the genetic complexity governing aging rate. *Proceedings of the National Academy of Sciences of the USA* 72: 4664.
Cutler, R. G. 1976. Evolution of longevity in primates. *Journal of Human Evolution* 5: 169.
Cutler, R. G. 1976. On the nature of aging and life maintenance processes. In R. G. Cutler, ed. *Interdisciplinary topics in gerontology*, Vol. 9, p. 83. S. Karger AG, Basel.
Dawkins, R. 1976. *The selfish gene.* Oxford University Press, New York.
Everitt, A. V., and J. A. Burgess, eds. 1976. *Hypothalamus, pituitary and aging.* Charles C Thomas, Publisher, Springfield, Ill.
Margulis, L. 1970. *Origin of eukaryotic cells.* Yale University Press, New Haven, Conn.
Medawar, P. B. 1952. *An unsolved problem of biology.* H. K. Lewis, London.
Monod, J. 1971. *Chance and necessity.* Alfred A. Knopf, Inc., New York.
Williams, G. C. 1957. Pleiotropy, natural selection, and the evolution of senescence. *Evolution* 11: 398.

GLOSSARY

Antibiosenescent processes — biological structures or reactions which act to counter the biosenescent effects of normal cellular biochemical reactions. Life-maintenance process indicates the same concept.
Bioage — the age of an organism in terms of its ability to function biologically.
Biosenescence — the state of having undergone a loss of functional capacity as the result of the passage of time. *Biosenescent* describes this state.
Chromatin — the complete genetic apparatus, composed of the complex of DNA and protein.
Chronoage — the chronological age of an organism due solely to the passage of time.

Continuously acting pleiotropic biosenescent processes — one class of pleiotropic biosenescent processes that act continuously throughout the life-span of an organism. They result from necessary metabolic processes that act in the organism regardless of its stage of development or growth.

Developmentally linked pleiotropic biosenescent processes — the other class of pleiotropic biosenescent processes that act according to the stage of development. These processes can result from a genetic program necessary for continued development. The changing internal environment of the organism can transform earlier useful processes into biosenescent processes, or the continuous expression of an earlier useful process can have a steadily increasing accumulation of harmful side effects.

Eukaryotic cells — cells with a true nucleus and nuclear membrane surrounding the genetic apparatus.

Extrinsic biosenescent processes — biosenescent processes caused by an agent outside the body.

Genes — the basic unit of genetic inheritance or information contained in the genome of a cell. Three major classes of genes are: structural, catalytic, and regulatory.

Genetic informational storage and flow — the central dogma of molecular biology. All information describing the structure and function of an organism is stored in the DNA of the genome and flows via RNA (DNA is transcribed to RNA) from the nucleus to the cytoplasm. The information in RNA is used to synthesize proteins. The flow of information is frequently denoted as: DNA → RNA → protein.

Genetic-programmed biosenescence — the condition where death is preceded by a long period of biosenescence. However, biosenescence has no known advantage, so this process is not expected to exist.

Genetic-programmed death — the conditions where the death of the organism occurs via a genetic process that kills the organism without an accompanying long biosenescent process.

Genome — the genetic apparatus of the cell that contains a haploid set of genes.

Germ plasm — the genetic information of an organism (a complete set) found in the germ cells (sex cells) of an organism.

Hominids — the organisms that, during the evolution of the primate species, formed the ancestral descendant sequence leading to modern man.

Intrinsic biosenescent processes — all the biosenescent processes that result from internal metabolic processes.

MCC — maximum life-span-potential calorie consumption — the number of calories an organism can consume on a per gram basis if it lives to its full MLP. Units are given as kcal/g (kilocalories per gram). MCC is the product of MLP and SMR (specific metabolic rate).

MLP — maximum life-span potential — the life-span of an organism under minimum environmental hazards. Generally, MLP is inversely proportional to the rate of biosenescence.

Nucleic acids — a polymer containing four different bases. Two main classes are DNA (deoxyribonucleic acid) and RNA (ribonucleic acid). Their main function is the storage and/or transfer of information through template-guided reactions.

Pleiotropic gene — a gene whose action has more than one effect. The effects could be of a beneficial or a detrimental nature, and could be continuously acting or dependent on the developmental stage of the organism. All genes have some degree of detrimental pleiotropic effects.

Prokaryotic cell — single-cell organisms that do not have their genetic material compartmentalized by a nuclear membrane, such as bacteria.

Protein — a polymer made up of about 20 different amino acids. There are three major classes of proteins: structural, catalytic (enzymes), and regulatory.

SMR (specific metabolic rate) — the number of calories (related to oxygen consumption) an animal consumes per unit body weight.

Soma cells — all the cells making up the body of the organism except the germ cells, which are solely used for reproduction.

21

The Origin, Evolution, Nature, and Causes of Aging

TRACY M. SONNEBORN

Aging is here defined as inherent, progressive, irreversible impairments of function. It occurs not only in higher organisms but also as far down the scale of beings as some — but not all — unicellular organisms. What does the whole array of organisms tell us about aging? Is there a good identifiable reason for its existence? Are there any *general* bases or causes of aging? In searching for answers to these questions, I shall be guided by evolutionary considerations.

THE RELATION OF AGING TO DEATH

First, we need to consider the relation of aging to death. Death can, of course, come at any stage of life, from conception on, as a result of accident, infectious disease, or heredity. For example, there is a gene called "drop-dead" that causes adult fruit flies of *any* age suddenly to turn on their backs and die instantly. In most organisms, there are critical periods in the life cycle. After the fertilized egg has divided into a number of cells forming a hollow ball, some of the cells begin to move into the hollow; at this stage, many embryos die. In organisms such as insects, that undergo a series of marked changes — from egg to larva, from larva to pupa, and then from pupa to adult — many die at each of these critical turning points. In humans, there are periods of high mortality before birth and at or shortly after birth. Nevertheless, in many organisms the chance of death from impairment of functions increases with age. In humans, this is the safely calculated basis of the insurance business. For many other organisms, predation and other environmen-

TRACY M. SONNEBORN, Ph.D. ● Department of Biology, Indiana University, Bloomington, Indiana 47401

tal factors may be more important lethal factors than aging. However, aging, when it occurs, by definition leads to death. Whether there are partial exceptions among some unicellular organisms (e.g., the ciliate *Tetrahymena*) in the sense of progressive impairment of one or more functions without leading to death is still not entirely clear.

THE MULTIPLICITY OF THEORIES

In the search for a general basis or cause of aging, the first thing one is struck with is the multiplicity of theories. A recent reviewer made the comment that there are almost as many ideas and theories about aging as there are researchers in the field — and that is a very large number. Why is that so? The answer seems quite obvious, and it tells us something very important. Each researcher is a specialist focusing his attention on a particular part of the whole subject, like the blind men who describe an elephant by the part they are next to and feel. The body of the elephant and any other organism — even the simplest — is an integrated machine of staggering complexity, with many parts that are interdependent and interact in a multiplicity of ways. Hence, deterioration or defect in almost *any* one part has repercussions on many other parts. Consequently, *all* parts of the body exhibit aging changes: an opthalmologist can look in your eye and estimate pretty well your physiological age; the same can be done be examination of your skin or your foot or your backbone. A research gynecologist once told me that he could tell a woman's age by the number of ova (or, more correctly, oocytes) in her ovary, not because of loss of the relatively small number that mature, usually one per month, but because of the enormously larger number that progressively decay and disappear without ever maturing. Practically every part of the body undergoes progressive age change. Each specialist, focusing on the aging changes in the part that is at the center of his attention, notes how the changes in this part affect or are correlated with changes seen in other parts. So, he thinks he has spotted the fundamental feature of aging and bases his theory on it. Hence, the multiplicity of theories.

Is it possible to discern in all of this highly specialized knowledge a fundamental, general basis or cause of aging? The trend of modern biology is clearly to seek fundamentals not in organ systems, organs, or tissues or in any gross part of the body, but in the basic building block of body structure and function, the cell; and, within the cell, in its organelles and especially in its molecules and their activities. There, if anywhere — it is widely believed — lie the ultimate clues to much of biology, including the biology of aging. So there has been in recent years a widespread and intense investigation of aging at the basic levels of the cell, its organelles and its molecules.

If you think this would automatically simplify and unify the study of aging, you are quite wrong. On the contrary, it turns out — as should really have been expected — that there is at least as great a diversity of ideas and theories of aging based on studies at cellular and molecular levels as on studies at higher structural levels. The reason is similar. There are many different kinds of cells in the body — at least 100 — and each type of cell is an enormously complex machine with organelles and with thousands of kinds of molecules acting and interacting. Not surprisingly, age changes in any one kind of cell — indeed, in any one kind of molecule — may have repercussions on many other kinds of cells and molecules. Again, we find specialists at the cellular or molecular level who focus on one kind or one class of cell or molecule and note their age changes and the effects or correlates of these changes in other cells or molecules. Various specialists have focused on changes with age in production of insulin or other hormones; or on age changes in the composition or properties of cellular membranes; or on age changes (including cross-linkage) in this or that enzyme or other protein (and there are hundreds or thousands of kinds of enzymes and other proteins); or on age changes in the neurotransmitters that mediate communication in the nervous system; or on cell products, such as collagen; or on the genetic material, DNA. Again, each specialist thinks that the kind of structure or molecule he focuses on is the basic or fundamental one because age changes in it appear to have such ramifying effects or correlates. So, again, there is a multiplicity of cellular and molecular theories of aging.

What is to be made of this confusing situation? Let's assume that all the facts and observations are correct. Are all the theories wrong? Is one of them *the* correct theory? How can we make a judgment? Before we can try to answer these questions, we must first seek a reliable guide to judgment. Where and how can we find it? As a biologist, I have to seek guidance in the evolutionary origin and development of aging. This is not the same as seeking evolutionary guidance, as some have done, to explain the more limited problem of the mere *existence* of aging in certain kinds of animals (e.g., how natural selection could operate either positively or by default in higher animals merely to permit the existence of aging in them). More comprehensive evolutionary guidance is needed to account for the fact that aging arose long before higher animals and has been maintained thereafter through the course of further evolutionary developments. The main contribution of this chapter to the field of aging is to point out a remarkable difference between organisms that age and those that do not and to indicate its significance.

ORGANISMS THAT AGE AND ORGANISMS THAT DO NOT

The first thing that strikes one is that aging is not an invariable correlate of life. It appeared rather late in the course of evolution. The simplest organisms now living — those we believe to be most like the first forms of life — are the bacteria. They simply do not grow old. They go on and on living and reproducing their kind, remaining vigorous, and dying only by accident — by unfavorable external conditions (including starving to death and being eaten by predators), or by an internal accident such as a lethal mutation. Bacteria appear to be potentially immortal. The same is true of many much more complex and highly evolved creatures: Many one-celled plants, such as algae, and one-celled animals, such as amebae, and certain flagellated or ciliated protozoans.

Natural aging and death first appear in the course of evolution among some — but by no means all — of the more complex unicellular organisms. For example, the ciliates *Paramecium* and *Euplotes* cannot simply go on growing and dividing into two and maintaining their full normal vigor. Eventually, their growth rate slows down little by little, longer and longer times intervening between successive cell divisions; then the cells stop dividing altogether and die. Along with declining reproductive vigor goes progressive impairment of other functions. Structural abnormalities appear with increasing severity, and cells die with increasing frequency until none remains alive.

IS AGING CORRELATED WITH SEX?

How, then, do such aging unicells escape extinction? The answer is clear: by sexual processes. *Paramecium, Euplotes,* and many other ciliated protozoans can undergo fertilization, usually by conjugation. For example, two paramecia come together, mouth to mouth, and form a temporary mating organelle through which reciprocal fertilization is accomplished, each mate sending one of its nuclei to unite with one of its partner's nuclei. After fertilization, the two cells separate, and each undergoes repeated growth and division in full vigor, initiating a new cycle that will lead again to aging and death unless the age clock is set back by another fertilization.

This age cycle of division in *Paramecium, Euplotes,* and other ciliates is marked by periods comparable to those in the life of higher animals. Cells produced by the early cell divisions after fertilization are immature; they are unable to conjugate. Immaturity is followed by a long period of maturity — many cell generations during which the cells, if they find a suitable partner, can mate and yield viable progeny by fertilization. If mature cells multiply too long without mating, senescence begins: The cells become more and more reluctant to undergo fertilization, and when they do, the proportion of their progeny that die soon after fertilization increases with parental age until eventually none can survive. At this point, the lineage is doomed; it dies whether mating occurs or not. So, to keep going, the organisms *have* to undergo fertilization before they are far into the senescent period.

Did senescence and death, then, evolve in invariable correlation with sexual processes? Are we to seek in sex the basis of senescence? Certainly not. Even bacteria, which do not age or die naturally, have the beginnings of sexuality. And some of the higher unicells (e.g., some algae) have fully developed sexuality and sexual reproduction, but they do not age and die. Under good cultural conditions, they can go on growing and dividing — so far as we know, forever — regardless of whether they indulge in sex or not. So species can have sex without aging, but cannot have aging without sex. These facts tell us that aging and death evolved *later* than sexuality; they are *not* — as some have held — the price paid for sex. There is another feature of organisms which evolved after sex and on which the origin and evolution of aging depend. Without this feature, aging is forestalled in organisms possessing sex. This critical, other feature is diploidy, as will now be argued.

THE RELATION OF AGING TO DIPLOIDY

Bacteria and most other unicellular organisms that lack aging differ from organisms that have aging in the *number* of each kind of chromosome per nucleus. *All organisms that have only one chromosome of each kind, that is, haploid organisms, lack natural aging. All organisms that have natural aging have two chromosomes of each kind; that is, they are diploid.*

Haploids and diploids differ in their ability to accumulate mutations, a difference of tremendous importance in the origin and evolution of aging. Genetic adaptation to variable and changing conditions involves accumulation in the same individual of favorable combinations of mutations. While single mutations usually are harmful or nonadaptive under existing conditions, certain combinations of mutations can be beneficial, especially under changing conditions. But mutations are rare events. Any one kind of gene mutates in something like 1 gene in 10 million. Since there is only one gene of each kind in haploids, mutations are quickly expressed and, because each mutation is likely to be individually harmful, it is likely to lead to death or be lost as possessors of it are overgrown by nonmutated cells. Accumulation of favorable combinations of mutations in sexless haploids is, therefore, a slow, precarious, and inefficient process.

The situation is somewhat, but not greatly, better in sexual haploids. They can usually bring together no more than one mutation in each of the two cells that mate because, as mentioned, one is likely to be lost before a second occurs in a cell bearing the first one. So the chance of getting favorable combinations of many mutations is still very small. Sex and fertilization are not entirely useless in haploids, but they are inefficient. So they remain optional, not mandatory.

The situation is very different in diploids with two genes of each kind (and here I include organisms such as certain fungi, for example, that can have or have had an extensive diploid phase as well as an extensive haploid phase). When one of the two genes mutates, the mutation usually is recessive; that is, it is not expressed because there is present also the other, normal, dominant gene. When a diploid organism starts life with both genes of a kind normal, the chance of *both* genes of the same kind mutating — and, therefore, being expressed — is the square of the chance for one of them (i.e., 1 in about 10 million times 10 million, or 1 in a hundred million million). That means practically never. When one gene is mutant and the other normal at the start of life, the chance of the mutation being expressed as a result of mutation of the normal gene is still about 1 in 10 million, so the single mutant gene can be safely carried along for very many cell generations without doing much, if any, harm, long enough for similarly unexpressed mutations of other genes to accumulate in the same individual.

In diploid organisms that have sex and fertilization, accumulated mutations can be assorted in all possible combinations, including some combinations in which *both* genes of one or more pairs are mutant and can, therefore, be expressed. The different combinations of mutations can then be tried out for their adaptive value or harm, and natural selection can preserve the adaptive ones. This is exactly what sex and fertilization are all about in diploids. They are mechanisms of recombining accumulated unexpressed genes on a large scale and bringing various combinations of them to expression. This it is that has made possible the explosive and highly diversified evolution of diploids. It took about 3 billion years to proceed from the origin of life to unicellular diploids with sex and fertilization and about one-sixth as long to go from there to all existing kinds of multicellular organisms. Although scholars still dispute the evolutionary value of recombining accumulated mutations, the evolutionary record speaks strongly for it.

Now recall — and this is the most significant point for the origin and evolution of aging — that it was among diploids that sex and fertilization ceased to be optional and became mandatory for survival of the species, as noted above in the examples of *Paramecium* and *Euplotes*. How could they become mandatory? By selective pressure for, or no selection against, mutations that assure a program consisting of a limited period in which fruitful fertilization is possible (whether it occurs or not) followed by aging and death after this period. This is how I believe aging and death evolved. The law is: Fertilize and recombine genes when you can, or tomorrow you will die. The key role of diploidy in the evolution of aging has, as far as I am aware, not heretofore been fully realized, if at all. Although the combination of diploidy, sex, and fertilization may have evolved independently in several lines of descent (e.g., in algae, ciliates, fungi, and perhaps other lines), once it appeared in any evolutionary line, it was usually retained through further stages of evolution.

MULTICELLULAR ORGANISMS

With the evolution of multicellular species, especially multicellular animals, came another important development which was already foreshadowed by certain unicellular organisms: the evolution of the distinction between body cells and germ cells. Body cells of multicellular animals do not develop into germ cells. They are doomed to age and die (unless they turn into cancer cells). Germ cells, on the contrary, give rise to sperm and egg cells, some of which unite and develop into a new body, with all its diverse cells *and* more germ cells. Here, then, is a basic fact about aging and death, one that is no less important because it is familar to us all. The succession of germ cells from generation to generation is *potentially* immortal. Given a chance, germ cell lines can go on forever, but body cells are doomed.

Not only is the germ line potentially immortal, but there is now very strong evidence to support the old faith that the germ cells (and body cells) of all living organisms are direct uninterrupted descendants of the original progenitors of the simplest currently living organisms, the bacteria. All living organisms that have been investigated, from lowest to highest, have very similar chemical components, including similar enzymes and other proteins, and most significantly, they all use exactly the same — or nearly the same — genetic code. So, the germ lines alive today have a history of more than 3 billion years without aging.

Whatever the differences between germ-line cells and body cells may be, they should provide clues to the nature and causes of aging. In searching for the clues, two basic questions present themselves: (1) What makes body cells age and die? (2) How do germ-line cells (and cancer cells) escape that fate? The two questions obviously are inter-related and occasionally will have to be considered together. I shall comment on a number of the main hypotheses that have been put forth in efforts to answer them. The hypotheses are not all mutually exclusive; a good deal of overlapping occurs among them. For this and other reasons, more than one may be valid.

CELLULAR DIFFERENTIATION, SPECIALIZATION, AND WEAR AND TEAR AS POSSIBLE CAUSES OF AGING

I consider first, but only briefly, two old, long-held, and related hypotheses and, in somewhat more detail, a third, related hypothesis. The first is that cellular differentiation leads inevitably to body cell mortality. To be sure, body cells, such as muscle, nerve, or skin cells, are highly differentiated in structure and function. But cells of certain lower organisms that do not undergo natural aging or death are more highly structured and differentiated than any kind of body cell in a higher organism. So, cellular differentiation per se can hardly be a general or fundamental cause of cellular aging and death.

The closely related second hypothesis is that specialization is the cause of body cell death. Body cells are, of course, specialized to perform certain functions and cannot or do not carry out other functions which are relegated to other body cells. This hypothesis is difficult to prove or disprove.

The third related hypothesis is that aging and death result simply from the wear and tear due to use, as happens sooner or later with any machine. At the cellular level, this hypothesis might be, and probably is, valid for cells such as nerve cells, which are all formed early in development and cannot be formed later. We, for example, are provided at the start with fantastic numbers of nerve cells, apparently far more than are needed for normal functioning. We depend on the initial surplus to keep us going, because large numbers are continually wearing out and dying.

Some lower organisms in particular seem to offer strong cir-cumstantial evidence in support of the wear-and-tear hypothesis. Small aquatic flatworms, such as *Stenostomum*, possess a fairly complicated head formed once and for all in early development, after which its cells do not divide. However, the trunk, that is, the rest of the body (includ-ing intestine, excretory apparatus, etc.), is composed of cells that can continue to divide, and the trunk grows continually longer and longer. When a certain length is reached, a new second head is formed some distance from the original head, and the worm divides into two, the front part having the old head, the rear part having the new head. The trunk attached to the original head then grows again, forms another head, and divides again. This goes on repeatedly for about a month, then begins to slow up — age creeps up on it. Growth and reproduction continue to slow up until they finally cease at age about 2 months, after less than 150 reproductions. Then the worm with the original head dies. Progressive retardation of growth ending in death does not occur in the line of descent composed of successive hind animals produced from the hind animal arising at each preceding division. In this case, obviously, a new head is developed for each worm in that line of descent. This comparison shows that aging and death in the worm are correlated with long possession of one and the same head, which presumably wears out while its trunk is continually growing. The production of a new head at each act of reproduction is correlated with an apparently immortal, ever youthful line of descent.

An extreme case is that of the rotifer, another type of small aquatic multicellular organism. In rotifers, all the body cells that will ever be formed arise during embryonic development. None divides later during the rest of life. In agreement with the idea of wear and tear of non-renewable cell types, these organisms have very short lives, usually only a matter of days or a few weeks.

Some interesting and possibly relevant facts are also shown by certain ciliated protozoans. The most complex structure in ciliates, such as *Paramecium* and *Tetrahymena*, is the oral or feeding apparatus. Like the head in the flatworm, it passes intact at each act of reproduction to the front product of division; the hind product develops a new oral apparatus. Curiously, however, *Paramecium* eventually ages, slowing up cell divisions and dying, whereas *Tetrahymena* does not. Can it be a mere coincidence that *Tetrahymena* can *at any time* discard its old oral apparatus and replace it by a new one, whereas *Paramecium* cannot? Only during fertilization does *Paramecium* replace its oral apparatus by a new one, and only after fertilization is the age clock set back to a new start. These correlations are impressive. *Tetrahymena* can replace its old oral apparatus in the absence of fertilization, and it can go on growing and dividing indefinitely without somatic senescence. Lacking this capacity for oral replacement in the absence of fertilization, *Paramecium* must undergo fertilization or eventually senesce and die.

LIMITED REPRODUCTIVE CAPACITY OF BODY CELLS AS A POSSIBLE CAUSE OF AGING

Our consideration of the wear-and-tear hypothesis has focused on structures, cells, and body parts that are not renewable resources. Some types of body cells (e.g., blood-forming cells or fibroblasts), unlike nerve cells, are not all formed early once and for all but can continue to multiply throughout life. Hence, cells of these types or their descendants that become defective or worn out can be overgrown and replaced by others that are normal. If such purging and replacement could go on forever, these cell types would be permanently renewable resources. In recent years, evidence has been presented that they are not permanently renewable but that, like *Paramecium,* these cells can undergo only a limited number of cell divisions. This, then, becomes the fourth hypothesis: that aging is due to the limited reproductive capacity of some kinds of body cells.

An interesting variation of this hypothesis is currently being considered. According to it, a given type of body cell, such as fibroblasts, initially consists of cells with unlimited reproductive capacity; but in the course of cell divisions, there arise with a certain probability cells committed to give rise exclusively to cellular descendants that will age and die. This hypothesis can be applied only with extreme assumptions to organisms such as rotifers in which very few body cells of any kind are formed before the organism ages and dies. In a general way, it is to be viewed as a postponement for varying numbers of body cell generations of the distinction others postulate between germ cells and all body cells in reproductive capacity.

The question arises whether body cells, either at the start or later, become inherently programmed for cessation of growth and division or whether such decline ("aging") is a cumulative effect of suboptimal conditions. For many years in the early decades of this century, this question was debated and experimented on with regard to the limited capacity for cell division in *Paramecium* and certain other ciliates. It was finally shown that fertilization, if not too long delayed, can cancel the degenerative age changes — to whatever they might be due — and that no ingenuity of man was able to prevent decline and death of lines of descent that failed to undergo fertilization. So it is generally agreed that the retardation and eventual cessation of growth and division in the absence of fertilization is inherently programmed in these cells. External conditions can hasten or retard cell aging and death but not prevent them.

As everyone knows, the limited reproductive capacity of body cells disappears when a body cell becomes a cancer cell. It then acquires the capacity to grow and divide endlessly, like germ cells. A number of theories have been put forth about the basis of this change from body cells to cancer cells, from limited to unlimited capacity for growth and division, but I think it is fair to say that the basic change is still not understood. If and when it is, we shall probably understand better the basis of the limited capacity of body cells for growth and reproduction, and hence know more about at least one aspect of aging.

THE ERROR-CATASTROPHE HYPOTHESIS OF AGING

The limited reproductive capacity of cells and the wear and tear of nonreplaceable cells or parts of cells, although together seeming to account for the decay or senescence of most cells (all but cancer cells and germ-line cells), do not really explain how reproductive capacity can be limited or of what wear and tear consist. The fifth hypothesis — that of error catastrophe — attempts to go deeper. The central idea is that certain special classes of intracellular accidents would have a cascade effect; that is, the initial accident leads progressively to more and more difficulties. Any accident or error in either the machinery or the process of making proteins would cascade into multiple effects. Enzymes are proteins; they are essentially involved in the whole complex of cell activities. Mistakes in the machinery or processes of making enzymes, yielding faulty enzymes, could thus interfere with virtually all cell activities, including actions on other cells. The difficulty would be compounded and spread, eventually leading to deterioration of the whole organism. Since no machine is perfect, error will eventually creep in. And in fact, age-correlated errors in enzyme and protein synthesis have been reported.

THE GENETIC BASIS OF AGING

Mutation is a special form of accident or error and, as a sixth hypothesis, aging has been ascribed by some to the accumulation of mutations in body cells. Such accumulation would be accentuated by the special class of mutations which bring about deficiencies in mechanisms known to be able to repair damage to the genetic material. Mutations affecting the machinery for making proteins would lead to the cascade effect mentioned above. Because the existence of two genes of each kind in diploid cells greatly reduces the probability that new mutations in body cells would have an appreciable effect, mutations would be effective mainly for those pairs of genes of which one was inherited in mutant form by the fertilized egg.

Most of the hypotheses thus far considered fail to explain why the germ line (and cancer cell lines) are free of the restraints and limitations presumed to affect body cells. Why doesn't the germ line wear out, have a limited number of possible cell divisions, undergo error catastrophe, accumulate mutations, or have its damage-repair mechanism decline in efficiency? More important, *all* the hypotheses fail to account for the fact that the rate of aging and the maximal length of life differ greatly in different strains and species. This variation is an old evolutionary acquisition. It exists even among unicellular organisms. Some species of *Paramecium* age and die in the course of 200 to 300 cell generations; other species of *Paramecium* and *Euplotes* age very slowly and live for thousands of cell generations. Among higher organisms, some live for only a few days, others for a hundred years or more. Even within a species, different families or strains differ greatly in rate of aging and length of life. These facts argue strongly for a genetic basis of aging, as was indeed implied when I developed the idea that aging arose in diploids by selection of mutations that determine aging and death after the period of sexual maturity or by default of selection in the postreproductive ages. In other words, aging is a part of the developmental program inscribed in the genetic material, DNA.

To recognize this fact is the first step, but only the first step, toward understanding how the hereditary developmental program works. Fundamentally, the problem is essentially the same as the problem of any other aspect of normal development. When we understand one, we should understand the other. At present, we have only very limited understanding of how the developmental program is inscribed in DNA, but knowledge of the structure and functioning of DNA is increasing with explosive speed. Enough is already known for us to imagine models to explain how one or another aspect of development — including the terminal development of aging and death — is programed. A DNA mechanism for counting cell divisions and undergoing a change of cell character after a definite number of divisions has recently been proposed. Something like this could, in principle, account for the fact that certain types of cells can divide only a limited number of times as well as for the very different hereditary rates of aging and lengths of life in different organisms.

Very recently, another suggestion has been made at the DNA level to account for some of the differences between body cell lines and germ cell lines. This suggestion is based on a certain known behavior of DNA at the time germ cells are being formed. At that time, the two chromosomes of a kind come together side by side. The suggestion is that at this time there is massive repair of genetic damage that has accumulated in preceding cell generations of the germ line, so that each new sexual generation is to a degree genetically rejuvenated.

These DNA mechanisms to which I have alluded are at present purely hypothetical. Whether any current model or some future model proves to be valid remains to be discovered. I intend only to point out that it is now *possible* to imagine models of aging at the basic DNA level which could (if valid) account for diverse rates of aging in different species and individuals, that it is possible to imagine DNA mechanisms that could account for differences between germ cells and body cells, and that it is, in fact, necessary to seek the explanation of aging in the structure and functioning of DNA. I realize that some may feel that our perspective on these matters is unduly affected by the recent and current brilliant, rapid successes of DNA research. I do not share that feeling. I am compelled to look to DNA for the solution of the aging problem because aging is a hereditary feature of normal development in organisms that age and because of the way in which I perceive the origin and evolution of aging.

SUMMARY

To recapitulate briefly the main points of this chapter, aging is not an inevitable event, characteristic of all life. Many lower forms of life — especially unicellular haploid organisms — do not undergo natural aging or death; they can go on living, growing, reproducing forever in full vigor. The same is true for the germ plasm of higher organisms, which traces its uninterrupted ancestry back more than 3 billion years, to the origin of life. Aging and death first appeared in the course of evolution *after* sex, fertilization, and diploidy had evolved but *before* multicellular organisms appeared. Sex and fertilization confer on diploids — far more than on haploids — the possibility of reaping the substantial advantage of genetic adaptability in the form of varied new combinations of genes. This appears, in spite of some arguments to the contrary, to have been so advantageous — both on the short- and long-term scales — that natural selection preserved or did not eliminate mutations that made sexual reproduction mandatory. As a result of mutations, some ciliates had to reproduce sexually to avoid growing old and dying. Mandatory fertilization, assured by eliminating all that live beyond the fruitful age, was retained by all higher organisms indicating that it probably played a key role in their relatively rapid evolution. In them, the body *has* to grow old and die, but germ-line cells can go on reproducing forever. Differences among organisms in rate of aging and length of life show clearly that the general basis of aging has to be hereditary. We do not know exactly how heredity determines aging and death of the body, but we can imagine reasonable and testable mechanisms based on known features of the structure and operation of DNA, the hereditary material. Eventually, the actual DNA mechanisms may be discovered.

ACKNOWLEDGMENT. The author acknowledges gratefully suggestions by Douglas Brash for improvement of the manuscript.

BIBLIOGRAPHY

General

Comfort, A. 1964. *Ageing, the biology of senescence*. Holt, Rinehart and Winston, Inc., New York.

Cutler, R. G., ed. 1976. Cellular ageing: Concepts and mechanisms. In Vols. 9 and 10 of *Interdisciplinary topics in gerontology*. S. Karger AG, Basel.

Harrison, D., ed. *Symposium on genetic effects on aging*. National Foundation–March of Dimes, Washington, D.C. In press.

Lansing, A. I., ed. 1952. *Cowdry's problems of ageing*. The Williams & Wilkens Company, Baltimore.

Shock, N. W., ed. 1962. *Biological aspects of ageing*. Columbia University Press, New York.

Strehler, B. L., ed. 1960. *The biology of aging*. Publication 6, American Institute of Biological Sciences, Washington, D.C.

References to Hypotheses and Particular Examples

On hypotheses about aging, see: L. Hayflick, *American Journal of Medical Science* 265: 433 (1973), for programmed number of possible cell divisions of fibroblasts; R. Holliday and J. E. Pugh, *Science* 187: 226 (1975), for a possible molecular mechanism controlling the number of cell divisions until a differentiative event occurs; R. Hart and R. B. Setlow, *Mechanisms of Aging and Development* 5: 67 (1976), for decline of repair of genetic damage in aging; R. Holliday, L. I. Huschtscha, G. M. Tarrant, and T. B. L. Kirkwood, *Science* 198: 366 (1977), for the hypothesis of commitment to aging as an event occurring among descendants of previously potentially immortal tissue cells; R. Martin, in Vol. 7 of *ICN–UCLA symposium proceedings (molecular human cytogenetics)* (D. E. Comings, R. S. Sparks, and C. F. Fox, eds.), Academic Press, Inc., New York (1977), for hypothesis of repair of genetic damage in germ-like cells during meiosis and for discussion and references concerning the genetic material as the primary seat of aging; D. L. Nanney, *Mechanisms of Aging and Development* 3: 81 (1974), for the hypothesis that cells may use random events to measure time, including aging, an idea related to the commitment and error hypotheses; L. E. Orgel, *Proceedings of the National Academy of Sciences of the USA* 67: 1476 (1970), for hypothesis of aging by accumulation of errors.

On aging in *Paramecium*, see H. S. Jennings, *Journal of Experimental Zoology* 99: 15 (1945); T. M. Sonneborn, *Journal of Protozoology* 7: 38 (1954); articles by T. M. Sonneborn and coworkers in Strehler's book listed among the general references above; articles by Joan Smith-Sonneborn and collaborators in *Radiation Research* 46: 64 (1971), *Journal of Cell Biology* 61: 591 (1974), and *Journal of Cell Science* 14: 691 (1974). On aging in *Euplotes*, see R. Katashima, *Journal of Science of Hiroshima University*, Ser. B, Div. 1, 23: 59 (1971). On aging in the flatworm, Stenostomum, see T. M. Sonneborn, *Journal of Experimental Zoology* 57: 157 (1930).

Index